图解 Java 并发编程

汪 建◎著

人民邮电出版社
北京

图书在版编目（CIP）数据

图解Java并发编程 / 汪建著. -- 北京：人民邮电出版社，2021.10
ISBN 978-7-115-56173-2

Ⅰ. ①图… Ⅱ. ①汪… Ⅲ. ①JAVA语言-程序设计 Ⅳ. ①TP312.8

中国版本图书馆CIP数据核字(2021)第050086号

内 容 提 要

本书采用图文并茂外加大量案例代码的方式讲解了Java并发编程机制的运行原理。

本书分为12章，内容涵盖了线程机制、线程I/O模型、Java内存模型、并发知识、AQS同步器、常见的同步器、原子类、阻塞队列、锁、任务执行器、其他并发工具等内容。此外，还在最后一章介绍了如何使用C++来模拟实现Java线程。

本书适合Java中高级开发人员、对Java并发编程机制感兴趣的人员以及Java架构师阅读。

◆ 著　　汪 建
责任编辑　傅道坤
责任印制　王 郁　焦志炜

◆ 人民邮电出版社出版发行　北京市丰台区成寿寺路11号
邮编　100164　电子邮件　315@ptpress.com.cn
网址　https://www.ptpress.com.cn
固安县铭成印刷有限公司印刷

◆ 开本：800×1000　1/16
印张：24.25
字数：544千字　　　　　　2021年10月第1版
　　　　　　　　　　　　　2025年2月河北第2次印刷

定价：109.90元

读者服务热线：(010)81055410　印装质量热线：(010)81055316
反盗版热线：(010)81055315

作者简介

汪建，笔名 seaboat，毕业于广东工业大学光信息科学与技术专业，毕业后从事各类业务系统、中间件、基础架构、人工智能系统等研发工作，目前致力于用 AI 提升企业业务系统效率以节约人力成本。擅长工程算法、人工智能算法、自然语言处理、计算机视觉、架构、分布式、高并发、大数据、搜索引擎等方面的技术，会使用大多数编程语言，更擅长 Java、Python 和 C++。平时喜欢看书、运动、写作、编程、绘画。崇尚开源，崇尚技术自由。已出版的图书有《图解数据结构与算法》《Tomcat 内核设计剖析》。

个人博客：blog.csdn.net/wangyangzhizhou。

个人公众号：远洋号。

远洋号

致　　　谢

　　七年前，我在博客上写过一个 Java 并发专栏，很多读者在阅读后反馈对他们的帮助很大，于是我基于该专栏编著了这本书。为了让读者更好地理解 Java 并发原理，我绘制了大量示意图，并提供了大量代码案例。

　　正是因为读者的反馈和建议，我才能进一步完善本书，在此向各位读者表示感谢。感谢公司提供的平台，让我得到了很多学习和成长的机会。感谢人民邮电出版社的编辑傅道坤老师不断给我提供帮助和机会，使我终能连续成书。最后，感谢一直默默支持我的家人，你们让我的世界更加丰富多彩。谨将本书献给我可爱的儿子和女儿，希望你们健康快乐成长。

前言

经过二十多年的发展，Java 语言已经成为互联网时代主要的编程语言之一。虽然这些年 Java 受到很多其他编程语言的挑战，但它仍然在编程语言排行榜中排名前三，这足以看出它的生命力之顽强。对于开发者来说，学会使用 Java 编程语言是非常有必要的。Java 在诞生之初是用于智能电器，然而并没有在该领域取得太大的成功。随着互联网的兴起，Java 因其跨平台和面向网络的优势而火了起来，当然这其中一个非常重要的原因是 Java 是开源的。

近几年，Java 又开始快速发展，相关的社区非常活跃且大版本的发布保持着很高的频率。目前 Java 主要用于移动端和服务端的开发，移动端主要针对 Android 系统，而服务端则针对企业的业务系统、中间件和大数据组件等。

就 Java 知识而言，并发模块属于高阶知识。为了应付高并发场景，我们必须彻底明白 Java 的并发原理。想要征服 Java 并发这座大山并非易事，实际上并发并非 Java 语言的问题，从本质上来说是并发操作本身的问题，此外还涉及计算机体系结构带来的问题。Java 并发涉及计算机的底层结构、CPU 指令的执行、程序的编译等方面的知识，我们要明白 Java 并发原理，就需要理解上述各方面的知识。

七年前，我在工作中经常遇到系统并发问题，于是决定深入学习 Java 并发知识。正是在学习的过程中，我写了一个 Java 并发主题的专栏并发布在博客上，这些年还在不断收到读者的反馈并得到了读者的肯定。很多读者经常问我该如何系统地学习 Java 并发知识，我在给他们解答的过程中萌生了创作一本书的想法，因此我对该专栏进行了扩展、优化，最终形成了读者手里的这本书。为了让读者能更好地理解 Java 并发原理，我绘制了两百余张示意图，并且提供了一百多个代码案例，这些都能极大地帮助读者理解其中的原理，以达到"一图胜千言"和"一例胜千言"的效果。

我尝试列出以下关键词来让读者了解本书的内容：线程调度、内存模型、阻塞唤醒、线程协作、指令重排、乐观锁、线程池、悲观锁、AQS、线程状态、synchronized、volatile、死锁、非阻塞、Lock、信号量、优先级、阻塞队列、线程饥饿、读写锁、竞争条件、互斥共享、数据竞争、可见性、CAS、CPU、中断、原子、同步、I/O。其实这些关键词也是 Java 并发的核心内容。如果我们能掌握这些内容，就可以说基本掌握了 Java 并发知识。当然，本书的定位是并发的实现原理层面，所以重点不在于 Java 并发包 API 的使用，而是对 Java 并发机制原理的讲解，以及对 JDK 相关核心源码的讲解。

本书组织结构

本书采用"一图胜千言"和"一例胜千言"的写作方法，使用了大量的示意图和代码来讲解 Java 并发相关的原理机制及具体实现，帮助读者更深入地理解相关知识点。本书共分为 12 章，具体内容如下。

- **第 1 章：线程机制**，介绍了什么是线程，具体包括线程的映射、线程的状态、线程的调度、线程优先级、线程的 CPU 时间及 yield 操作、sleep 操作、interrupt 操作、join 操作、阻塞唤醒操作。
- **第 2 章：线程 I/O 模型**，介绍了计算机中的线程与 I/O 模型。阻塞 I/O 模型包括单线程阻塞 I/O 模型和多线程阻塞 I/O 模型；非阻塞 I/O 模型包括应用层 I/O 多路复用、内核 I/O 多路复用和内核回调事件驱动 I/O。多线程虽然提升了任务执行效率和用户体验，但也使编码更复杂且带来了更多的资源开销。
- **第 3 章：Java 内存模型**，讲解了 Java 内存模型和计算机运行模型。CPU、主存和外设通过总线进行通信，CPU 内部包含了算术逻辑单元、控制单元和寄存器。为了能屏蔽底层硬件和操作系统的差异从而实现跨平台的内存描述，Java 在虚拟机中定义了一种统一的内存模型。本章也分析了 volatile 的可见性问题和指令重排问题，此外还讲解了 happens-before 的八大原则。
- **第 4 章：并发知识**，介绍了 Java 并发知识，详细分析了 synchronized 互斥锁的使用及实现原理，讲解了自旋锁（包括原始自旋锁、排队自旋锁、CLH 锁、MCS 锁）及线程饥饿（包括 synchronized 饥饿、优先级饥饿、线程自旋饥饿、等待唤醒饥饿、公平性解决饥饿），最后分别深入分析了数据竞争、竞争条件和死锁。
- **第 5 章：AQS 同步器**，讲解了 Java 并发的基石——AQS 同步器，包括 AQS 的等待队列与状态转换、AQS 的互斥锁与共享锁、AQS 独占锁的获取与释放、AQS 共享锁的获取与释放、AQS 的阻塞与唤醒、AQS 的中断机制、AQS 的超时机制、AQS 的原子性、AQS 的自旋锁、AQS 的公平性、AQS 的条件队列以及 AQS 自定义同步器。
- **第 6 章：常见的同步器**，介绍了常见同步器（包括闭锁、信号量、循环屏障、相位器和交换器）的使用与实现原理。
- **第 7 章：原子类**，介绍了并发中常见的原子工具，包括原子整型、原子引用、原子数组以及原子变量更新器等。
- **第 8 章：阻塞队列**，讲解了 JDK 中常见阻塞队列的使用及实现原理，包括数组阻塞队列、链表阻塞队列、优先级阻塞队列、延迟阻塞队列以及链表阻塞双向队列。
- **第 9 章：锁**，介绍了可重入锁和读写锁的使用及实现原理，它们都包含公平和非公平两种模式。对于读写锁，本章分别对读锁和写锁进行了详细分析，此外还讲解了锁的 Condition 机制。

- **第 10 章：任务执行器**，介绍了任务执行器的 Executor 接口和 ExecutorService 接口的设计思想，并且介绍了同步执行器、一对一执行器、线程池执行器和串行执行器等自定义执行器，旨在帮助读者更好地理解任务执行器的原理。最后，本章深入分析了 JDK 的线程池任务执行器 ThreadPoolExecutor 的实现原理。
- **第 11 章：其他并发工具**，介绍了 JDK 中线程本地变量和写时复制数组列表这两种并发工具的使用及实现原理。
- **第 12 章：C++模拟实现 Java 线程**，讲解了如何通过 C++ 来模拟实现 Java 线程，包括模拟实现 Java 线程、yield 语义、sleep 操作、synchronized 语义以及 Interrupt 操作等。

本书特色

本书具备如下特色。
- 本书通篇大量采用图解的方式进行介绍，作者绘制了两百多张示意图来帮助读者理解相关知识点，并针对每个关键点和难点给出了相应的图示，使读者能轻松理解 Java 并发相关工具和概念的思想。
- 本书提供了大量代码案例，总共编写了一百多个代码案例来讲解 Java 并发工具和问题，让读者能从代码角度去理解并发是如何实现的。
- 本书讲解的 Java 并发知识在实际工程中很常见，因此能够很好地帮助读者在实际的项目开发中理解相关的实现原理。
- 本书的主题是 Java 并发原理，重点是并发问题和工具的讲解与分析，而不是如何使用 Java 并发 API。
- 本书脉络结构清晰，由基础概念到高层工具，循序渐进，且各知识点的连贯性较强，读者只要具备基本的 Java 编程知识就能阅读。

读者对象

- 如果你是 Java 开发人员，可以通过本书学习到 Java 并发相关的知识。
- 如果你是资深的 Java 工程师，对 Java 并发机制感兴趣，可以通过本书深入学习到 Java 并发的原理。
- 如果你是 Java 架构师，可以通过本书巩固 Java 底层知识。

反馈

在本书交稿时,我仍在担心是否遗漏了某些知识点,本书的内容是否翔实齐备,是否能让读者有更多收获,是否会因为自己理解的偏差而误导读者。由于写作水平和写作时间有限,书中难免存在不妥之处,恳请读者评判、指正。

读者可将意见及建议发送到邮箱 wyzz8888@foxmail.com,本书相关的勘误也会发布在我的个人博客 blog.csdn.net/wangyangzhizhou 上。欢迎读者通过邮件或博客与我交流。

资源与支持

本书由异步社区出品，社区（https://www.epubit.com/）为您提供相关资源和后续服务。

配套资源

本书提供如下资源：
- 本书源代码；
- 书中彩图文件。

要获得以上配套资源，请在异步社区本书页面中单击 配套资源 ，跳转到下载界面，按提示进行操作即可。注意：为保证购书读者的权益，该操作会给出相关提示，要求输入提取码进行验证。

如果您是教师，希望获得教学配套资源，请在社区本书页面中直接联系本书的责任编辑。

提交勘误

作者和编辑尽最大努力来确保书中内容的准确性，但难免会存在疏漏。欢迎您将发现的问题反馈给我们，帮助我们提升图书的质量。

当您发现错误时，请登录异步社区，按书名搜索，进入本书页面，单击"提交勘误"，输入勘误信息，单击"提交"按钮即可。本书的作者和编辑会对您提交的勘误进行审核，确认并接受后，您将获赠异步社区的 100 积分。积分可用于在异步社区兑换优惠券、样书或奖品。

扫码关注本书

扫描下方二维码,您将会在异步社区微信服务号中看到本书信息及相关的服务提示。

与我们联系

我们的联系邮箱是 fudaokun@ptpress.com.cn。

如果您对本书有任何疑问或建议,请您发邮件给我们,并请在邮件标题中注明本书书名,以便我们更高效地做出反馈。

如果您有兴趣出版图书、录制教学视频,或者参与图书翻译、技术审校等工作,可以发邮件给我们。

如果您所在的学校、培训机构或企业,想批量购买本书或异步社区出版的其他图书,也可以发邮件给我们。

如果您在网上发现有针对异步社区出品图书的各种形式的盗版行为,包括对图书全部或部分内容的非授权传播,请您将怀疑有侵权行为的链接发邮件给我们。您的这一举动是对作者权益的保护,也是我们持续为您提供有价值的内容的动力之源。

关于异步社区和异步图书

"异步社区" 是人民邮电出版社旗下 IT 专业图书社区,致力于出版精品 IT 技术图书和相关学习产品,为作译者提供优质出版服务。异步社区创办于 2015 年 8 月,提供大量精品 IT 技术图书和电子书,以及高品质技术文章和视频课程。更多详情请访问异步社区官网 https://www.epubit.com。

"异步图书" 是由异步社区编辑团队策划出版的精品 IT 专业图书的品牌,依托于人民邮电出版社近 30 年的计算机图书出版积累和专业编辑团队,相关图书在封面上印有异步图书的 LOGO。异步图书的出版领域包括软件开发、大数据、AI、测试、前端、网络技术等。

异步社区

微信服务号

目录

第1章 线程机制 ··············· 1
- 1.1 线程是什么 ··············· 1
- 1.2 线程的映射 ··············· 3
  - 1.2.1 多对一映射 ··············· 4
  - 1.2.2 一对一映射 ··············· 4
  - 1.2.3 多对多映射 ··············· 5
  - 1.2.4 Java 层到内核层 ··············· 6
- 1.3 Java 线程的状态 ··············· 7
- 1.4 Java 线程的调度 ··············· 10
- 1.5 Java 线程的优先级与执行机制 ····· 11
- 1.6 Java 线程的 CPU 时间 ··············· 14
- 1.7 Java 线程的 yield 操作 ··············· 15
- 1.8 Java 线程的 sleep 操作 ··············· 17
- 1.9 Java 线程的 Interrupt 操作 ········· 20
  - 1.9.1 可运行状态的中断 ··············· 23
  - 1.9.2 阻塞/等待状态的中断 ········ 23
  - 1.9.3 经典中断实现方式 ··············· 24
  - 1.9.4 park 的特殊中断 ··············· 26
- 1.10 Java 线程的阻塞与唤醒 ··············· 26
- 1.11 Java 线程的 join 操作 ··············· 34

第2章 线程 I/O 模型 ··············· 38
- 2.1 线程与阻塞 I/O ··············· 38
  - 2.1.1 单线程阻塞 I/O 模型 ··············· 39
  - 2.1.2 多线程阻塞 I/O 模型 ··············· 40
- 2.2 线程与非阻塞 I/O 模型 ··············· 41
  - 2.2.1 应用层 I/O 多路复用 ··············· 43
  - 2.2.2 内核 I/O 多路复用 ··············· 43
  - 2.2.3 内核回调事件驱动 I/O ········· 44
- 2.3 Java 多线程非阻塞 I/O 模型 ········· 46
- 2.4 多线程带来了什么 ··············· 48
  - 2.4.1 提升执行效率 ··············· 49
  - 2.4.2 提升用户体验 ··············· 50
  - 2.4.3 让编码更难 ··············· 50
  - 2.4.4 资源开销与上下文切换开销 ··············· 51

第3章 Java 内存模型 ··············· 53
- 3.1 计算机的运行 ··············· 53
- 3.2 Java 内存模型 ··············· 56
- 3.3 volatile 能否保证线程安全 ········· 59
- 3.4 happens-before 原则 ··············· 62
  - 3.4.1 单线程原则 ··············· 65
  - 3.4.2 锁原则 ··············· 66
  - 3.4.3 volatile 原则 ··············· 67
  - 3.4.4 线程 start 原则 ··············· 67
  - 3.4.5 线程 join 原则 ··············· 68
  - 3.4.6 线程 interrupt 原则 ··············· 68
  - 3.4.7 finalize 原则 ··············· 69
  - 3.4.8 传递原则 ··············· 70
- 3.5 Java 指令重排 ··············· 70

第4章 并发知识 ··············· 75
- 4.1 synchronized 互斥锁 ··············· 75
  - 4.1.1 作用在对象方法上 ··············· 76
  - 4.1.2 作用在类静态方法上 ··············· 79
  - 4.1.3 作用在对象方法里面 ··············· 80
  - 4.1.4 作用在类静态方法里面 ··············· 81
- 4.2 乐观的并发策略 ··············· 85
- 4.3 自旋锁 ··············· 87

4.3.1　UMA 架构与 NUMA 架构 ·········· 88
4.3.2　原始自旋锁 ·········· 90
4.3.3　排队自旋锁 ·········· 91
4.3.4　CLH 锁 ·········· 93
4.3.5　MCS 锁 ·········· 95
4.4　线程饥饿 ·········· 97
4.4.1　synchronized 饥饿 ·········· 97
4.4.2　优先级饥饿 ·········· 99
4.4.3　线程自旋饥饿 ·········· 100
4.4.4　等待唤醒饥饿 ·········· 101
4.4.5　公平性解决饥饿 ·········· 102
4.5　数据竞争 ·········· 104
4.6　竞争条件 ·········· 108
4.6.1　线程执行顺序的不确定性 ·········· 110
4.6.2　并发机制 ·········· 111
4.7　死锁 ·········· 113
4.7.1　锁的顺序化 ·········· 116
4.7.2　资源合并 ·········· 116
4.7.3　避免锁嵌套 ·········· 117
4.7.4　锁超时机制 ·········· 117
4.7.5　抢占资源机制 ·········· 118
4.7.6　撤销线程机制 ·········· 118
4.7.7　死锁的检测 ·········· 119

第 5 章　AQS 同步器 ·········· 120
5.1　什么是 AQS 同步器 ·········· 120
5.2　AQS 的等待队列与状态转换 ·········· 120
5.3　AQS 的独占锁与共享锁 ·········· 125
5.4　AQS 独占锁获取与释放 ·········· 127
5.4.1　获取独占锁的逻辑 ·········· 128
5.4.2　尝试获取独占锁 ·········· 130
5.4.3　入队操作逻辑 ·········· 130
5.4.4　入队后的操作 ·········· 132
5.4.5　虚节点可能消失 ·········· 133
5.4.6　取消锁获取操作 ·········· 134
5.4.7　唤醒后继节点 ·········· 136
5.4.8　释放独占锁的逻辑 ·········· 137
5.5　AQS 共享锁获取与释放 ·········· 138
5.5.1　获取共享锁的逻辑 ·········· 138
5.5.2　入队操作 ·········· 140
5.5.3　入队后的操作 ·········· 140
5.5.4　引入 PROPAGATE 状态 ·········· 143
5.5.5　释放共享锁的逻辑 ·········· 146
5.6　AQS 的阻塞与唤醒 ·········· 146
5.6.1　许可机制 ·········· 148
5.6.2　LockSupport 示例 ·········· 149
5.6.3　park 与 unpark 的顺序 ·········· 150
5.6.4　park 对中断的响应 ·········· 151
5.6.5　park 是否会释放锁 ·········· 152
5.6.6　LockSupport 的实现 ·········· 154
5.7　AQS 的中断机制 ·········· 156
5.7.1　synchronized 不支持中断 ·········· 157
5.7.2　AQS 独占模式的中断 ·········· 158
5.7.3　AQS 共享模式的中断 ·········· 160
5.8　AQS 的超时机制 ·········· 161
5.8.1　synchronized 不支持超时 ·········· 161
5.8.2　AQS 独占模式的超时 ·········· 162
5.8.3　AQS 共享模式的超时 ·········· 163
5.9　AQS 的原子性如何保证 ·········· 164
5.10　AQS 的自旋锁 ·········· 175
5.11　AQS 的公平性 ·········· 177
5.12　AQS 的条件队列 ·········· 179
5.12.1　await 方法 ·········· 179
5.12.2　signal 方法 ·········· 182
5.13　AQS 自定义同步器 ·········· 183
5.13.1　AQS 设计思想 ·········· 183
5.13.2　独占模式 ·········· 184
5.13.3　共享模式 ·········· 186

第 6 章 常见的同步器 189
6.1 常见的同步器 189
6.2 闭锁 192
6.3 信号量 197
6.3.1 非公平模式的实现 198
6.3.2 公平模式的实现 201
6.3.3 信号量的使用示例 202
6.4 循环屏障 204
6.5 相位器 210
6.5.1 相位器的主要概念及方法 211
6.5.2 相位器的 3 个例子 212
6.5.3 相位器的状态示意图 215
6.5.4 相位器的实现原理 216
6.6 交换器 220
6.6.1 交换器的实现原理 223
6.6.2 交换器的单槽模式 226
6.6.3 交换器的多槽模式 228

第 7 章 原子类 233
7.1 原子整型 233
7.1.1 一行代码等于原子性吗 233
7.1.2 volatile 能保证原子性吗 234
7.1.3 synchronized 能解决问题吗 235
7.1.4 AtomicInteger 237
7.1.5 实现原理 238
7.2 原子引用 239
7.3 原子数组 243
7.3.1 AtomicIntegerArray 244
7.3.2 AtomicLongArray 247
7.3.3 AtomicReferenceArray 248
7.4 原子变量更新器 249

第 8 章 阻塞队列 254
8.1 阻塞队列概述 254
8.2 数组阻塞队列 258
8.3 链表阻塞队列 263
8.4 优先级阻塞队列 271
8.5 延迟阻塞队列 279
8.5.1 优先级队列 281
8.5.2 DelayQueue 的阻塞与唤醒 283
8.5.3 DelayQueue 的实现原理 284
8.6 链表阻塞的双向队列 288

第 9 章 锁 298
9.1 可重入锁 298
9.1.1 非公平模式的实现 299
9.1.2 公平模式的实现 301
9.1.3 公平模式的 3 个示例 302
9.2 读写锁 305
9.2.1 读写锁的性质 306
9.2.2 简单的实现版本 307
9.2.3 读写锁的升级与降级 308
9.2.4 读写锁的实现思想 309
9.2.5 读写锁的共用状态变量 312
9.2.6 读写锁的公平/非公平模式 313
9.2.7 写锁的实现 314
9.2.8 读锁的实现 315
9.2.9 读写锁的使用示例 317
9.3 锁的条件机制 318
9.3.1 wait/notify 模式 318
9.3.2 Condition 320

第 10 章 任务执行器 322
10.1 任务执行器接口 322
10.1.1 同步执行器 322
10.1.2 一对一执行器 323
10.1.3 线程池执行器 323
10.1.4 串行执行器 325
10.2 任务执行器的 ExecutorService 接口 326
10.3 线程池任务执行器 331

10.3.1 线程池任务执行器的运行状态 332
10.3.2 线程池任务执行器的使用示例 334
10.3.3 线程池任务执行器的实现原理 335

第 11 章 其他并发工具 344
11.1 线程本地变量 344
11.1.1 线程本地变量的使用示例 344
11.1.2 线程本地变量的 3 个主要方法 346
11.1.3 JDK 中线程本地变量的实现思想 347
11.1.4 JDK 中线程本地变量的实现源码 348
11.1.5 线程本地变量的内存泄漏 352
11.2 写时复制数组列表 353

第 12 章 C++模拟实现 Java 线程 359
12.1 模拟实现 Java 线程 359
12.2 模拟实现 yield 语义 365
12.3 模拟实现 sleep 操作 367
12.4 模拟实现 synchronized 语义 369
12.5 模拟实现 Interrupt 操作 372

第 1 章

线程机制

1.1 线程是什么

对于语言层面的线程，Java 开发人员再熟悉不过了。Java 线程类为 java.lang.Thread，当任务不能在当前线程中执行时，我们会创建一个 Thread 对象，然后启动该线程去工作。Java 有两种创建线程的方式：直接继承 java.lang.Thread 类；实现 java.lang.Runnable 接口。代码清单 1.1 简单地展示了线程的两种不同创建方式。其实这两种方式都与 Thread 类有关，它们都创建了 Thread 对象。代码清单 1.1 在运行后分别输出 3 个线程的名称：main、Thread-0 和 Thread-1。

代码清单 1.1　线程的不同创建方式

```java
 1.  public class JavaThreadTest {
 2.
 3.      public static void main(String[] args) {
 4.          System.out.println(Thread.currentThread().getName());
 5.          new MyThread().start();
 6.          new Thread(new MyThread2()).start();
 7.      }
 8.
 9.      static class MyThread extends Thread {
10.          public void run() {
11.              System.out.println(Thread.currentThread().getName());
12.          }
13.      }
14.
15.      static class MyThread2 implements Runnable {
16.          public void run() {
17.              System.out.println(Thread.currentThread().getName());
18.          }
19.      }
20.
21.  }
```

在早期的操作系统中，执行任务被抽象为进程，进程也是操作系统运行和调度的基本单元。然而随着计算机技术的不断发展，以进程为调度单元的方式逐渐产生了弊端，因为它的资源开销较大。于是人们在进程的基础上提出了线程。线程是进程里面的运行单位，可以把线程看成轻量级的进程。CPU 会按某种策略为每个线程分配一定的时间片去执行。比如图 1.1 中有 3 个线程，它们分别定义了 3 个执行任务，CPU 会轮着去执行这 3 个线程。

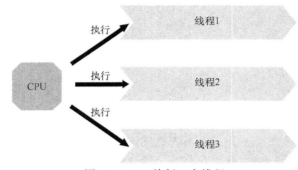

图 1.1　CPU 执行 3 个线程

进程是指程序的一次动态执行过程，计算机中正在执行的程序就是进程，每个程序都会各自对应着一个进程。一个进程包含了从代码加载到执行完成的一个完整过程，它是操作系统中资源分配的最小单元。

线程则是比进程更小的执行单位，是 CPU 调度和分配的基本单位。每个进程至少有一个线程，而一个线程却只能属于一个进程。线程可以对所属进程的所有资源进行调度和运算。线程既可以由操作系统内核来控制调度，也可以由用户程序来控制调度。

在图 1.2 中，假设某台计算机有 4 个 CPU，那么这 4 个 CPU 将按照某种策略去执行进程 1、进程 2 和进程 3。可以看到每个进程都包含至少一个线程，其中进程 1 包含 4 个线程，进程 2 和进程 3 只包含 1 个线程。此外，每个进程都有自己的资源，进程内的所有线程都共享进程所包含的资源。

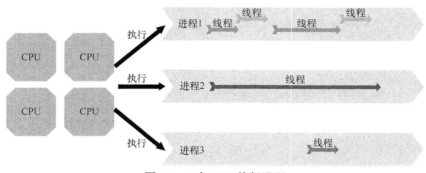

图 1.2　4 个 CPU 执行进程

1.2 线程的映射

现代计算机大体可以分为硬件和软件两大块。硬件是基础，而软件则是运行在硬件之上的程序。其中软件又可以分为操作系统和应用程序：操作系统专注于对硬件的交互管理并提供一个运行环境给应用程序使用；应用程序则是能实现若干功能且运行在操作系统环境中的软件。现代计算机的结构如图 1.3 所示。

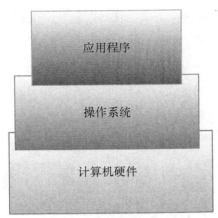

图 1.3 现代计算机的结构

Java 语言编译后的字节码运行在 JVM（Java 虚拟机）上，而 JVM 其实又是一个进程，所以 Java 属于应用程序层。我们在 Java 层通过 new 关键词创建一个 Thread 对象，然后调用 start 方法启动该线程，那么从线程角度来看其实就涉及 Java 层线程、JVM 层线程、操作系统层线程。JVM 主要由 C/C++实现，所以 Java 层线程终究还是要映射到 JVM 层线程，而 Java 层线程到操作系统层线程的映射就要看 JVM 的具体实现了。

线程按照操作系统和应用程序两个层次可以分为内核线程（Kernel Thread）和用户线程（User Thread）。所谓内核线程，就是直接由操作系统内核支持和管理的线程，线程的建立、启动、同步、销毁、切换等操作都由内核完成。而用户线程则是线程的管理工作在用户空间完成，它完全建立在用户空间的线程库上，由内核提供支持但不由内核管理，内核也无法感知到用户线程的存在。用户线程的建立、启动、同步、销毁、切换都在用户空间完成，无须切换到内核。

我们可以将用户线程看成更高层面的线程，而内核线程则向用户线程提供支持。这样一来，用户线程与内核线程之间必然存在着一定的映射关系，不同的操作系统可能采取不同的映射方式，一般包括多对一映射（用户级方式）、一对一映射（内核级方式）和多对多映射（组合方式）这 3 种映射方式。

1.2.1 多对一映射

多对一映射就是多个用户线程被映射到一个内核线程上。每个进程都对应着一个内核线程，进程内的所有线程也都对应着该内核线程。在图 1.4 中，UT 表示用户线程，KT 表示内核线程。处于用户空间的进程 1 包含 3 个用户线程，而进程 2 只包含 1 个用户线程，这些用户线程由库调度器进行调度。内核空间一共有 4 个内核线程，内核线程的调度由操作系统调度器来完成。进程 1 包含的 3 个用户线程都映射到一个内核线程上，这 3 个用户线程按一定策略轮流执行，具体的调度算法由库调度器完成，任意时刻每个进程中都只有一个用户线程被执行。

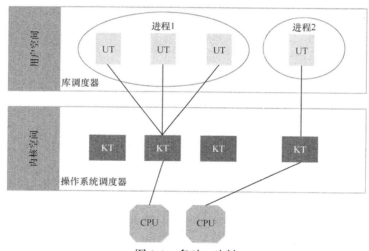

图 1.4 多对一映射

多对一映射可在不支持线程的操作系统中由库来实现线程机制，用户线程创建、销毁、切换的代价比内核线程的小。但它也存在一个较大的风险，那就是进程内的某个线程发生系统阻塞时将导致该进程中的所有线程都被阻塞。

1.2.2 一对一映射

一对一映射就是每个用户线程都被映射到一个内核线程上，用户线程的整个生命周期都绑定到所映射的内核线程上。在图 1.5 中，用户空间中的进程 1 包含 3 个用户线程，而进程 2 只包含 1 个用户线程，这些用户线程由库调度器进行调度。内核空间一共有 4 个内核线程，它们都与用户线程一一对应，内核线程的调度由操作系统调度器来完成。

在这种映射方式下，多个 CPU 能够并行执行同一个进程内的多个线程。如果进程内的某个线程被阻塞，就可以切换到该进程的其他线程继续执行，并且还能够切换执行其他进程的线程。但是，因为每个用户线程都需要对应一个内核线程，所以内核开销比多对一映射方式大。

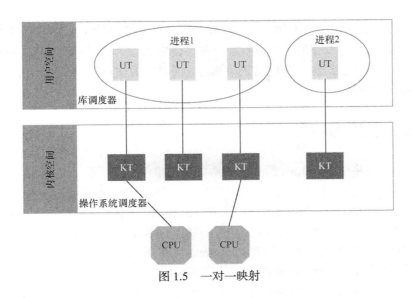

图 1.5　一对一映射

1.2.3　多对多映射

多对多映射也称为组合方式，它是将多对一和一对一两种方式组合起来，通过综合两者的优点所形成的一种方式。该方式在用户空间创建、销毁、切换、调度线程，但进程中的多个用户线程会被映射到若干个内核线程上。在图 1.6 中，用户空间包含进程 1 和进程 2，进程 1 中的 3 个用户线程不再分别对应一个内核线程或者全部都对应同一个内核线程，而是 2 个用户线程对应一个内核线程，剩下的 1 个用户线程对应另一个内核线程。这也就形成了三对二的映射，此即多对多映射。

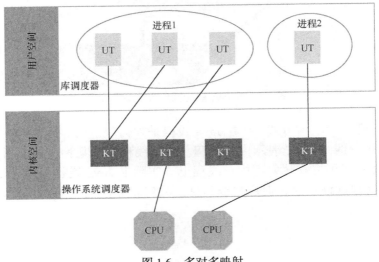

图 1.6　多对多映射

多对多映射方式综合了多对一映射和一对一映射的优点,每个内核线程负责与之绑定的若干用户线程,进程中某个线程发生系统阻塞时并不会导致整个进程阻塞,而是阻塞该内核线程所对应的若干用户线程,其他线程仍然正常执行。同时因为用户线程数量比内核数量多,所以也能有效减少内核开销。

1.2.4　Java 层到内核层

根据图 1.7,我们将线程从 Java 层到操作系统层的整个脉络串起来,从中可以看到整个 Java 及 JVM 都处于用户空间。Java 层通过 new 创建 java.lang.Thread 对象,该对象实际上在 JVM 中被封装成 JavaThread(C++)对象。而 JavaThread 对象又会关联到一个 OSThread 对象,后者是对操作系统内核线程的抽象表示,它统一维护了操作系统线程的句柄,包括 Linux、Windows 和其他操作系统的线程。这里以 Linux 操作系统为例,OSThread 对象其实是使用 pthread 库来创建线程,至于 pthread 库创建的线程属于用户线程还是内核线程,就得看操作系统和库的实现了。Linux 内核的早期版本不支持线程,而是由库实现线程,后期版本则提供了内核线程。Java 层到内核层的整体流程大致为 java.lang.Thread → JavaThread → OSThread → pthread → 内核线程。

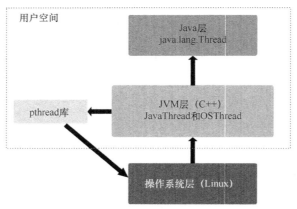

图 1.7　从 Java 层到操作系统层的线程

我们重新回到 3 种映射方式,看看 Java 线程是如何映射的。在图 1.8 中,JT 表示 Java 线程,UT 表示用户线程,KT 表示内核线程。Java 线程会以某种方式映射到用户线程上,而用户线程又会以某种方式映射到内核线程上。实际上 Java 线程与用户线程是一一对应的,JVM 中会处理这种映射,但用户线程到内核线程之间的映射则由具体的库和操作系统决定。

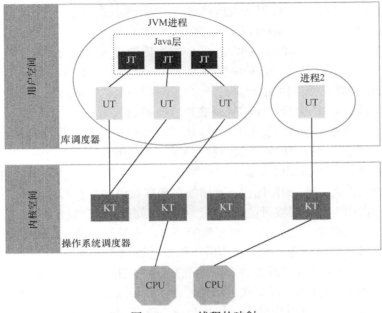

图 1.8 Java 线程的映射

1.3 Java 线程的状态

与人类一样，线程也拥有自己的生命周期，一条线程从创建到死亡的过程就是线程的生命周期。在整个生命周期内，线程在不同时刻可能处于不同的状态。有些线程的任务简单，涉及的状态就少。而有些线程的任务复杂，涉及的状态就多。那么线程到底有多少种状态，不同状态之间又是如何转化的呢？

对于线程状态的分类实际上并没有严格的规定，只要能正确表示状态即可。我们来看一种常见的状态分类，如图 1.9 所示。一个线程从创建到死亡期间可能会经历若干个状态，但在任意一个时间点上线程只能处于其中一种状态。线程总共包含 5 个状态：新建（new）、可运行（runnable）、运行（running）、不可运行（not runnable）、死亡（dead）。

图 1.9 常见的线程状态分类

- **新建状态**：一个线程被创建但未被启动，则处于新建状态。在程序中使用 new MyThread() 创建的线程实例在调用 start() 方法之前都处于此状态。
- **可运行状态**：创建的线程实例调用 start() 方法后便进入可运行状态。处于此状态的线程并不是说处于运行状态，只是说它是可以运行的。可运行线程会被加入到队列中等待 CPU 的执行时间。我们可以想象有一个可运行线程池，start() 方法把线程放进可运行线程池中，而后 CPU 会按一定的规则去执行池里的线程。
- **运行状态**：当可运行线程获取到 CPU 执行时间后则进入了运行状态。
- **不可运行状态**：运行中的线程暂时放弃 CPU 的使用权，可能是因为执行了挂起、睡眠或等待等操作。比如在执行 I/O 操作时，外部设备的速度远低于处理器速度，就会导致线程暂时放弃 CPU 的使用权。又比如在获取对象同步锁的过程中，如果同步锁先被别的线程占用，那么同样可能会导致线程暂时放弃 CPU 的使用权。
- **死亡状态**：线程执行完 run() 方法实现的任务，或因异常情况导致停止执行并退出，均处于死亡状态。线程进入死亡状态后将无法再转换成其他状态。

Java 线程的某些方法会导致线程状态发生改变，例如 Thread 类的 start、stop、sleep、suspend、resume、wait 和 notify 等方法。需要注意的是，stop、suspend 和 resume 等方法因为容易引起死锁问题而已被弃用。Java 将线程分为 6 种状态：NEW、RUNNABLE、BLOCKED、WAITING、TIMED_WAITING 和 TERMINATED。

```
1.    public enum State {
2.        NEW, RUNNABLE, BLOCKED, WAITING, TIMED_WAITING, TERMINATED;
3.    }
```

图 1.10 所示为 Java 线程的相关状态以及它们之间的转换。其实主要的几个状态与前面介绍的 5 个状态是相同的，不同的地方在于 Java 将不可运行（not runnable）状态进行了进一步的细分，引申出了 3 个状态：阻塞（BLOCKED）、等待（WAITING）、计时等待（TIMED_WAITING）。需要注意的是，图 1.10 中的 RUNNABLE 和 RUNNING 两个状态实际上都被 Java 归类到 RUNNABLE 状态中，但为了说明 CPU 分配和 yield 操作，这里将它们分开了。

Java 线程创建后处于 NEW 状态，调用 start() 方法后变为 RUNNABLE 状态，即表示该线程已经就绪，可以由调度器调度运行了。这里先看 RUNNING 状态，线程在分配到 CPU 执行时间后就处于真正的运行状态（RUNNING），但当 CPU 时间使用完后或者线程调用 yield() 方法后，都将使线程重新进入到 RUNNABLE 状态以等待下一次的调度。注意，RUNNABLE 和 RUNNING 状态其实是无法从 Java 层获知的，这里分开讲解是为了让大家能更好地理解线程的调度概念。

如果 RUNNING 状态的线程遇到同步（synchronized）块且其锁已经被其他线程获得，那么此时该线程就会进入 BLOCKED 状态，处于该状态的线程将暂停运行且放弃参与 CPU 时间片的分配。处于该状态的线程会被扔到一个等待队列中，当持有锁的线程释放锁后，队列中的线程再继续竞争锁，得到锁的线程则变为 RUNNABLE 状态，并参与 CPU 时间片的分配。

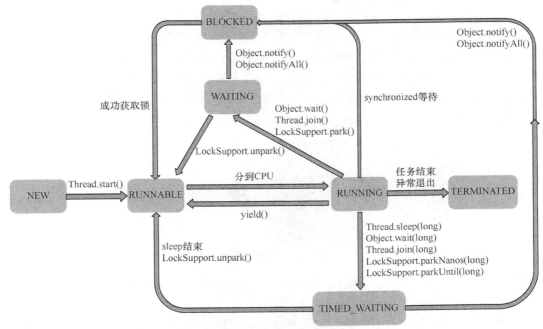

图 1.10　线程状态的转换

如果 RUNNING 状态的线程调用了 Object.wait()、Thread.join()和 LockSupport.park()，就会进入 WAITING 状态。如果被 park 的线程被其他线程执行 LockSupport.unpark()，就会被唤醒，并直接进入到 RUNNABLE 状态。这里重点关注 Object.wait()，它与 Object.notify()一起使用，且都是在 synchronized 块中，所以通过 Object.notify()和 Object.notifyAll()唤醒的线程还会进入 BLOCKED 状态重新参与锁竞争。一个线程执行了 wait()方法后将进入所指定对象的等待队列中，同时它还会释放对象锁。只有其他线程调用 notify()或 notifyAll()方法后才会唤醒等待队列中的线程。一般认为 notify()是随机唤醒等待队列中的一个线程，而 notifyAll()则是唤醒等待队列中的所有线程，被唤醒的线程将重新进行锁竞争。

如果处于 RUNNING 状态的线程调用了拥有超时机制的等待方法，则会进入 TIMED_WAITING 状态，比如 Thread.sleep(long)、Object.wait(long)、Thread.join(long)、LockSupport.parkNanos(long) 和 LockSupport.parkUntil(long)等。如果 sleep()结束或者调用 LockSupport.unpark()，则会使相关线程进入 RUNNABLE 状态；而如果被 Object.notify()或 Object.notifyAll()唤醒，则会进入 BLOCKED 状态重新参与锁竞争。

最后，正常执行完毕或因异常退出的线程会进入 TERMINATED 状态。

1.4　Java 线程的调度

在 Java 多线程环境中，为了保证所有线程都能按照一定的策略执行，JVM 需要有一个线程调度器，如图 1.11 所示。这个调度器定义了线程调度的策略，通过特定的机制为多个线程分配 CPU 的使用权。线程调度器中一般包含多种调度策略算法，由这些算法来决定 CPU 的分配。此外，每个线程还有自己的优先级（比如有高、中、低级别），调度算法会通过这些优先级来实现优先机制。

图 1.11　线程调度器

这里不深入讨论调度的具体算法，主要关注两种常见的线程调度模式：抢占式调度和协同式调度。

- **抢占式调度**：每个线程的执行时间和线程的切换都由调度器控制，调度器按照某种策略为每个线程分配执行时间。调度器可能会为每个线程都分配相同的执行时间，也可能为某些特定线程分配较长的执行时间，甚至在极端情况下还可能不给某些线程分配执行时间片，从而导致某些线程得不到执行。在抢占式调度机制下，一个线程的堵塞不会导致整个进程堵塞。
- **协同式调度**：某一线程执行完后会主动通知调度器切换到下一个线程上执行。这种调度模式就像接力赛一样，一个人跑完自己的路程后就把接力棒交接给下一个人，下一个人继续往下跑。在这种模式下，线程的执行时间由线程本身控制，也就是说线程的切换点是可以预先知道的。但是它有一个致命弱点，即如果某一个线程的逻辑存在问题，则可能导致系统运行到一半就一直阻塞了，最终可能导致整个系统崩溃。

我们通过图 1.12 来理解这两种模式。在图 1.12 中，左边为抢占式线程调度，现在假如存在 3 个待运行的线程，在抢占式调度下处理器的执行路径为先在线程 1 执行一段时间，然后强制切换到线程 2 执行一段时间，最后再切换到线程 3 执行一段时间。接着下一轮又重新回到线程 1；如此循环往复，直至 3 个线程都执行完。而在图 1.12 右边的协同式线程调度下，执行的策略则不同，调度器会先将执行时间分配给线程 1 并且一次性执行完。然后线程 1 再通知线程 2，线程 2 也一次性执行完。最后线程 2 再通知线程 3，线程 3 同样一次性执行完。

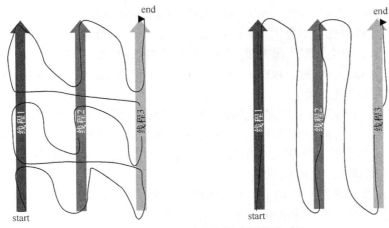

图 1.12　抢占式调度和协同式调度的对比

在了解了两种线程调度模式后,再来看看 Java 使用的是哪种线程调度模式。Java 线程使用的是抢占式调度。Java 的线程调度涉及 JVM 的实现,JVM 规范中规定每个线程都有各自的优先级,且优先级越高,越优先执行。但优先级高并不代表能独自占用执行时间,可能是优先级越高得到的执行时间越多。反之,优先级低的线程所分到的执行时间少,但不会不分配执行时间。

1.5　Java 线程的优先级与执行机制

Java 线程的调度机制由 JVM 实现。假如有若干个线程,我们想让一些线程拥有更多的执行时间或者少分配点执行时间,那么就可以通过设置线程的优先级来实现。所有处于可执行状态的线程都在一个队列中,且每个线程都有自己的优先级,JVM 线程调度器会根据优先级来决定每次的执行时间和执行频率。

但是,优先级高的线程一定会先执行吗?我们能否在 Java 程序中通过优先级值的大小来控制线程的执行顺序呢?答案是不能。这是因为影响线程优先级语义的因素有很多,具体如下:
- 不同版本的操作系统和 JVM 都可能会产生不同的行为;
- 优先级对于不同的操作系统调度器来说可能有不同的语义;
- 有些操作系统的调度器不支持优先级;
- 对于操作系统来说,线程的优先级存在"全局"和"本地"之分,不同进程的优先级一般相互独立;
- 不同的操作系统对优先级定义的值不一样,Java 只定义了 1~10;
- 操作系统常常会对长时间得不到运行的线程给予增加一定的优先级;
- 操作系统的线程调度器可能会在线程发生等待时有一定的临时优先级调整策略。

JVM 线程调度器的调度策略决定了上层多线程的运行机制,每个线程执行的时间都由它分配管理。调度器将按照线程优先级对线程的执行时间进行分配，优先级越高得到的 CPU 执行时间越长，执行频率也可能更大。Java 把线程优先级分为 10 个级别，线程在创建时如果没有明确声明优先级，则使用默认优先级。Java 定义了 Thread.MIN_PRIORITY、Thread.NORM_PRIORITY 和 Thread.MAX_PRIORITY 这 3 个常量，分别代表最小优先级值（1）、默认优先级值（5）和最大优先级值（10）。

此外，由于 JVM 的实现是以宿主操作系统为基础的，所以 Java 各优先级与不同操作系统的原生线程优先级必然存在着某种映射关系，这样才能够封装所有操作系统的优先级来提供统一的优先级语义。例如优先级 1～10 在 Linux 中可能要与–20～19 之间的优先级值进行映射，而 Windows 系统则有 9 个优先级要映射。

代码清单 1.2 是一个简单的例子，里面创建了两个线程并设置了不同的优先级值。每次运行的结果可能都不相同，比如输出可能为 110000000000011111111，所以并非优先级高就先执行。

代码清单 1.2　创建两个具有不同优先级的线程

```java
1.  public class ThreadPriorityTest {
2.
3.      public static void main(String[] args) {
4.          Thread t = new MyThread();
5.          t.setPriority(10);
6.          t.setName("00");
7.          Thread t2 = new MyThread();
8.          t2.setPriority(8);
9.          t2.setName("11");
10.         t2.start();
11.         t.start();
12.     }
13.
14.     static class MyThread extends Thread {
15.         public void run() {
16.             for (int i = 0; i < 5; i++)
17.                 System.out.print(this.getName());
18.         }
19.     }
20. }
```

为了更好更深入地理解线程的优先级，下面继续从 JVM 源码的角度来对优先级的处理进行讲解。我们先从 Java 层对优先级的设置与获取开始，然后一层层往下看 JVM 的处理。对于 Java 层，其实操作很简单，getPriority 和 setPriority 两个方法即可以完成任务，如代码 1.2 所示。在执行代码清单 1.3 后，输出分别为 5 和 1，其中 5 是线程默认优先级的值。

代码清单 1.3　设置与获取线程的优先级

```java
1.  public class ThreadPriorityTest2 {
2.      public static void main(String[] args) {
3.          Thread t = Thread.currentThread();
```

```
4.          System.out.println(t.getPriority());
5.          Thread.currentThread().setPriority(1);
6.          System.out.println(t.getPriority());
7.      }
8.  }
```

Java 层的优先级值需要转换成操作系统的优先级值，这中间存在一个映射操作。JVM 通过 java_to_os_priority 数组进行映射，这个数组一共有 12 个元素。下面看一下 Linux 和 Windows 操作系统各自的优先级值。

Java 层的 1 和 10 分别对应 Linux 系统的 4 和 -5，Linux 系统不同版本的线程优先级值的范围可能会不一样，一般认为是 -20～19，其中 -20 为最高优先级，19 为最低优先级，Java 则使用了其中的 4～-5 来映射 1～10 的优先级，如代码清单 1.4 所示。

代码清单 1.4　优先级在 Linux 系统上的显示

```
1.  int os::java_to_os_priority[CriticalPriority + 1] = {
2.    19,           // 0 Entry should never be used
3.    4,            // 1 MinPriority
4.    3,            // 2
5.    2,            // 3
6.    1,            // 4
7.    0,            // 5 NormPriority
8.    -1,           // 6
9.    -2,           // 7
10.   -3,           // 8
11.   -4,           // 9 NearMaxPriority
12.   -5,           // 10 MaxPriority
13.   -5            // 11 CriticalPriority
14. };
```

对于 Windows 系统，Java 层的 1 和 2 都映射到 THREAD_PRIORITY_LOWEST，其他优先级值也都进行类似的映射，如代码清单 1.5 所示。需要注意的是，Windows 系统实际包含的优先级不止下面这些，Java 层并没有全部映射。

代码清单 1.5　优先级在 Windows 系统上的显示

```
1.  int os::java_to_os_priority[CriticalPriority + 1] = {
2.    THREAD_PRIORITY_IDLE,             // 0 Entry should never be used
3.    THREAD_PRIORITY_LOWEST,           // 1 MinPriority
4.    THREAD_PRIORITY_LOWEST,           // 2
5.    THREAD_PRIORITY_BELOW_NORMAL,     // 3
6.    THREAD_PRIORITY_BELOW_NORMAL,     // 4
7.    THREAD_PRIORITY_NORMAL,           // 5 NormPriority
8.    THREAD_PRIORITY_NORMAL,           // 6
9.    THREAD_PRIORITY_ABOVE_NORMAL,     // 7
10.   THREAD_PRIORITY_ABOVE_NORMAL,     // 8
11.   THREAD_PRIORITY_HIGHEST,          // 9 NearMaxPriority
12.   THREAD_PRIORITY_HIGHEST,          // 10 MaxPriority
```

```
13.    THREAD_PRIORITY_HIGHEST                         // 11 CriticalPriority
14. };
```

1.6　Java 线程的 CPU 时间

　　Java 线程的执行由 JVM 进行管理，每个线程在从启动到结束的过程中都可能经历多种状态。多个线程执行则意味着线程的并发和并行，也就涉及 CPU 的执行时间。图 1.13 是 3 个线程分配到的 CPU 执行时间的示意图，3 个线程除了真正执行阶段，从启动到结束的过程中还包含了等待阶段。

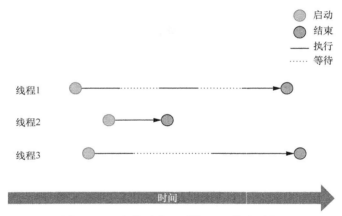

图 1.13　3 个线程分配到的 CPU 执行时间

　　在一个线程从启动到结束的过程中，有两个时间概念需要理解：线程的真正执行时间和总消耗时间。总消耗时间等于真正执行时间与等待时间的和。在图 1.14 中可以很清晰地看到这两者之间的关系，T1 得到了多个 CPU 执行时间，即真正的执行时间，而 T1 的总消耗时间还包括 T1 执行期间 CPU 分配给其他线程执行的时间，所以总消耗时间总是大于等于真正执行时间。

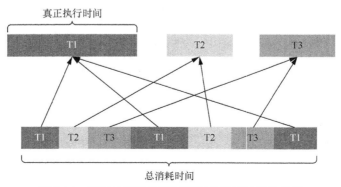

图 1.14　线程的真正执行时间和总消耗时间的关系

在 Java 中，线程会按其优先级来分配 CPU 时间，那么在执行过程中线程何时会放弃 CPU 的使用权呢？其实可以归类成以下 3 种情况。

- 线程死亡，即线程运行结束，也就是运行完 run() 方法中的任务后整个线程的生命周期结束。在这种情况下，任务执行完了自然就要放弃 CPU 的使用权。
- 线程主动放弃 CPU。需要注意的是，基于时间片轮转调度的操作系统不会让线程永久放弃 CPU，也就是说只是放弃本轮 CPU 时间片的执行权。比如在调用 yield() 方法时，线程将放弃参与当前 CPU 时间片的分配。
- 因等待放弃 CPU，指线程因为进入阻塞等待状态，从而放弃 CPU 执行时间。进入等待状态的原因可能有很多种，比如磁盘 I/O、网络 I/O、主动睡眠、锁竞争和执行等待等。

1.7 Java 线程的 yield 操作

实际上，yield 操作语义在不同操作系统中的实现机制可能不一样。我们以基于时间片的线程调度器为例来理解 yield 的语义。在讨论 yield 之前，必须先了解时间片这个概念。时间片是指在抢占式操作系统中线程或进程在被抢占前所能持续运行的时间。通俗地说就是每个线程轮流分配到的 CPU 执行时间。每个线程或进程都执行一定的时间，然后再切换到另一个线程或进程去执行一定的时间。在图 1.15 中，假设只有一个 CPU 并发地执行 3 个线程任务，线程 1 先执行一会儿，然后轮到线程 2，然后再到线程 3，这样不断继续下去。时间片的长短由操作系统控制，它也可能与线程优先级相关。时间片太长会导致并发效果差，影响系统的交互表现。反之，又会带来大量的切换成本，从而浪费 CPU，导致真正用在线程或进程的执行时间变少。

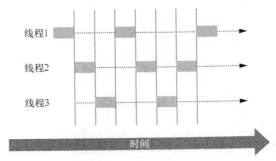

图 1.15　CPU 并发执行 3 个线程

操作系统通过时间片的机制来执行线程，一般会通过一个就绪队列来维护待执行的线程，新创建的线程就绪后会加入到就绪队列中，然后 CPU 会依次执行就绪队列中的线程。图 1.16 是一个常规的流程，原来就绪队列有 4 个线程，当线程 5 创建后被加入到就绪队列中，然后就绪队列中一共有 5 个待执行线程。CPU 将队列头部的线程 1 拿出来执行一个完整的时间片，然后将线程 1 放到队列尾部，轮到线程 2 执行；以此类推，不断循环下去。

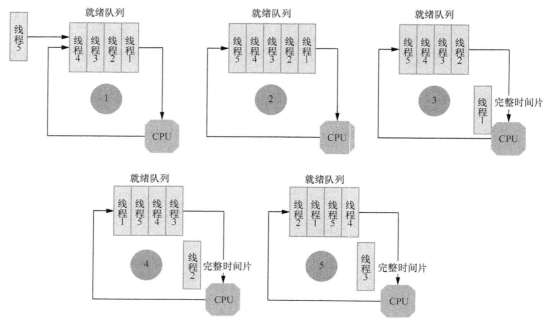

图 1.16　线程的常规执行流程

实际上,除了完整地使用完时间片之外,还可以明确指定放弃时间片。放弃时间片就意味着将 CPU 的执行时间让给其他线程,放弃时间片的线程只有在下一轮才能再次分配到时间片。这个放弃 CPU 时间片的操作称为 yield,Java 中的线程对象提供了 yield()方法。在图 1.17 中可以看到,线程 1 执行了完整的时间片,而线程 2 则直接放弃时间片,此时线程 2 被加入到队列尾部,而线程 3 则继续执行。

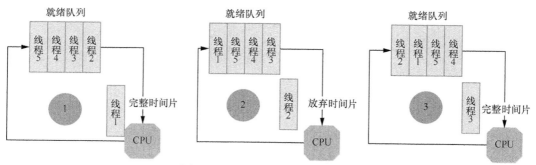

图 1.17　放弃时间片的 yield 操作

yield 操作的示例如代码清单 1.6 所示。其中 MyThread 线程定义的任务是让出自己的 CPU 时间,该例子会输出 100 次"主线程",但只有额外的一两次"让出线程 CPU 时间"(注意,我们只关注主线程输出"主线程"从 0 次到 100 次之间的这个过程)。整个过程可以理解为:在运行该类后,主线程会创建一个 MyThread 对象并启动它,然后就一直输出"主线程",而 MyThread 对象作为线程也会分到 CPU 时间片,它每次分到时间片后就输出"让出线程 CPU

时间"并放弃该轮 CPU 的使用权。

代码清单 1.6　yield 操作示例

```
1.   public class YieldThreadTest {
2.       public static void main(String[] args) {
3.           MyThread mt = new MyThread();
4.           mt.setDaemon(true);
5.           mt.start();
6.           for (int i = 0; i < 100; i++) {
7.               System.out.println("主线程");
8.           }
9.       }
10.
11.      static class MyThread extends Thread {
12.          public void run() {
13.              while (true) {
14.                  System.out.println("让出线程CPU时间");
15.                  Thread.currentThread().yield();
16.              }
17.          }
18.      }
19.  }
```

Java 中 yield 操作的具体语义取决于 JVM 的实现，而 JVM 的实现又依赖于不同的操作系统。所以不同的操作系统语义可能不相同，即使相同操作系统的不同版本也可能不同。我们没有必要去深究每种操作系统、每个版本的线程调度算法的实现，而是可以通过经典的基于时间片的调度策略来理解 yield 操作的含义。实际上，现代操作系统已经发展出很多不同的调度算法，实现更加复杂，考虑的因素也更多。

1.8　Java 线程的 sleep 操作

sleep 操作是一个经常使用的操作，特别是在开发环境中进行调试的时候，我们为了模拟长时间的执行而使用 sleep。该操作对应着 java.lang.Thread 类的 sleep 本地方法，它能使当前线程睡眠指定的时间，如图 1.18 所示。

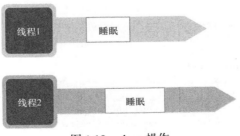

图 1.18　sleep 操作

sleep 操作的简单示例如代码清单 1.7 所示。这里让主线程睡眠 3 000ms。主线程先输出"当前线程睡眠 3 000ms",然后暂停 3s,最后再输出"睡眠结束"。

代码清单 1.7　sleep 操作示例

```
1.   public class TestSleep2 {
2.
3.       public static void main(String[] args) {
4.           System.out.println("当前线程睡眠 3000ms");
5.           try {
6.               Thread.sleep(3000);
7.           } catch (InterruptedException e) {
8.               e.printStackTrace();
9.           }
10.          System.out.println("睡眠结束");
11.      }
12.
13.  }
```

在使用 sleep 方法时,应注意以下事项。
- 该方法只针对当前线程,即让当前线程进入休眠状态。也就是说,哪个线程调用 Thread.sleep,则哪个线程睡眠。
- sleep 方法传入的睡眠时间并不精准,这取决于操作系统的计时器和调度器。
- 如果在 synchronized 块内进行 sleep 操作,或在已获得锁的线程中执行 sleep 操作,都不会让线程失去锁,这点与 Object.wait 方法不同。
- 当前线程执行 sleep 操作进入睡眠状态后,其他线程能够中断当前线程,使其解除睡眠状态并抛出 InterruptedException 异常。

前面提到,sleep 操作不会使得当前线程释放锁,现在我们看代码清单 1.8。该示例通过 synchronized 同步块来实现锁机制。主线程创建 thread1 和 thread2 两个线程,其中 thread1 比 thread2 先启动。thread1 获取锁后输出"thread1 gets the lock."并开始睡眠 3s。接着 thread2 启动,但它会因为获取不到锁而阻塞,直到 thread1 睡眠结束输出"thread1 releases the lock."并释放锁,thread2 才能获得锁往下执行,输出"thread2 gets the lock 3 second later."。

代码清单 1.8　通过 synchronized 同步块实现锁机制

```
1.   public class TestSleep3 {
2.       public static void main(String[] args) throws InterruptedException {
3.           Object lock = new Object();
4.           Thread thread1 = new Thread(() -> {
5.               synchronized (lock) {
6.                   System.out.println("thread1 gets the lock.");
7.                   try {
8.                       Thread.sleep(3000);
9.                   } catch (InterruptedException e) {
10.                  }
```

```
11.             System.out.println("thread1 releases the lock.");
12.         }
13.     });
14.     Thread thread2 = new Thread(() -> {
15.         synchronized (lock) {
16.             System.out.println("thread2 gets the lock 3 second later.");
17.         }
18.     });
19.     thread1.start();
20.     Thread.sleep(100);
21.     thread2.start();
22. }
23. }
```

sleep 操作支持中断机制。如果某个线程处于睡眠状态，那么其他线程就可以对其执行中断操作。在代码清单 1.9 中，主线程创建 thread1 和 thread2 两个线程，thread1 启动后输出"thread1 sleeps for 30 seconds."，然后开始睡眠 30s。主线程休眠 2s 后开始启动 thread2，它输出 "thread2 interrupts thread1." 后对 thread1 进行中断操作。最终 thread1 还没休眠够 30s 就被中断，捕获异常后输出 "thread1 is interrupted by thread2."。

代码清单 1.9　中断机制的演示

```
1.  public class TestSleep4 {
2.
3.      public static void main(String[] args) throws InterruptedException {
4.          Thread thread1 = new Thread(() -> {
5.              System.out.println("thread1 sleeps for 30 seconds.");
6.              try {
7.                  Thread.sleep(30000);
8.              } catch (InterruptedException e) {
9.                  System.out.println("thread1 is interrupted by thread2.");
10.             }
11.         });
12.         Thread thread2 = new Thread(() -> {
13.             System.out.println("thread2 interrupts thread1.");
14.             thread1.interrupt();
15.         });
16.         thread1.start();
17.         Thread.sleep(2000);
18.         thread2.start();
19.     }
20. }
```

sleep 方法传入的时间参数值必须大于等于 0，正常情况下我们都会传入大于 0 的值，但有时也会有传入 0 值。0 是不是表示不睡眠呢？如果不睡眠又为什么要执行 sleep 操作呢？实际上 sleep(0) 并非指睡眠 0s，它的意义是让出该轮 CPU 时间。也就是说它的意义与 yield 方法相同，而 JVM 在实现时也可以用 yield 操作来代替。在代码清单 1.10 中，MyThread 会不断让出自己的 CPU 时间，主线程则得到更多的执行时间，这个过程一共输出 100 次 "main thread"，却仅输出几次 "yield cpu time"（注意，我们只关注主线程输出 "main thread" 从 0 次到 100 次之间的这个过程）。

代码清单 1.10 sleep(0) 使用示例

```
1.   public class TestSleep5 {
2.       public static void main(String[] args) {
3.           MyThread mt = new MyThread();
4.           mt.setDaemon(true);
5.           mt.start();
6.           for (int i = 0; i < 100; i++) {
7.               System.out.println("main thread");
8.           }
9.       }
10.
11.      static class MyThread extends Thread {
12.          public void run() {
13.              while (true) {
14.                  System.out.println("yield cpu time");
15.                  try {
16.                      Thread.sleep(0);
17.                  } catch (InterruptedException e) {
18.                      e.printStackTrace();
19.                  }
20.              }
21.          }
22.      }
23.  }
```

本节介绍了 Java 线程中 sleep 方法的使用及要点，其中最重要的是睡眠不会释放当前线程已获得的锁，并且 sleep 方法支持中断。本节还介绍了比较少见的 sleep(0) 操作，其实它的语义与 yield 方法相同，也就是说它等同于 yield 方法。

1.9 Java 线程的 Interrupt 操作

中断（Interrupt）通常被定义为一个事件或者信号，它表示发生了某个事件需要进行关注。一个中断系统一般包括中断源、中断信号、中断控制器和中断处理器。中断源指发起中断信号的对象，中断信号就是一个信号标识，中断控制器则负责对中断信号进行管理，中断处理器负责在接收到中断信号后进行处理。

Java 的中断机制也包含了上述 4 个对象，如图 1.19 所示。其中线程 2 为中断源，线程 1 是中断处理器，它负责中断处理，中断信号是线程 1 中的一个标识位（也称为中断变量、中断位），JVM 作为中断控制器来管理中断信号及触发工作，比如 JVM 能够以抛出 InterruptedException

异常的形式来触发睡眠线程的中断。

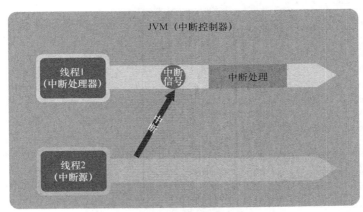

图 1.19　Java 中断机制

Java 的中断机制使用一个中断变量作为标识，假如某个线程的中断变量被标记为 true，那么该线程在适当的时机会抛出异常，我们在捕获异常后进行相应的处理。要实现 Java 中断机制，必须明白中断标识是如何检测和触发的。这并非直接在 Java 层写个 for/while 循环去不断地轮询检测，这样效率太低，而且 JVM 层一般也不执行这类的轮询检测。实际上最高效的方式就是使用硬件层提供的信号触发机制。我们可以在需要中断机制的方法中通过本地调用来实现硬件层信号触发。比如 Java 的 Thread 线程类的 sleep() 方法被定义为本地方法，可直接使用 C 库提供的信号触发来实现中断检测触发。

可以使用一个布尔类型变量来表示中断标识，那么这个中断标识应该放在哪里呢？由于中断是针对线程实例而言的，所以把标识变量放到线程中就再合适不过了。从 Java 应用开发的角度来看，我们可以自定义中断标识变量。而从 JDK 的角度看，由于线程是由 JVM 维护的，所以中断标识也可以在 JVM 层的线程中定义。

下面先看看如何在 Java 层中自定义中断标识变量，如代码清单 1.11 所示。首先定义一个线程类，接着再定义一个布尔类型变量 isInterrupted（为了保证其可见性，应该声明为 volatile），最后在 run 方法里面通过 while 循环不断检测 isInterrupted 变量，一旦该变量被设为 true，就立即停止执行。该示例中创建的新线程会一直运行，直到主线程调用 customInterrupt 方法将 isInterrupted 变量设置为 true 后才停止运行。

代码清单 1.11　在 Java 层中自定义中断标识变量

```
1.   public class InterruptThreadDemo extends Thread {
2.
3.       private volatile boolean isInterrupted = false;
4.
5.       public void customInterrupt() {
6.           isInterrupted = true;
7.       }
```

```
8.
9.      public void run() {
10.         while (!isInterrupted) {
11.             System.out.println("Thread is running....");
12.         }
13.         System.out.println("Interrupt thread.... ");
14.     }
15.
16.     public static void main(String args[]) {
17.         InterruptThreadDemo thread = new InterruptThreadDemo();
18.         thread.start();
19.         try {
20.             Thread.sleep(10);
21.         } catch (InterruptedException e) {
22.         }
23.         thread.customInterrupt();
24.     }
25. }
```

再看 JVM 提供的中断标识，如代码清单 1.12 所示。它内部也使用了一个中断标识来实现中断操作，具体的实现原理可参考第 12 章。新线程会不断检查线程的中断标识变量，如果没有中断则一直执行。而一旦主线程调用 interrupt 方法将新线程的中断标识变量设为 true，则停止执行。

代码清单 1.12　使用 JVM 提供的中断标识

```
1.  public class InterruptThreadDemo2 extends Thread {
2.
3.      public void run() {
4.          while (!Thread.interrupted()) {
5.              System.out.println("Thread is running....");
6.          }
7.          System.out.println("Interrupt thread....");
8.      }
9.
10.     public static void main(String args[]) {
11.         InterruptThreadDemo2 thread = new InterruptThreadDemo2();
12.         thread.start();
13.         try {
14.             Thread.sleep(10);
15.         } catch (InterruptedException e) {
16.         }
17.         thread.interrupt();
18.     }
19. }
```

JDK 预留了 3 个方法用于操作中断标识，如代码清单 1.13 所示。这 3 个方法依次用于设置线程为中断状态、判断线程状态是否中断、清除当前线程的中断状态并返回它之前的值。通过 interrupt() 方法设置中断标识，设置中断状态后线程并非立即停止，除非我们自己对变量进行检查处理。而如果线程正在执行 sleep()、wait()、join()等方法，则会抛出 InterruptedException 异常（这个由 JVM 实现）。需要注意的是，interrupted()不仅能让线程变成中断状态，它还会清除中断标识。

代码清单 1.13　JDK 预留的 3 个方法

```
1.  public class Thread {
2.      public void interrupt() {...}
3.      public Boolean isInterrupted() {...}
4.      public static Boolean interrupted() {...}
5.  }
```

1.9.1　可运行状态的中断

线程最常见的状态就是可运行（RUNNABLE）状态，在此状态下需要我们自己来检测中断标识，从而提供中断操作。比如在代码清单 1.14 中，通过 while (!Thread.currentThread().isInterrupted())来不断检测中断标识。当 thead1 启动后就一直输出"running..."，主线程睡眠 2s 后调用 thread1 的 interrupt()方法将其线程中的中断标识标为 true，此时 thread1 便跳出 while 循环并停止运行。

代码清单 1.14　检测中断标识

```
1.  public class RunnableInterrupt {
2.      public static void main(String[] args) {
3.          Thread thread1 = new Thread(() -> {
4.              while (!Thread.currentThread().isInterrupted()) {
5.                  System.out.println("running...");
6.              }
7.          });
8.          thread1.start();
9.          try {
10.             Thread.currentThread().sleep(2000);
11.         } catch (InterruptedException e) {
12.             e.printStackTrace();
13.         }
14.         thread1.interrupt();
15.     }
16. }
```

1.9.2　阻塞/等待状态的中断

对于处于可运行状态的线程，我们需要通过 while 的形式来检测中断标识，而对于处于阻塞或等待状态的线程，则无须自己检测中断标识，我们要做的就是捕获中断异常并进行处理。代码清单 1.15 所示为 Java 线程睡眠与等待的中断情况。thread1 和 thread2 启动后分别输出

"thread1 is running..." "thread2 is running..."，然后 thread1 睡眠 200s 而 thread2 直接进入等待状态。当两个线程都调用 interrupt()方法后，两个线程都抛出 InterruptedException 异常，捕获异常后分别输出 "thread1 has stoped!" 和 "thread2 has stoped!"。假如不对 thread2 进行中断，则 thread2 会一直等待下去。

代码清单 1.15　Java 线程睡眠与等待的中断情况

```
1.   public class WaitingInterrupt {
2.       public static void main(String[] args) {
3.           Object lock = new Object();
4.           Thread thread1 = new Thread(() -> {
5.               try {
6.                   System.out.println("thread1 is running...");
7.                   Thread.currentThread().sleep(200000);
8.               } catch (InterruptedException e) {
9.                   System.out.println("thread1 has stoped!");
10.              }
11.          });
12.          Thread thread2 = new Thread(() -> {
13.              try {
14.                  System.out.println("thread2 is running...");
15.                  synchronized (lock) {
16.                      lock.wait();
17.                  }
18.              } catch (InterruptedException e) {
19.                  System.out.println("thread2 has stoped!");
20.              }
21.          });
22.          thread1.start();
23.          thread2.start();
24.          try {
25.              Thread.currentThread().sleep(2000);
26.          } catch (InterruptedException e) {
27.              e.printStackTrace();
28.          }
29.          thread1.interrupt();
30.          thread2.interrupt();
31.      }
32.  }
```

1.9.3　经典中断实现方式

前面说到，对于可运行状态的线程我们需要自己维护中断状态标识，而对于阻塞和等待状态的线程则需要捕获中断异常并处理。通常情况下，线程既可能处于可执行状态又可能处于阻塞/等待状态，我们要如何处理呢？

在代码清单 1.16 中，通过 while(!Thread.currentThread().isInterrupted()) 来不断检测中断标

识。在这个循环中，线程除了可运行状态还会进入睡眠阻塞状态，也就是说在这两种状态下都要支持中断。睡眠操作是由 JVM 层面支持中断的，所以当 thread1 调用 interrupt()方法时，它会中断睡眠并抛出 InterruptedException 异常。需要注意的是，当捕获异常后要再次调用 Thread.currentThread().interrupt()，因为 sleep()方法被中断后，它会将中断标识清理掉，然后抛出中断异常。因此需要再次设置中断标识，以便能跳出 while 循环，结束整个 thread1 线程的执行。

代码清单 1.16　检测中断标识

```
1.   public class RunningWaitingInterrupt {
2.       public static void main(String[] args) {
3.           Thread thread1 = new Thread(() -> {
4.               while (!Thread.currentThread().isInterrupted()) {
5.                   System.out.println("running...");
6.                   try {
7.                       Thread.currentThread().sleep(200);
8.                   } catch (InterruptedException e) {
9.                       System.out.println(Thread.currentThread().isInterrupted());
10.                      Thread.currentThread().interrupt();
11.                      System.out.println("thread1 has stoped!");
12.                  }
13.              }
14.          });
15.          thread1.start();
16.          try {
17.              Thread.currentThread().sleep(2000);
18.          } catch (InterruptedException e) {
19.              e.printStackTrace();
20.          }
21.          thread1.interrupt();
22.      }
23.  }
```

执行代码清单 1.16 后，某次输出结果如下。

```
1.   running...
2.   running...
3.   running...
4.   running...
5.   running...
6.   running...
7.   running...
8.   running...
9.   false
10.  thread1 has stoped!
```

1.9.4　park 的特殊中断

实际上通过 park 方式来阻塞线程时，它的中断方式比较特殊。在这种方式阻塞下的中断并不会抛出 InterruptedException 异常，但中断标识会被设置为 true。我们来看代码清单 1.17，thread1 启动后输出 "thread1 is running..."，然后调用 park() 方法阻塞，2s 后主线程调用 thread1 的 interrupt() 方法来中断 thread1 的阻塞，此时 thread1 结束阻塞，而且 thread1 的中断标识已经被设置为 true。

代码清单 1.17　使用 park 方式进行阻塞

```
1.   public class ParkInterrupt {
2.       public static void main(String[] args) {
3.           Thread thread1 = new Thread(() -> {
4.               System.out.println("thread1 is running...");
5.               LockSupport.park();
6.           });
7.           thread1.start();
8.           try {
9.               Thread.currentThread().sleep(2000);
10.          } catch (InterruptedException e) {
11.              e.printStackTrace();
12.          }
13.          thread1.interrupt();
14.          System.out.println(thread1.isInterrupted());
15.      }
16.  }
```

1.10　Java 线程的阻塞与唤醒

线程的阻塞和唤醒在多线程并发过程中是一个关键点，当很多线程参与并发时可能会带来很多隐蔽的问题。如何正确地暂停一个线程，暂停后又如何在某个运行节点恢复运行，这些都是需要仔细思考的细节。Java 提供了多种方式来对线程进行阻塞和唤醒操作，比如 suspend 与 resume、wait 与 notify 以及 park 与 unpark 等，如图 1.20 所示。

先来看一下为什么需要阻塞和唤醒操作。简单来说，主要是为了控制线程在某些关键节点的先后执行顺序。线程之间是独立的，它们之间需要某种机制才能相互通信，Java 为此引入了共享变量。某个线程将信息写到变量中，另一个线程通过该变量读取信息，这样便能实现线程之间的通信。比如在代码清单 1.18 中，线程 1 通过共享变量 message 与线程 2 进行通信。但代码清单 1.18 并不总是正确的，因为两个线程是独立执行的，无法控制线程 2 在线程 1 读取共享变量前先对共享变量进行赋值，从而导致线程 1 接收到的 message 可能是 null。

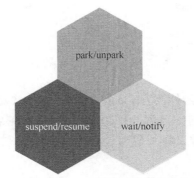

图 1.20　对线程进行阻塞和唤醒操作的各种方式

代码清单 1.18　线程之间的通信

```
1.  public class ThreadCommunicationDemo {
2.      private static String message;
3.
4.      public static void main(String[] args) {
5.          Thread thread1 = new Thread(() -> {
6.              System.out.println(message);
7.          });
8.          Thread thread2 = new Thread(() -> {
9.              message = "i am thread2";
10.         });
11.         thread1.start();
12.         thread2.start();
13.     }
14. }
```

为了解决上述问题，可以在线程 1 中增加一个 while 判断，当 message 变量为 null 时就一直循环，直到线程 2 将信息赋值到 message，如代码清单 1.19 所示。这样做确实能解决问题，但这种做法最多只能算二流方法。借助于 Java 提供的阻塞与唤醒操作，可使线程之间的通信更加高效。

代码清单 1.19　通过 while 判断解决通信问题

```
1.  public class ThreadCommunicationDemo2 {
2.      private static String message;
3.
4.      public static void main(String[] args) {
5.          Thread thread1 = new Thread(() -> {
6.              while (message == null) {
7.              }
8.              System.out.println(message);
9.          });
10.         Thread thread2 = new Thread(() -> {
11.             message = "i am thread2";
```

```
12.            });
13.            thread1.start();
14.            thread2.start();
15.        }
16.    }
```

在 Java 发展的过程中，曾经使用 suspend 和 resume 方法对线程进行阻塞和唤醒，它们能够在代码中控制阻塞和唤醒的时间节点。比如线程 1 启动后在某个时间点需要让它挂起，这时可以调用 suspend 方法，而线程 2 则可以通过调用 resume 方法唤醒线程 1 继续往下执行，如图 1.21 所示。

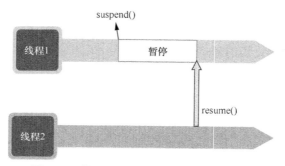

图 1.21　使用 suspend 和 resume 唤醒线程

下面通过 suspend 与 resume 的组合来解决上述两个线程的通信问题，如代码清单 1.20 所示。线程 1 和线程 2 都启动后，如果 thread1 先执行并发现 message 为 null，那么它就会调用 suspend 阻塞自己。后面 thread2 对 message 变量赋值后会调用 thread1 的 resume 方法来唤醒 thread1，然后 thread1 正确输出 "i am thread2"。

代码清单 1.20　使用 suspend 和 resume 的组合解决线程通信问题

```
1.  public class SuspendResumeDemo {
2.
3.      private static String message;
4.      static Thread thread1 = new Thread1();
5.      static Thread thread2 = new Thread2();
6.
7.      public static void main(String[] args) {
8.          thread1.start();
9.          thread2.start();
10.     }
11.
12.     static class Thread1 extends Thread {
13.         public void run() {
14.             if (message == null)
15.                 suspend();
16.             System.out.println(message);
17.         }
```

```
18.        }
19.
20.        static class Thread2 extends Thread {
21.            public void run() {
22.                message = "i am thread2";
23.                thread1.resume();
24.            }
25.        }
26.
27.    }
```

然而，suspend 与 resume 的组合存在很多问题，比较典型的是死锁问题。例如，在代码清单 1.21 中，本意是由主线程创建并启动 mt 线程，100ms 后通过 suspend 方法挂起线程，最后通过 resume 方法恢复线程。但现实并不是这样，程序执行到 suspend 时将一直卡住，永远等不来 "can you get here?" 的输出。

代码清单 1.21   suspend 与 resume 组合中的死锁问题

```
1.    public class ThreadSuspend {
2.        public static void main(String[] args) {
3.            Thread mt = new Thread(() -> {
4.                while (true) {
5.                    System.out.println("running....");
6.                }
7.            });
8.            mt.start();
9.            try {
10.               Thread.currentThread().sleep(100);
11.           } catch (InterruptedException e) {
12.           }
13.           mt.suspend();
14.           System.out.println("can you get here?");
15.           mt.resume();
16.       }
17.   }
```

为什么会出现上面的现象呢？这是死锁导致的。乍一看没什么问题，线程的任务只是简单地打印字符串。但问题就出现在 System.out.println 上，由于 println 被声明为一个同步方法，执行时将对 System 类的 out（PrintStream 类的一个实例）单例属性加同步锁。而 suspend 方法挂起线程但并不释放锁，当 mt 线程被挂起后，主线程调用 System.out.println 时同样需要获取 System 类 out 对象的同步锁才能打印 "can you get here?"。于是主线程一直在等待同步锁而 mt 线程又不释放锁，这就导致了死锁。

接着看另外一种阻塞唤醒方案——wait 与 notify 组合，即利用 Object 类的 wait() 和 notify() 方法实现线程阻塞。在使用这种阻塞唤醒方式时，需要稍微转变一下思维，它与面向线程阻塞思维有较大差异。前面的 suspend 与 resume 只需在线程内直接调用就能完成阻塞和唤醒，这很好理解。而如果改用 wait 与 notify 形式，则是以某个 object 为信号，线程 1 和线程 2 调用

object 的 wait 方法后都将阻塞,接着线程 3 调用 object 的 notify 方法将会唤醒线程 1 或线程 2,如图 1.22 所示。

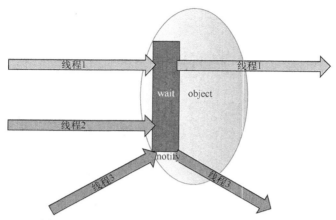

图 1.22　使用 wait 与 notify 组合的阻塞唤醒方案

使用 wait 与 notify 组合能在一定程度上避免死锁问题,但并不能完全避免,因此必须在编程过程中避免死锁。在使用过程中需要注意下面几点。

- wait 与 notify 方法是针对对象的,调用任意对象的 wait 方法都将导致线程阻塞,阻塞的同时也将释放该对象的锁。相应地,调用任意对象的 notify 方法则随机解除该对象阻塞的线程,但它需要重新获取该对象的锁,直到获取成功才能往下执行。
- wait 与 notify 方法必须在 synchronized 块或方法中调用,并且要保证同步块或方法的锁对象与调用 wait 与 notify 方法的对象是同一个。如此一来,当前线程在调用 wait 之前就已经成功获取某对象的锁,执行 wait 阻塞后就将之前获取的对象锁释放。当然,假如不按照上面规定约束编程,程序也能通过编译,但运行时将抛出 IllegalMonitorStateException 异常,因此在编程时必须保证用法正确。
- notify 用于随机唤醒一个阻塞中的线程并让其获取对象锁,进而往下执行,而 notifyAll 则是唤醒阻塞中的所有线程,让它们去竞争该对象锁,获取到锁的那个线程才能往下执行。

下面使用 wait 与 notify 改造代码清单 1.21 中存在的死锁问题(见代码清单 1.22)。改造的思想就是在 MyThread 中添加一个标识变量,一旦变量改变就相应地调用 wait 和 notify 阻塞和唤醒线程。由于在执行 wait 后将释放 synchronized(this)锁住的对象锁,此时 System.out.println ("running...")的锁也释放完毕,从而避免了死锁问题。

代码清单 1.22　改造代码清单 1.21 中的死锁问题

```
1.    public class ThreadWaitNotify {
2.
3.        public static void main(String[] args) throws InterruptedException {
```

```java
4.         MyThread mt = new MyThread();
5.         mt.start();
6.         Thread.sleep(100);
7.         mt.suspendThread();
8.         System.out.println("can you get here?");
9.         Thread.sleep(3000);
10.        mt.resumeThread();
11.    }
12.
13.    static class MyThread extends Thread {
14.        public boolean isSuspend = false;
15.
16.        public void run() {
17.            while (true) {
18.                synchronized (this) {
19.                    System.out.println("running...");
20.                    if (isSuspend)
21.                        try {
22.                            wait();
23.                        } catch (InterruptedException e) {
24.                        }
25.                }
26.            }
27.        }
28.
29.        public void suspendThread() {
30.            this.isSuspend = true;
31.        }
32.
33.        public void resumeThread() {
34.            synchronized (this) {
35.                this.isSuspend = false;
36.                notify();
37.            }
38.        }
39.    }
40.
41. }
```

下面再使用 wait 与 notify 组合来实现代码清单 1.18 中两个线程通信的例子，如代码清单 1.23 所示。注意必须使用 synchronized 对 wait 和 notify 进行加锁，否则会产生异常。假如线程 1 先得到锁，那么 message 肯定为 null，这时线程 1 会调用 wait 方法进入阻塞状态。而如果线程 2 先获得锁，那么 message 则已经被赋值，肯定不为空。这两种情况都没问题，因此保证了程序的准确性。

代码清单 1.23　使用 wait 与 notify 实现线程通信

```java
1. public class WaitNotifyDemo {
2.     private static String message;
```

```java
3.
4.     public static void main(String[] args) {
5.         Object lock = new Object();
6.         Thread thread1 = new Thread(() -> {
7.             synchronized (lock) {
8.                 if (message == null)
9.                     try {
10.                        lock.wait();
11.                    } catch (InterruptedException e) {
12.                    }
13.                }
14.                System.out.println(message);
15.        });
16.
17.        Thread thread2 = new Thread(() -> {
18.            synchronized (lock) {
19.                message = "i am thread2";
20.                lock.notify();
21.            }
22.        });
23.
24.        thread1.start();
25.        thread2.start();
26.    }
27. }
```

wait 与 notify 组合的方式看起来不错，但仍然存在一点不足，那就是 wait 必须在 notify 之前执行，假如先执行 notify，再执行 wait，将导致线程永远无法被唤醒。另外，它面向的主体是 object，阻塞的是执行 wait 方法的线程，而唤醒的是某个随机的线程或所有线程。接下来介绍第三种方式——park 与 unpark 组合。该方式能很好地规避死锁和竞态条件，从而很好地代替 suspend 与 resume 组合，如图 1.23 所示。

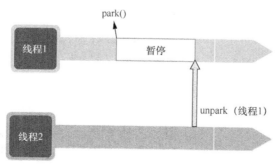

图 1.23　使用 park 与 unpark 组合的阻塞唤醒方案

下面通过 park 与 unpark 组合来实现代码清单 1.18 中两个线程通信的例子，如代码清单 1.24 所示。可以看到代码简洁且容易理解。假如线程 1 执行得更快，则该线程会执行 park 操作进入阻塞，等到线程 2 对 message 赋值后执行 unpark 操作才能唤醒线程 1 继续往下执行。假如线程 2 执行得更快，则对 message 赋值后会对线程 1 进行 unpark 操作，那么线程 1 在执行 park 操作时就能直接通过。相较于 wait 和 notify 组合对执行顺序的敏感性，park 与 unpark 则通过使用许可机制避免了这个问题，这就是 park 与 unpark 使用许可机制的优势。

代码清单 1.24　使用 park 与 unpark 实现线程通信

```
1.  public class ParkUnparkDemo {
2.      private static String message;
3.
4.      public static void main(String[] args) {
5.          Thread thread1 = new Thread(() -> {
6.              LockSupport.park();
7.              System.out.println(message);
8.          });
9.
10.         Thread thread2 = new Thread(() -> {
11.             message = "i am thread2";
12.             LockSupport.unpark(thread1);
13.         });
14.
15.         thread1.start();
16.         thread2.start();
17.     }
18. }
```

另外，需要注意的是，许可是一次性的。也就是说不管是先 unpark 一次还是 100 次，只要使用一次 park 许可就没有了，下次要得到许可就得重新 unpark。比如，在代码清单 1.25 中，先对当前线程 unpark 了 5 次，然后第一次 park 就将许可使用掉了，那么第二个 park 就阻塞了，主线程被阻塞无法往下执行。

代码清单 1.25　许可机制的使用

```
1.  public class ParkUnparkDemo2 {
2.
3.      public static void main(String[] args) {
4.          LockSupport.unpark(Thread.currentThread());
5.          LockSupport.unpark(Thread.currentThread());
6.          LockSupport.unpark(Thread.currentThread());
7.          LockSupport.unpark(Thread.currentThread());
8.          LockSupport.unpark(Thread.currentThread());
9.          LockSupport.park();
10.         LockSupport.park();
11.     }
12. }
```

LockSupport 类为线程阻塞唤醒提供了基础，它的实现原理在后面相关章节中会深入分析。park 与 unpark 组合在竞争条件问题上具有 wait 和 notify 无可比拟的优势。在使用 wait 和 notify 组合时，某一线程被另一线程 notify 之前必须要保证此线程已经执行到 wait 等待点，若错过

notify 则可能永远都在等待，另外 notify 也不能保证唤醒指定的某个线程。反观 LockSupport，由于 park 与 unpark 引入了许可机制，所以能很好解决该问题。许可机制的逻辑为：
- park 操作将在许可为 0 的时候阻塞，而如果许可为 1 则直接返回并将许可置为 0；
- unpark 操作则将许可置为 1，并尝试唤醒线程。

根据这两个逻辑，对于同一个线程，park 与 unpark 操作的顺序并不影响程序正确地执行。假如先执行 unpark 操作，则许可被置为 1，之后再执行 park 操作，此时因为许可等于 1 而直接返回并往下执行，不会进入阻塞。

park 与 unpark 组合真正解耦了线程之间的同步，不需要考虑同步锁。而 wait 与 notify 要保证有锁才能执行，而且执行 notify 操作释放锁后当前线程会重新进入等待队列来等待获取锁，LockSupport 则完全不用考虑锁、等待队列等问题。

1.11 Java 线程的 join 操作

计算机为了提升 CPU 使用效率和交互性而引入了并发机制，执行的任务也抽象成了线程，从宏观上看，多个线程就像是同时执行一样。但并发同样也使得线程的执行顺序不容易控制，实际工程中的很多场景都会涉及某个线程需要依赖另外一个或几个线程的执行结果，这就要求被依赖的线程必须先执行完毕，此时就需要 join 操作。比如在图 1.24 中，假如要计算 A+B 的结果且 A 和 B 的计算都比较耗时，那么我们将 B 的计算分给线程 2，而线程 1 则负责 A 的计算。如果线程 1 先执行完，则它要等待线程 2，直到线程 2 计算出 B 的结果后，线程 1 才继续往下执行，去计算 A+B。

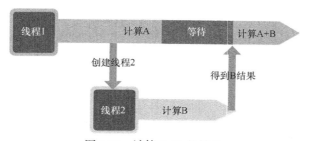

图 1.24　计算 A+B 的结果

join 操作类似于前面讲解的线程的 wait 与 notify 功能，某个线程可以通过调用 join 操作来等待另外一个线程的执行，直到另外一个线程执行完毕。我们根据图 1.25 看一下 join 的过程。线程 t1 首先创建了线程 t2 并启动该线程，接着线程 t1 继续创建线程 t3，然后线程 t1 调用 t2.join() 和 t3.join() 后进入等待状态。自此线程 t1 进入等待状态，而线程 t2 和线程 t3 则一直执行，等它们都执行完毕后线程 t1 才会继续往下执行。

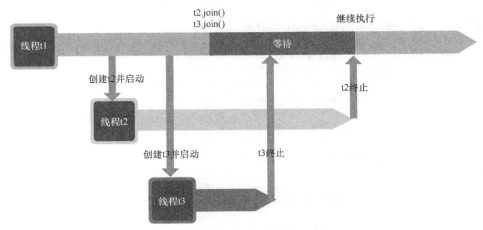

图 1.25 join 的过程

代码清单 1.26 是一个简单的例子，主线程创建了线程 t2 并启动它，t2 使用睡眠来模拟耗时计算，主线程调用了 t2.join()表示要等到 t2 执行完毕后才往下执行，也就是 3s 后主线程才输出 "got t2's result."。

代码清单 1.26　join 操作示例

```
1.   public class JoinTest {
2.       public static void main(String[] args) {
3.           Thread t2 = new Thread(() -> {
4.               try {
5.                   //模拟耗时计算
6.                   Thread.sleep(3000);
7.               } catch (InterruptedException e) {
8.                   e.printStackTrace();
9.               }
10.          });
11.          t2.start();
12.          try {
13.              t2.join();
14.              System.out.println("got t2's result.");
15.          } catch (InterruptedException e) {
16.              e.printStackTrace();
17.          }
18.      }
19.  }
```

join 操作具有中断机制，从代码清单 1.26 中可以看到，主线程调用 t2.join()后会一直处于等待状态。假如 t2 一直执行不完，则主线程会一直等待下去。然而 join 操作支持中断，因此可以通过中断来解除 join 的阻塞。在代码清单 1.27 中，t2 启动后会睡眠 60s，随后 t3 启动后主线程就进入等待状态。t3 在睡眠 3s 后就将主线程的中断标识设置为 true，即进行中断操作。主线程解除阻塞，并输出 "t3 has interrupted main thread."。

代码清单 1.27　通过中断解除 join 的阻塞

```
1.  public class JoinInterruptTest {
2.      public static void main(String[] args) {
3.          Thread mainThread = Thread.currentThread();
4.          Thread t2 = new Thread(() -> {
5.              try {
6.                  //模拟耗时计算
7.                  Thread.sleep(60000);
8.              } catch (InterruptedException e) {
9.              }
10.         });
11.         Thread t3 = new Thread(() -> {
12.             try {
13.                 Thread.sleep(3000);
14.                 mainThread.interrupt();
15.             } catch (InterruptedException e) {
16.             }
17.         });
18.         t2.setDaemon(true);
19.         t2.start();
20.         t3.start();
21.         try {
22.             t2.join();
23.         } catch (InterruptedException e) {
24.             System.out.println("t3 has interrupted main thread.");
25.         }
26.     }
27. }
```

join 操作也具有超时机制。由于 join 操作默认会无限等待，也就是说，不管另一个线程执行多久，都将等待其运行完。但如果希望等待的时间有期限，则可以传入超时时间，一旦等待超过该指定时间就会解除阻塞。代码清单 1.28 与前面不同的地方在于其调用了 t2.join(3000)，也就是 join 的超时为 3s，t2 会睡眠 60s，但主线程只会等待 3s 就解除阻塞，然后输出"join timeout."。

代码清单 1.28　带有超时机制的 join 操作

```
1.  public class JoinTimeoutTest {
2.      public static void main(String[] args) {
3.          Thread t2 = new Thread(() -> {
4.              try {
5.                  //模拟耗时计算
6.                  Thread.sleep(60000);
7.              } catch (InterruptedException e) {
8.                  e.printStackTrace();
9.              }
10.         });
11.         t2.setDaemon(true);
12.         t2.start();
```

```
13.         try {
14.             t2.join(3000);
15.             System.out.println("join timeout.");
16.         } catch (InterruptedException e) {
17.             e.printStackTrace();
18.         }
19.     }
20. }
```

最后看一下 join 操作的实现原理，对应的核心源码在 java.lang.Thread 类中。不带参数的 join 方法实际上间接调用了 join(0)，所以主要逻辑在 join(long millis) 方法中，如代码清单 1.29 所示。如果传入的超时时间为负数，则会抛出非法参数异常；如果超时时间为 0，则调用 wait(0) 方法，该方法会使当前线程 1 一直等待，直到其他线程进行了通知（notify）。这里需要注意的是，并没有其他线程显式地调用 notify 来通知阻塞的线程，实际上线程在终止时会调用自身的 notifyAll 方法。也就是说，JVM 会负责在线程退出前去进行通知操作，从而让 join 解除等待状态。如果超时时间大于 0，则计算最长的等待时间，然后调用 wait(delay) 使线程进入等待状态，传入的参数使得在等待超时后能解除等待状态。

代码清单 1.29 join 操作的实现原理

```
1.  public final void join() throws InterruptedException {
2.      join(0);
3.  }
4.
5.  public final synchronized void join(long millis) throws InterruptedException {
6.      long base = System.currentTimeMillis();
7.      long now = 0;
8.
9.      if (millis < 0) {
10.         throw new IllegalArgumentException("timeout value is negative");
11.     }
12.
13.     if (millis == 0) {
14.         while (isAlive()) {
15.             wait(0);
16.         }
17.     } else {
18.         while (isAlive()) {
19.             long delay = millis - now;
20.             if (delay <= 0) {
21.                 break;
22.             }
23.             wait(delay);
24.             now = System.currentTimeMillis() - base;
25.         }
26.     }
27. }
```

第 2 章
线程 I/O 模型

2.1 线程与阻塞 I/O

I/O 模型是指计算机在涉及 I/O 操作时使用到的模型。为了解决各种问题，人们提出了很多不同的 I/O 模型，与之相关的概念有线程、阻塞、非阻塞、同步以及异步等。I/O 可以分成阻塞 I/O 与非阻塞 I/O 两大类型。阻塞 I/O 在进行 I/O 操作时会使当前线程进入阻塞状态，而非阻塞 I/O 则不进入阻塞状态。以服务器处理客户端连接为例，在单线程情况下由一个线程负责所有客户端连接的 I/O 操作，而在多线程情况下则由若干线程共同处理所有客户端连接的 I/O 操作。

此外，需要注意的是，计算机的 I/O 其实包含了各种设备的 I/O，比如网络 I/O、磁盘 I/O、键盘 I/O 和鼠标 I/O 等。我们以经典的网络 I/O 场景（见图 2.1）为例来讲解单线程阻塞 I/O 模型和多线程阻塞 I/O 模型。

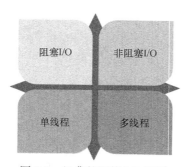

图 2.1 经典的网络 I/O 场景

程序在执行 I/O 时一般需要从内核空间复制数据，但内核空间的数据可能需要较长时间进行准备，由此导致用户空间产生阻塞。在图 2.2 中，应用程序处于用户空间，一个应用程序对应着一个进程，而进程则包含了缓冲区。当要进行 I/O 操作时，需要通过内核来执行相应的操作，比如由内核负责与键盘、磁盘、网络等控制器进行通信。当内核得到不同设备的控制器发送过来的数据后，会将数据复制到用户空间供应用程序使用。

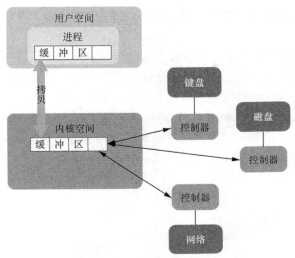

图 2.2 应用程序的 I/O 操作

网络 I/O 产生阻塞的过程如图 2.3 所示。应用程序首先发起读取操作,然后进入阻塞状态,接下来由操作系统内核完成 I/O 操作。内核刚开始时未准备好数据,它需要不断读取网络数据,一旦数据准备好,就将数据复制到用户空间供应用程序使用。应用程序在从发起读取操作到继续往下处理的这段时间就处于阻塞状态。

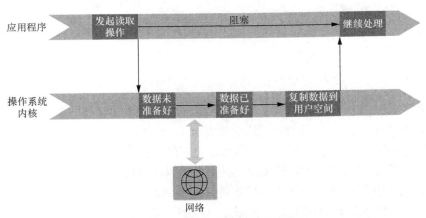

图 2.3 网络 I/O 的阻塞过程

2.1.1 单线程阻塞 I/O 模型

单线程阻塞 I/O 模型是最简单的一种服务器模型,几乎所有程序员在刚开始接触网络编程时都从这种模型开始。这种模型只能同时处理一个客户端访问,并且在 I/O 操作上是阻塞的,线程会一直处于等待状态而不会做其他事情。对于多个客户端访问的情况,必须要等到前一个客户端访问结束后才能进行下一个访问的处理。也就是说，请求一个一个排队，且只提供一问一答服务。

图2.4所示为单线程阻塞服务器响应客户端访问的时间节点。首先，服务器必须初始化一个套接字（socket）服务器，并绑定某个端口号使之监听客户端的访问。接着，客户端1调用服务器的服务，服务器接收到请求后对其进行处理，处理完后写数据回客户端1。最后，处理客户端2的请求并写数据回客户端2，期间即使客户端2在服务器处理完客户端1之前就进行请求，也要等服务器对客户端1响应完后才会对客户端2进行响应处理。

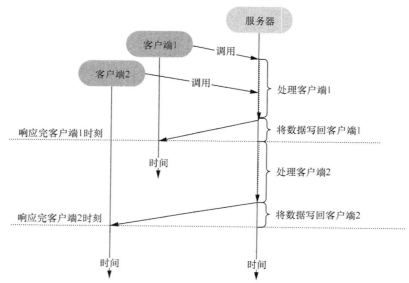

图2.4　单线程阻塞服务器响应客户端访问的时间节点

这种模型的特点在于单线程和阻塞I/O。单线程即服务器端只有一个线程处理客户端的所有请求，客户端连接与服务器端的处理线程比是 $n:1$，它无法同时处理多个连接，只能串行处理连接。而阻塞I/O是指服务器在读写数据时是阻塞的，在读取客户端数据时要等待客户端发送数据并且把操作系统内核中的数据复制到用户进程中，这时才解除阻塞状态。将数据写回客户端时要等待用户进程将数据写入内核后才解除阻塞状态。

单线程阻塞I/O模型是最简单的一种服务器模型，整个运行过程都只有一个线程，只能同时处理一个客户端的请求（如果有多个客户端访问，就必须排队等待）。服务器系统资源消耗较小，但并发能力低，容错能力差。

2.1.2　多线程阻塞I/O模型

针对单线程阻塞I/O模型的缺点，最简单的改进方式就是将其多线程化，使之能对多个客户端进行并发响应。多线程模型的核心就是利用多线程机制为每个客户端分配一个线程。如图2.5

所示，服务器端开始监听客户端的访问，假如有两个客户端发送请求过来，服务器端在接收到客户端请求后将创建两个线程分别对它们进行处理。每个线程负责一个客户端连接，直到响应完成。期间两个线程并发地为各自对应的客户端处理请求，包括读取客户端数据、处理客户端数据、将数据写回客户端等操作。

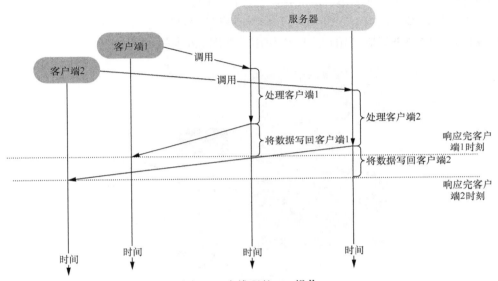

图 2.5　多线程的 I/O 操作

这种模型的 I/O 操作也是阻塞的，因为每个线程执行到读取或写入操作时都将进入阻塞状态，直到成功读取客户端的数据或数据成功写入内核后才解除阻塞状态。尽管此时的 I/O 操作还是会阻塞，但这种模式比单线程模式的性能明显提高，它不用等到第一个请求处理完才处理第二个，而是并发地处理客户端请求，客户端连接与服务器端处理线程的关系是一对一的。

多线程阻塞 I/O 模型的特点如下所示：
- 支持对多个客户端并发响应，处理能力得到大幅提高，有较强的并发能力；
- 服务器系统资源消耗量较大，而且多线程之间会产生线程切换成本，同时拥有较复杂的结构。

2.2　线程与非阻塞 I/O 模型

前面讲到，多线程阻塞 I/O 模型通过引入多线程的方法来提升服务器端的并发处理能力，确实能够达到一定的效果。但它还是存在一个严重的问题，那就是每个连接都需要一个线程负责 I/O 操作。当连接数量较多时将导致机器线程数量太多，而这些线程在大多数时间内都处于等待状态，线程之间的切换成本非常高。对于多线程阻塞 I/O 模型的这个缺点，有没有可能只用一个线程就可以维护多个客户端连接并且不会阻塞在读写操作呢？这就是下面要介绍的单线程非阻塞 I/O 模型。

就单线程非阻塞 I/O 模型来说，与阻塞 I/O 模型相同的地方是，程序在执行 I/O 时一般需要从内核空间和用户空间复制数据，不同之处在于非阻塞 I/O 模型不会一直等到内核准备好数据，而是直接返回去做其他的事，也就是说并不产生阻塞。应用程序进程包含一个缓冲区，单个线程会不断循环遍历所有客户端，尝试对它们进行读写操作。如果内核已准备好数据，那么应用层线程就会将数据复制到用户空间供使用，如图 2.6 所示。

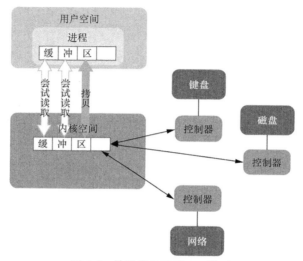

图 2.6　单线程非阻塞 I/O 模型

非阻塞 I/O 模型最重要的一个特点是在调用读或写接口后会立即返回，而不会进入阻塞状态。网络的非阻塞 I/O 的过程如图 2.7 所示。应用程序首先发起读取操作，它会告知操作系统

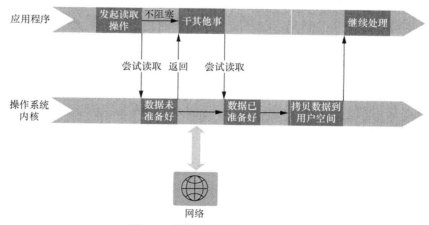

图 2.7　网络的非阻塞 I/O 的过程

内核去执行 I/O 操作。刚开始时由于内核未准备好数据，所以会马上返回而不是阻塞。此时应用程序可以做其他事，过一段时间后再尝试读取操作。如果发现数据已经准备好，内核就将数据复制到用户空间供应用程序使用。

非阻塞 I/O 模型可分为应用层 I/O 多路复用、内核 I/O 多路复用、内核回调事件驱动 I/O 这 3 种。下面进行详细介绍。

2.2.1 应用层 I/O 多路复用

应用层 I/O 多路复用如图 2.8 所示。当多个客户端向服务器发出请求时，服务器端会将每一个客户端连接保存到一个 socket 列表中，而应用层的线程则轮着对 socket 列表的客户端连接进行尝试读写。对于读取操作，如果成功读取到若干数据，则对读取到的数据进行处理；如果读取失败，则到下一个循环再继续尝试。对于写入操作，先尝试将数据写入指定的某个 socket，写入失败则到下一个循环再继续尝试。

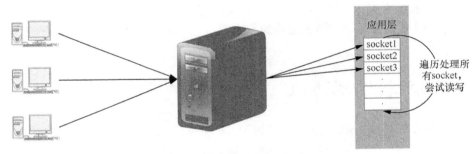

图 2.8　应用层 I/O 多路复用

这样一来，不管有多少个 socket 连接，它们都可以由一个线程来管理。这个线程负责遍历 socket 列表，不断地尝试读取或写入数据。这种方式很好地利用了阻塞的时间，使处理能力得到提升。但这种模型需要在应用程序中遍历所有的 socket 列表，同时需要处理数据的拼接。

2.2.2 内核 I/O 多路复用

在应用层执行遍历的成本较高，能不能把遍历工作下移到内核层呢？实际上完全可以将 socket 的遍历工作交给操作系统内核，然后把对 socket 遍历的结果组织成一系列的事件列表并返回应用层处理，应用层需要处理的对象就是这些事件。这是一种事件驱动的非阻塞方式，也就是内核 I/O 多路复用。

如图 2.9 所示，服务器端有多个客户端连接，应用层向内核请求读写事件列表。内核遍历所有 socket 并生成对应的可读列表 readList 和可写列表 writeList。readList 标明了每个 socket 是否可读，例如 socket1 的值为 1，则表示可读；socket2 的值为 0，则表示不可读。writeList

则标明了每个 socket 是否可写。应用层遍历读写事件列表 readList 和 writeList，并执行相应的读写操作。

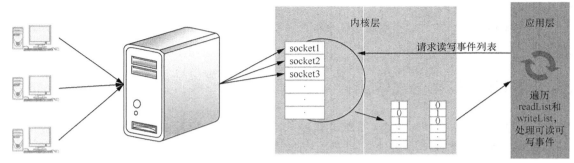

图 2.9　事件驱动的非阻塞方式

由于不用在应用层对所有 socket 进行遍历，而将遍历工作下移到内核层，因此这种方式有助于提高检测效率。然而，它需要将所有连接的可读事件列表和可写事件列表都传到应用层。假如 socket 连接数量变大，将列表从内核复制到应用层也是不小的开销。另外，当活跃连接较少时，内核与应用层之间会存在很多无效的数据副本，因为内核将活跃和不活跃的连接状态都复制到应用层中。

2.2.3　内核回调事件驱动 I/O

通过遍历的方式检测 socket 是否可读可写是一种效率比较低的方式，不管是在应用层中遍历还是在内核中遍历。所以需要另外一种机制来优化遍历的方式，那就是回调函数。内核中的 socket 都对应一个回调函数，当客户端往 socket 发送数据时，内核从网卡接收数据后就会调用回调函数。回调函数中维护一个事件列表，应用层获取此事件列表即可得到所有感兴趣的事件。这就是内核回调事件驱动 I/O。

内核回调事件驱动方式有两种。

方式 1 是用可读列表 readList 和可写列表 writeList 标记读写事件。socket 的数量与 readList 和 writeList 两个列表的长度一样，readList 的第一个元素标为 1 则表示 socket1 可读；同理，writeList 的第二个元素标为 1 则表示 socket2 可写。

在图 2.10 中，多个客户端连接服务器端，当客户端发送数据过来时，内核从网卡成功复制数据后会调用回调函数将 readList 的第一个元素置为 1，应用层发送读、写事件列表的请求，内核返回包含了事件标识的 readList 和 writeList 事件列表，应用层进而分别遍历读事件列表 readList 和写事件列表 writeList，然后对置为 1 的元素所对应的 socket 进行读或写操作。这样就避免了遍历 socket 的操作，但仍然有大量无用的数据（状态为 0 的元素）会从内核复制到应用层中。

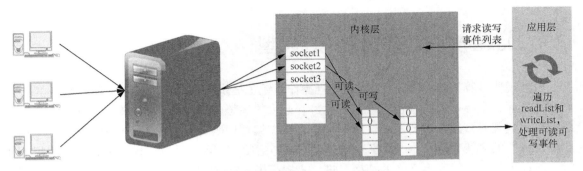

图 2.10　内核回调事件驱动 I/O 方式 1

方式 2 如图 2.11 所示。服务器端有多个客户端 socket 连接。首先，应用层告诉内核每个 socket 感兴趣的事件。接着，当客户端发送数据过来时，会对应一个回调函数，内核从网卡成功复制数据后即调用回调函数将 socket1 作为可读事件 event1 加入事件列表。同样，内核发现网卡可写时就将 socket2 作为可写事件 event2 加入事件列表。最后，应用层向内核请求读、写事件列表，内核将包含了 event1 和 event2 的事件列表返回应用层，应用层通过遍历事件列表得知 socket1 有数据待读取，于是进行读操作，而 socket2 则用于写入数据。

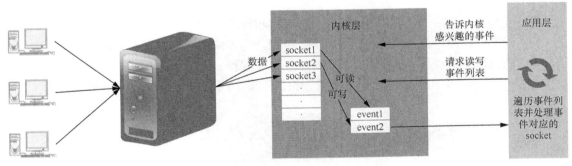

图 2.11　内核回调事件驱动 I/O 方式 2

上面两种方式由操作系统内核维护客户端的所有连接，并通过回调函数不断更新事件列表，应用层线程只需遍历这些事件列表即可知道可读或可写的连接，进而对这些连接进行读写操作，这就极大地提高了检测效率，自然也进一步增强了处理能力。

对于 Java 来说，非阻塞 I/O 的实现完全是基于操作系统内核的非阻塞 I/O，它将操作系统的非阻塞 I/O 的差异屏蔽并提供统一的 API，让我们不必关心操作系统。JDK 会帮我们选择非阻塞 I/O 的实现方式。例如对于 Linux 系统来说，在支持 epoll 的情况下，JDK 会优先选用 epoll 实现 Java 的非阻塞 I/O。这种非阻塞方式的事件检测机制就是效率最高的"内核回调事件驱动 I/O"中的方式 2。

2.3　Java 多线程非阻塞 I/O 模型

虽然现代计算机都是多 CPU 的，而且操作系统也提供了多线程机制，但并不是说单线程完全被抛弃，实际上单线程也有自己的优势。最大的优势就是一个 CPU 只负责一个线程，因此可以完全规避多线程中的所有疑难杂症，这样在编写代码时就简单多了，如图 2.12 所示。比如在以前的多线程环境中，共享变量的操作要考虑很多问题，而单线程则不必考虑这些。同时单线程也免去了线程上下文的切换，进一步提高了 CPU 的真正使用率。

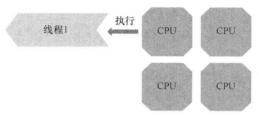

图 2.12　一个 CPU 只负责一个线程

在一个线程对应一个 CPU 的情况下，如果多核计算机中只执行一个线程，那么就只有一个 CPU 工作，这样也就无法充分利用 CPU 资源。为了解决这个问题，我们的程序可以根据 CPU 的数量来创建线程数，N 个 CPU 分别对应 N 个线程，如图 2.13 所示。这种模型充分利用了多个 CPU，同时也保持了单线程的优点，相当于多个线程并行执行而不是并发执行。

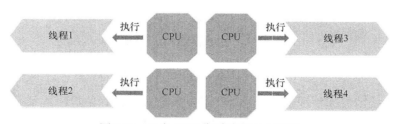

图 2.13　N 个 CPU 分别对应 N 个线程

在多核的机器时代，多线程和非阻塞都是提升服务器处理性能的利器。那么如何将它们结合起来呢？最常规的做法就是将客户端连接按组分配给若干线程，每个线程负责处理对应组内的连接。如图 2.14 所示，有 4 个客户端访问服务器，服务器将 socket1 和 socket2 交由线程 1 管理，而线程 2 则管理 socket3 和 socket4，通过事件检测及非阻塞读写就可以让每个线程都能高效运行。

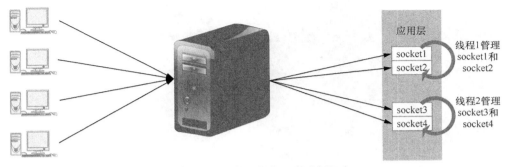

图 2.14　将多线程与非阻塞进行组合

在实际的工程中，最经典的多线程非阻塞 I/O 模式是 Reactor 模式。首先看单线程下的 Reactor。Reactor 将服务器端的整个处理过程分成若干个事件，例如分为接收事件、读事件、写事件、执行事件等。Reactor 通过事件检测机制将这些事件分发给不同的处理器去处理。如图 2.15 所示，若干客户端连接访问服务器端，Reactor 负责检测各种事件并分发到处理器，这些处理器包括接收连接的 accept 处理器、读数据的 read 处理器、写数据的 write 处理器以及执行逻辑的 process 处理器。在整个过程中只要有待处理的事件存在，就可以让 Reactor 线程不断往下执行，而不会阻塞在某处，所以处理效率很高。

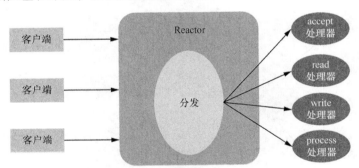

图 2.15　单线程下的 Reactor

基于单线程的 Reactor 模式在实际中很少使用，我们更多地是将它改进为多线程模式。常见的有下面两种：

- 多线程 Reactor 模式，即在耗时的 process 处理器中引入多线程，例如使用线程池；
- 多 Reactor 实例模式，即直接使用多个 Reactor 实例，每个 Reactor 实例对应一个线程。

Reactor 模式的第一种改进方式是多线程 Reactor 模式，如图 2.16 所示。多线程 Reactor 模式的整体结构基本与单线程的 Reactor 类似，只是额外引入了一个线程池。由于对连接的接收、对数据的读取和对数据的写入等操作基本上都耗时较少，因此把它们都放到 Reactor 线程中处理。然而，对于可能比较耗时的逻辑处理工作，则在 process 处理器中引入线程池。process 处理器自己不执行任务，而是交给线程池，从而在 Reactor 线程中避免了耗时的操作。将耗时的操作转移到线程池中后，尽管 Reactor 只有一个线程，也能保证 Reactor 的高效性。

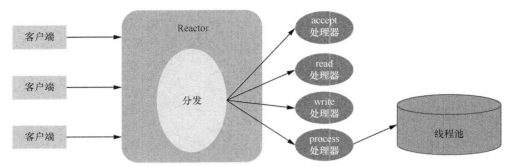

图 2.16 多线程 Reactor 模式

Reactor 模式的第二种改进方式是多 Reactor 实例模式,如图 2.17 所示。其中有多个 Reactor 实例,每个 Reactor 实例对应一个线程。因为接收事件是相对于服务器端而言的,所以客户端的连接接收工作统一由一个 accept 处理器负责,accept 处理器会将接收的客户端连接均匀分配给所有 Reactor 实例。每个 Reactor 实例负责处理分配到该 Reactor 上的客户端连接,包括连接的读数据、写数据和逻辑处理。这就是多 Reactor 实例的原理。

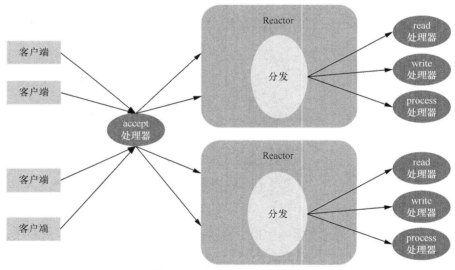

图 2.17 多 Reactor 实例模式

2.4 多线程带来了什么

并发和并行都是相对于进程或线程来说的。并发是指一个或若干个 CPU 对多个进程或线程进行多路复用,用简单的语言来说就是 CPU 交替执行多个任务,每个任务都执行一小段时间,从宏观上看,就像是全部任务都在同时执行一样。并行则是指多个进程或线程在同一时刻执行，这是真正意义上的同时执行，它必须有多个 CPU 的支持。图 2.18 是并发和并行的执行时间图。对于并发来说，线程 1 先执行一段时间，然后线程 2 再执行一段时间，接着线程 3 再执行一段时间。每个线程都轮流得到 CPU 的执行时间，这种情况下只需要一个 CPU 就能实现。对于并行来说，线程 1、线程 2 和线程 3 是同时执行的，这种情况下需要 3 个 CPU 才能实现。

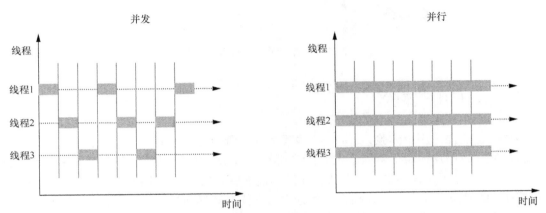

图 2.18　并发和并行的执行时间图

2.4.1　提升执行效率

多线程技术可以实现并发和并行执行，所以多线程能提升总体的执行效率。如果不支持多线程，那么当某个执行任务进入等待阻塞状态时，则可能因为阻塞而导致运行效率低下。在图 2.19 中，一个请求任务发起服务器请求后则开始等待响应，在整个运行线上可以看到，真正运行的时间（方块）很少，这便是运行效率低下。但在多线程的情况下，则可以在等待阻塞时去做其他的工作。

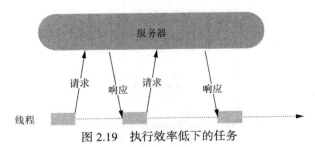

图 2.19　执行效率低下的任务

此外，在多 CPU 环境下，如果一个任务能够分解成多个小任务，那么就能够用多个 CPU 同时执行它，这样就能以更加快的速度完成任务（毕竟单个 CPU 运行能力有限）。在图 2.20 中，一旦将任务分解成 3 个小任务，这 3 个小任务在多 CPU 环境下就能够并行执行，从而大大减少了整体执行时间。

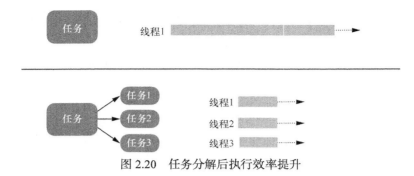

图 2.20　任务分解后执行效率提升

2.4.2　提升用户体验

多线程也能提升用户体验，如果一个线程的任务既包含耗时的任务，又包含用户交互的任务，就可能导致用户体验很糟糕。大家在看到图 2.21 所示的窗口时（一直在打转，且无法对其进行操作），是不是很难受？一个线程在发起请求后开始等待请求结果，用户界面则一直卡着没响应，此时便可以通过多线程将任务分为请求任务和界面操作两部分，这样就能在请求后保持对界面操作的响应，从而提供更好的用户体验。

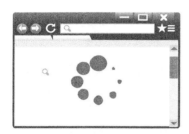 

图 2.21　一直打转的窗口

2.4.3　让编码更难

"天下没有免费的午餐"，多线程也是需要付出代价的。从代码编写的角度来看，多线程使编码变得更加复杂,从本质上来说这是由多线程机制与现代计算机结构带来的。纵然在编程语言设计专家的努力下,现在产生了很多可以简化多线程编程的语言和模型,但相较于单线程来说,多线程的编写仍然很复杂。数据在主存储到 CPU 之间有若干层缓存和寄存器(见图 2.22),而且多个线程可能访问共享内存,这就涉及数据同步问题,从而增加了多线程编程的复杂性。此外,线程与线程之间的通信也比较麻烦,这也增加了多线程编码的复杂性。总之,尽管很多编程语言尝试提供更便捷的多线程编程,但在语言层面仍然无法完全屏蔽多线程与计算机结构的复杂性,所以不管使用什么语言,在进程多线程的编码时都需要考虑很多。

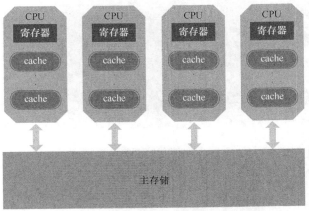

图 2.22 缓存与寄存器的存在带来了数据同步问题

2.4.4 资源开销与上下文切换开销

除了增加编码难度外,多线程还在执行过程中带来了实际的损耗,包括资源开销和上下文切换开销。资源开销主要包括其本身占用的内存资源、执行时线程本地栈的开销以及对这些线程进行管理的开销。而上下文切换开销则是因为 CPU 在从一个线程切换到另外一个线程时因为需要进行现场保护和现场恢复工作而引入的开销。现场保护和现场恢复涉及线程标识、寄存器内存、线程状态、线程优先级、线程资源清单等。在图 2.23 中,假设线程 1 和线程 2 由某个 CPU 执行。线程 1 执行一段时间后将相关信息保存到现场数据结构中,现场数据结构存放在主存储中,然后从线程 2 对应的现场数据结构中恢复线程 2 的相关信息,完成现场恢复后线程 2 开始执行。接下去的过程反过来,由线程 2 切换到线程 1。可以看到由虚线分割的两部分被标为切换开销,从完整的时序来看,这两部分并非用于执行线程的任务,而是消耗在了现场的保护和恢复上,这便是上下文切换的开销。

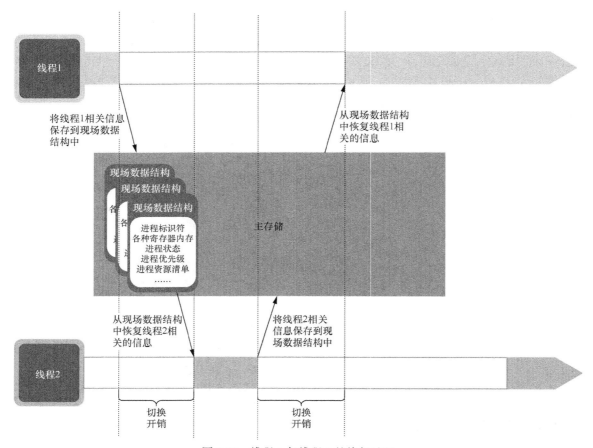

图 2.23　线程 1 与线程 2 的执行过程

在实践中要综合考虑多线程的优缺点，不能一味地去追求多线程。在使用多线程之前，必须衡量多线程所带来的好处与付出的代价。

第 3 章
Java 内存模型

3.1 计算机的运行

在分析 Java 内存模型之前，需要先学习一下计算机的运行模型。在信息时代，CPU 是一个耳熟能详的概念，大家都知道 CPU 就是计算机的大脑。计算机中一连串复杂的指令就是由它负责执行的，这些指令通常称为程序。那么 CPU 到底是什么呢？它是计算机中控制数据操控的电路，主要由 3 部分构成：控制单元、算术逻辑单元和寄存器单元，如图 3.1 所示。

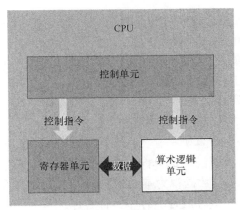

图 3.1 CPU 的构成

控制单元是整个 CPU 的指挥控制中心，它的主要职责就是指挥 CPU 工作，通过向其他两个单元发送控制指令来达到控制效果。算术逻辑单元主要的职责是执行运算，包括算术运算和逻辑运算，它对控制单元发送过来的指令执行相应的运算操作。寄存器单元主要用于存储临时数据，它保存着待处理的或已处理的数据。寄存器单元的出现是为了减少 CPU 对内存的访问次数，提升读取数据性能，最终提升 CPU 的整个工作效率。

CPU 中的寄存器分为通用寄存器和专用寄存器。通用寄存器用于临时存放 CPU 正在使用的数据，而专用寄存器则用于 CPU 的专有用途，比如指令寄存器和程序计数器。CPU 与主存（主存储器）通过总线进行通信，CPU 通过控制单元能够操作主存中的数据（见图 3.2）。

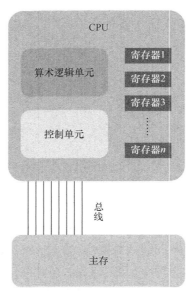

图 3.2　CPU 与主存的通信

CPU 执行两个数值相加的过程大致如下所示。
1. 从主存读取第一个值放到寄存器 1。
2. 从主存读取第二个值放到寄存器 2。
3. 两个寄存器保存的值作为输入送到加法电路。
4. 将加法结果保存到寄存器 3。
5. 控制单元将寄存器 3 的结果放到主存中。

原始的计算机并不像现代计算机一样将程序保存起来。在以前，人们只对数据进行保存，而设备执行的步骤被当作计算机的一部分内置在控制单元中。这样很不灵活，最多只能通过重新布线来提升灵活性。将程序与数据视为本质相同的事物是很大的思想突破，因为人们一直认为它们是不同的事物，数据应该存放在主存中，而程序应该属于 CPU 的一部分。

将程序当作数据保存在主存中大有好处，这样一来控制单元能够从主存读取程序，然后对它们解码并执行。当要修改执行程序时，可以在计算机的主存中修改，而不必对 CPU 进行更改或重新布线。

每个程序都包含着大量的机器指令，CPU 对这些指令进行解码并执行。CPU 分为两类体系：精简指令集计算机（RISC）和复杂指令集计算机（CISC）。RISC 提供了最小的机器指令集，计算机效率高，速度快且制造成本低。而 CISC 提供了强大、丰富的指令集，能更方便地实现复杂的软件。

机器指令分为以下 3 类。
- **数据传输类**：用于将数据从一个地方移动到另一个地方。比如将主存单元的内容加载到寄存器的 LOAD 指令，将寄存器的内容保存到主存的 STORE 指令，都属于数据传输类指令。此外，CPU 与其他设备（键盘、鼠标、打印机、显示器、磁盘等）进行通信的指令称为 I/O 指令。
- **算术逻辑类**：用于让控制单元请求在算术逻辑单元内执行运算。这些运算包括算术、与、或、异或和位移等。
- **控制类**：用于指导程序执行，比如转移（JUMP）指令（包括无条件转移和条件转移）。

CPU 将主存的指令加载进来后对其进行解码并执行，其中涉及指令寄存器与程序计数器这两个重要的寄存器，如图 3.3 所示。指令寄存器用于存储正在执行的指令，而程序计数器则保存下一个待执行的指令地址。

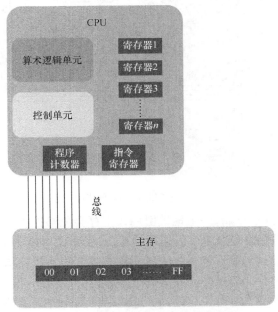

图 3.3　CPU 加载并执行指令的过程

那么指令是如何执行的呢？比如要计算 11+22，过程如下所示。
1. 将主存地址为 00 的内容加载到寄存器 1 中。
2. 将主存地址为 01 的内容加载到寄存器 2 中。
3. 将寄存器 1 和寄存器 2 的数据相加并将结果保存到寄存器 3 中。
4. 将寄存器 3 的结果存储到主存地址为 02 的位置。
5. 停止。

在这个过程中，CPU 涉及 4 个操作：加载（load）、存储（store）、加法（add）和停止（halt）。可以对这些操作进行编码，比如可以分别用 1、2、3、0000 来表示这 4 个操作。然后就可以用 1100、1201、3312、2302、0000 编码来表示 11+22 的执行过程，其中 1100 表示加载主存中 00 地址的内容到寄存器 1，1201 则表示加载主存中 01 地址的内容到寄存器 2，其余编码类似。

此外，CPU 与其他设备的通信一般通过控制器来实现。控制器可能在主板上，也可能以电路板的形式插到主板中。控制器本身可以看成是小型计算机，它也有自己简单的 CPU。以前每连接一种外设都需要购买对应的控制器，而现在随着通用串行总线（USB）成为通用的标准，很多外设都可以直接用 USB 控制器作为通信接口。每个控制器都连接在总线上并通过总线进行通信，如图 3.4 所示。

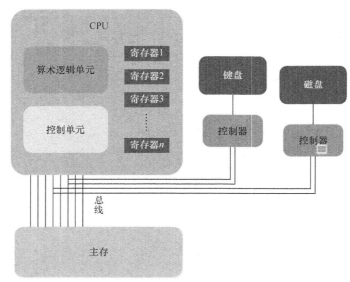

图 3.4　控制器与总线相连

每个控制器都被设计成对应着一组地址引用，主存会忽略这些地址引用。当 CPU 向这些地址发送消息时，其实是直接穿过主存传到控制器，因此操作的是控制器而非主存。这种模式称为存储映射 I/O。此外，这种模式的另外一种实现方式可以在机器指令中提供特定的操作码，专门用于与控制器通信，这样的操作码称为 I/O 指令。

最后再来了解直接存储器存取（DMA），这是一种提升外设通信性能的措施。由于 CPU 并非总是需要使用总线，因此当总线空闲的时候控制器可以将其充分利用起来。因为控制器都与总线相连接，而且控制器有执行指令的能力，所以可以将 CPU 的一些工作分给控制器来完成。比如要在磁盘中检索数据，CPU 可以告知控制器，然后由控制器找到数据并放到主存上，期间 CPU 可以去执行其他任务，这样便能节省 CPU 资源。不过 DMA 会使总线通信更加复杂，而且会导致总线竞争的问题。

3.2　Java 内存模型

内存模型可以看成系统底层与编程语言层面之间的规范，主要是多线程程序对共享存储器访问的行为表现。一方面，系统底层希望能对程序层面进行更多的优化策略（包括处理器和编译器），从而能更好地提升运行性能。另一方面，更多的优化工作也给编程语言层面带来了可编程性问题，因为复杂的内存模型将带来更多约束，这会增加多线程的编程难度。

对于单线程程序而言，编译器和处理器对程序的优化都能够做到对编程人员透明，也就是说不管怎么优化都不会影响程序最终结果的准确性。但对于多线程程序来说却完全不同，编译器和处理器的优化工作没有办法做到对编程人员完全透明。为了在提升运行性能的同时又能兼顾运行结果的准确性，只能让编程人员参与进来，而为了规范编程人员，又需要提供内存模型规范。

在多核时代，如何提高 CPU 的使用率成为一个永恒的话题。面对多核系统，我们需要建立一个能在多线程模式下工作的模型。这个话题讨论的主要是如何定义一个高效且安全的内存模型。内存模型用于定义处理器的各层缓存（Cache）与共享内存的同步机制，以及线程和内存交互的规则，如图 3.5 所示。

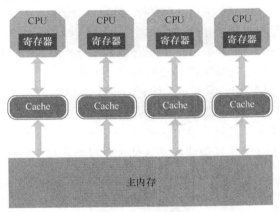

图 3.5　内存模型

在多核多线程环境中，为了提升运行时数据的访问性能，经常会使用多层缓存策略。这种架构带来了共享变量的可见性问题，即每个核或线程都有自己的本地副本，对共享变量的读写操作可能不会被其他核或线程知晓（也就是不可见）。比如在图 3.6 中，主存中的 x=1，而其他线程的本地副本却可能为 0 或 3，而且它们彼此之间也不可见。

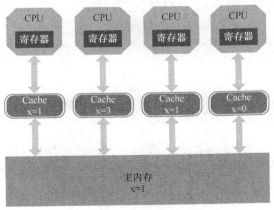

图 3.6　变量可见性问题

Java 也有属于它自己的内存模型，即 Java 内存模型（Java Memory Model，JMM）。由于 Java 被定义成一种跨平台的语言，所以在内存的描述方面也是跨平台的。Java 虚拟机试图定义一种统一的内存模型，将各种底层硬件及操作系统的内存访问差异进行封装，使 Java 程序在不同硬件及操作系统上都能达到相同的并发效果。Java 内存模型描述了程序中各个变量之间的关系，包括实例域、静态域、数据元素，还描述了在实际计算机系统中将变量存储到内存和从内存中取出变量的底层细节，包括某个线程对主存中的共享变量进行写操作时，如何和何时让其他线程知道，还描述了多个线程对主存中共享资源的安全访问。

我们通过图 3.7 来更好地理解 JMM 的工作机制。从整体上看，有几个比较重要的概念：主内存、线程本地内存、共享变量、变量副本、线程等。首先看主内存与线程本地内存以及它们的关系。主内存保存了 Java 程序的所有共享变量，而线程本地内存则保存了这些共享变量的副本。当然主内存不保存局部变量和方法参数，因为这些都不是共享变量。再看一下线程与线程本地内存的关系。每个线程都有一个属于自己的本地内存，不同线程之间的本地内存是互相不可见的，而且线程对变量的操作也只能针对自己的本地内存。最后看一下线程之间的通信机制（线程之间不可直接通信）。假如某个线程对一个共享变量进行修改，那么该如何让另一个线程知道呢？实际上只有线程 A 先将变量的更改反映到主存中，然后再由线程 B 从主存中读取变量，这样才能完成线程之间的通信。通信过程中也涉及可见性问题，即一个线程对变量的修改什么时候同步到主存中，其他线程又在什么时候与主存进行同步。

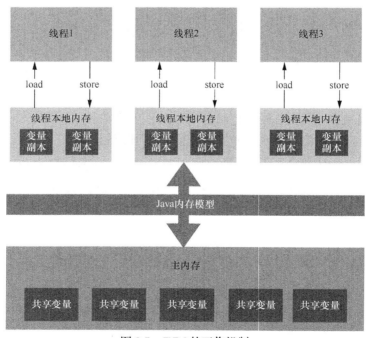

图 3.7　JVM 的工作机制

在 Java 内存模型中，如果一个线程更改了共享变量的值，其他线程能立刻知道这个更改，则说这个变量具有可见性。一般来说有 4 种常见的方式能保证变量的可见性，分别为 volatile、synchronized、final 和锁。

被 volatile 关键词声明的变量，每当有任何更改时都将立即同步到主存中，而每个线程在使用这个变量时，都要重新从主存刷新到工作内存，这样就确保了变量的可见性。当然，普通变量最终也会同步到主存，再由主存同步到每个线程的工作内存，只是这个最终可能比较"久"，不能保证可见性。

就 synchronized 来讲，由于 synchronized 的底层也通过锁来实现，所以 synchronized 和锁的本质是一样的。当一个线程释放一个锁时，将会强制刷新工作内存中的变量值到主存中。而当另一个线程获取此锁时将会强制重新装载此变量值。当然这两个线程获取释放的必须是同一个锁，所以 synchronized 也保证了变量的可见性。

被 final 声明的变量一旦完成初始化，其他线程就能看到这个 final 变量。

可以说 JMM 是 Java 的基础，也是 Java 多线程的基础。JMM 的定义直接影响 JVM 及 Java 多线程实现的机制。要想深入了解多线程并发中的相关问题和现象，则有必要对 Java 内存模型进行深入研究。在定义 JMM 时必须考虑下面几个方面：首先是如何更好地提高线程的性能效率；然后是如何屏蔽底层物理硬件及操作系统的差异并提供统一的对外概念；最后是如何使它的模型既严谨又宽松，保证语义不会产生歧义且能做一些优化扩展。

3.3 volatile 能否保证线程安全

为了提高执行性能，JMM 引入了工作内存和主存两个概念。在继续讨论之前必须先了解 4 种存储介质：寄存器、高级缓存（Cache）、随机存取存储器（RAM）和只读存储器（ROM）。寄存器是处理器中的一部分，而高级缓存是 CPU 设计人员为提高性能所引入的一个缓存，也可以说是处理器的一部分。

在利用 CPU 进行运算时必定涉及操作数的读取，假如 CPU 直接读取 ROM，那么这个读取速度将是无法忍受的，于是便引入了 RAM。这样做确实让速度提高了很多，但由于 CPU 的发展十分迅猛，RAM 受到技术及成本的限制而发展缓慢，此时就产生了一个很难调和的问题：CPU 的运算速度比从 RAM 读取数据的速度快了几个数量级。"木桶原理"告诉我们，桶的容量大小取决于最短的那块。由于存在这个问题，它必将影响处理器的效率，于是又引入了高级缓存。通过直接在 CPU 中添加几个不同级别的缓存，虽然它们的速度无法与寄存器相比，但是速度已经提升了很多，基本能跟 CPU 的计算速度相匹配。

由上可知，为了解决 CPU 运算速度与读取速度的问题，引入了多种存储机制。读取速度快慢的排序为寄存器 > Cache > RAM > ROM，如图 3.8 所示。可以用一个比较好理解但不完全正确的概念来解释，即寄存器距离 CPU 最近，所以读取最快；Cache 次之；RAM 第三；ROM 离得最远，自然速度最慢。当然不能完全用距离来说明这个问题，但用距离来解释比较好理解。另外的影响因素还涉及硬件设计、工作方式等。

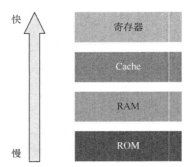

图 3.8 不同存储机制的读取速度

机器的这 4 种存储介质都是有联系的，程序在运行时一般会将 ROM 相关的程序数据读到 RAM 中，而需要运算的数据或运算过程中即将要用到的数据则会被读到 Cache 或寄存器中。假如要进行的运算所需要的所有数据及指令都在寄存器和 Cache 中，则这个运算过程会相当迅速，因为此时不存在性能瓶颈，运算速度与读取速度基本匹配。

CPU 读取数据的顺序则是先尝试读寄存器，如果不存在则尝试读 Cache。如果还不存在则读 RAM，最后才是读 ROM，如图 3.9 所示。假如 CPU 有三级 Cache，那么在读取时是一级一级往下，直到找到需要的操作数。设计精良的 CPU 的 3 级缓存能让命中率达到 95%以上。

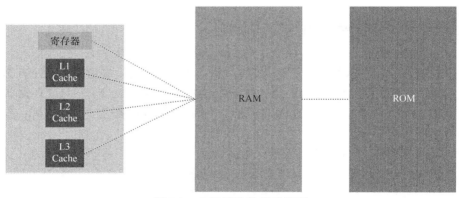

图 3.9 CPU 读取数据的顺序

有了上面的知识铺垫后再往下探索就水到渠成了。如果把 Java 内存模型与多级存储机制类比，我们能够发现 Java 为了提高性能而引入了工作内存的概念。可以把 Java 模型中的主存和工作内存分别与 RAM 和 Cache 或寄存器对应起来，每个线程的工作内存预先把需要的数据复制到 Cache 或寄存器（但是不保证所有的工作内存的变量副本都放在 Cache 中，也可能在 RAM 中，具体还要看 JVM 是如何实现的），这样就提高了线程执行时数据的读取速度（见图 3.10）。当然，寄存器和 Cache 由于成本原因存在容量大小限制的问题，这也是一个考验 JVM 实现的难题。

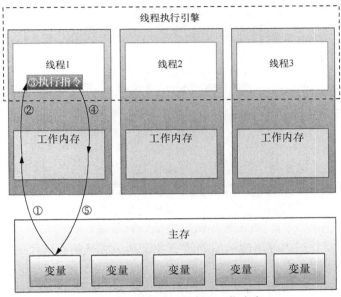

图 3.10 Java 中的主存和工作内存

一般来说，在引入一种机制解决了某个问题时，也会带来另外一些问题。数据同步就是带来的一个问题，即能否保证当前运算使用的变量值总是当前时刻最新的值。如果变量值并非最新值，将会导致数据的脏读，从而可能导致计算结果是错误的。这时可能有人会想到 Java 中有个 volatile 关键词，毫无疑问它能保证可见性，能让每个线程得到的都是主存中最新的变量值。但它是否足以保证多线程下执行结果的准确性呢？

一个典型的示例如代码清单 3.1 所示。在执行完所有线程任务，我们期望的结果会是 10*10 000。但实际却是一个小于 10*10 000 的数，乍一看没什么问题，但仔细一想就清楚了。count++编译后最终并非一个原子操作（后面章节会讲到），它由几个指令组合而成。在 Java 内存模型中，count++被分割成多个步骤，这些步骤不是一个具备原子性的整体。假如在执行的过程中其他线程读取了主存的 count 变量，这时就会产生脏读现象，从而造成结果错误。

代码清单 3.1　数据同步的问题

```
1.  public class VolatileTest {
2.
3.      private volatile static int count = 0;
4.
5.      public static void increase() {
6.          count++;
7.      }
8.
9.      public static void main(String[] args) {
10.         for (int i = 0; i < 10; i++)
```

```
11.         new Thread(() -> {
12.             for (int j = 0; j < 10000; j++)
13.                 increase();
14.         }).start();
15.     try {
16.         Thread.sleep(10000);
17.     } catch (InterruptedException e) {
18.     }
19.     System.out.println(count);
20. }
21. }
```

要解决这个问题其实不难,就是将这些操作变为原子操作。即保证某个线程在执行完 increase 操作之前不能有其他线程读取 count 变量。要达到这个目的,只需对 count 变量的访问加一个互斥锁即可。某个线程在执行 increase 操作之前需要先对 count 加锁,加锁成功后其他线程将无法对 count 进行访问。当该线程执行完后就释放锁,此时其他线程才能访问该变量。

volatile 可见性是通过内存屏障(后面章节会讲到)实现的。内存屏障的另一个作用就是强制刷新各种 CPU 缓存数据,因此任何 CPU 上的线程都可以读取到这些数据的最新版本。

3.4 happens-before 原则

happens-before(先发生)原则实际上是一种一致性模型,它主要规定了 Java 内存在多线程下操作的顺序性。如果某个操作先发生于另外一个操作,那么先发生的操作的执行结果将对后发生的操作可见。正式一点的描述就是:假设操作 A 和操作 B 由多线程执行,如果 A 操作先发生于 B 操作,那么 A 操作对内存产生的影响将对 B 操作可见,即执行 B 操作时能看到 A 操作对内存的改变。在图 3.11 中,线程 1 执行了 A 操作并对内存做了一些修改,此时如果 A 和 B 符合 happens-before 原则,那么线程 2 中的 B 操作就能看到 A 操作对内存的修改。

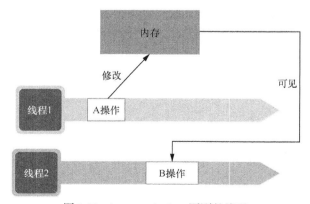

图 3.11　happens-before 原则的演示

为什么要制定 happens-before 原则呢？总体来说，happens-before 原则用于保证某个线程执行的某个操作能被其他线程中的某个操作可见。对于 Java 体系而言，它希望通过某些规范来将复杂的物理计算机的底层封装到 JVM 中，从而向上提供一种统一的内存模型语义。通过 happens-before 原则能够规范 JVM 的实现，同时又能向上层的 Java 开发人员描述多线程并发中的可见性问题，而且不必关注 JVM 复杂的实现，如图 3.12 所示。

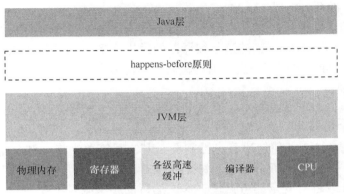

图 3.12　happens-before 原则的作用

下面通过几个例子来说明可见性问题以及如何解决可见性问题。在代码清单 3.2 中，有 x 和 y 两个共享变量，线程 2 负责更改 x 和 y 的值，而线程 1 则通过 while 循环检测 x 和 y 是否被线程 2 更改过，主线程启动线程 1 后睡眠 1s 再启动线程 2。原本我们期望的结果是当线程 2 启动后线程 1 应该输出 "thread1 可以看到变量改变"，但事实却是永远都不会输出该提示。

代码清单 3.2　可见性问题的演示

```
1.   public class VisibilityDemo {
2.       static int x = 0;
3.       static int y = 1;
4.
5.       public static void main(String[] args) throws InterruptedException {
6.           Thread thread1 = new Thread(() -> {
7.               while (true)
8.                   if (x == 2 && y == 3) {
9.                       System.out.println("thread1 可以看到变量改变");
10.                  }
11.          });
12.          Thread thread2 = new Thread(() -> {
13.              x = 2;
14.              y = 3;
15.          });
16.          thread1.start();
17.          Thread.sleep(1000);
```

```
18.            thread2.start();
19.        }
20. }
```

由于 JVM 体系非常复杂，我们写的 Java 层代码可能会在编译时被编译器改动，也可能会在执行时被改动。所以实际上无法百分之百地保证代码清单 3.2 就一定是严格的可见性示例，因为其中的 Java 代码可能会被编译器优化。而上面对可见性的分析也只是从 Java 层源码进行的，可能会发生下面两种情况。

- **情况 1**：编译器将线程 1 中的 while(true)if(x==2&&y==3)的顺序反过来了，变成 if(x==2&&y==3)while(true)。当然这是编译器的优化策略，我们无法控制。与此同时也导致了线程 1 只会去内存读一次 x 和 y，后面不再读内存，也就读不到被线程 2 修改的 x 和 y 的新值。
- **情况 2**：编译器不会改变 Java 层代码逻辑，线程 1 无限循环，处于繁忙状态，而不去读取内存中最新的 x 和 y 值。

接着看代码清单 3.3 如何解决可见性问题。其实很简单，将 x 和 y 都声明为 volatile 便可。关于 volatile，我们在前面分析过，每个线程对 volatile 变量的读写相当于直接对主内存读写，所以能保证线程对变量的可见性。这次运行的结果则是不断地输出"thread1 可以看到变量改变"。

代码清单 3.3　可见性问题的解决

```
1.  public class VisibilityDemo2 {
2.      static volatile int x = 0;
3.      static volatile int y = 1;
4.
5.      public static void main(String[] args) throws InterruptedException {
6.          Thread thread1 = new Thread(() -> {
7.              while (true)
8.                  if (x == 2 && y == 3) {
9.                      System.out.println("thread1 可以看到变量改变");
10.                 }
11.         });
12.         Thread thread2 = new Thread(() -> {
13.             x = 2;
14.             y = 3;
15.         });
16.         thread1.start();
17.         Thread.sleep(1000);
18.         thread2.start();
19.     }
20. }
```

最后，再提一下上面所讲的编译器优化问题，这里给大家抛出一些现象，大家可以自己去深入探索。我们在代码清单 3.2 的基础上分别增加 3 种情况的代码：调用 System.out.println()、调用 new Random().nextInt()、调用 Thread.sleep()，如代码清单 3.4 所示。最终可以发现这 3 种情况都能使线程 1 输出"thread1 可以看到变量改变",这说明线程 1 能看到线程 2 对 x 和 y 的修改。这是因为这些代码增加后阻止了编译器的优化操作,还是这些代码增加后使得 while 循环会去读取内存中的 x 和 y 的最新值呢?感兴趣的读者可以深入研究一下。

代码清单 3.4　在代码清单 3.2 的基础上分别增加 3 种情况的代码

```
1.  public class VisibilityDemo3 {
2.      static int x = 0;
3.      static int y = 1;
4.  
5.      public static void main(String[] args) throws InterruptedException {
6.          Thread thread1 = new Thread(() -> {
7.              while (true) {
8.                  if (x == 2 && y == 3)
9.                      System.out.println("thread1 可以看到变量改变");
10. //                 System.out.println("x,y = " + x + "," + y);//第一种
11.                  new Random().nextInt();//第二种
12. //                 try {
13. //                     Thread.sleep(2000);//第三种
14. //                 } catch (InterruptedException e) {
15. //                 }
16.              }
17.          });
18.          Thread thread2 = new Thread(() -> {
19.              x = 2;
20.              y = 3;
21.          });
22.          thread1.start();
23.          Thread.sleep(1000);
24.          thread2.start();
25.      }
26.  
27.  }
```

Java 并发常见的 happens-before 原则一般分为 8 个,其中很多都是我们非常熟悉的,只是我们很少从 happens-before 的角度去理解它。下面分别介绍这 8 个原则。

3.4.1　单线程原则

单线程原则是最简单的 happens-before 规则,就是说在单个线程内,前面的代码先发生于后面的代码。比如在代码清单 3.5 中,主线程内的 step-1 比 step-2 输出更早,4 个输出操作按照代码顺序执行。

代码清单 3.5　单线程原则的演示

```
1.  public class VisibilityDemo4 {
2.  
```

```
3.     public static void main(String[] args) {
4.         System.out.println("step-1");
5.         System.out.println("step-2");
6.         System.out.println("step-3");
7.         System.out.println("step-4");
8.     }
9.
10. }
```

3.4.2 锁原则

锁原则是指某个锁解锁前的操作先发生于接下来获取该锁后的其他操作。以 synchronized 锁为例，我们都知道进入和离开 synchronized 大括号分别对应的是加锁和解锁操作，假如线程 1 先获取锁，则在解锁前的所有操作都先发生于线程 2 获取该锁后的其他操作；反之亦然。

由于 happens-before 原则保证了可见性，代码清单 3.6 可能存在以下两种输出情况。

- 情况 1：线程 1 解锁前的操作对线程 2 获取锁后的操作可见，所以线程 1 的 x=3 而线程 2 的 x=7。
- 情况 2：线程 2 解锁前的操作对线程 1 获取锁后的操作可见，所以线程 2 的 x=4 而线程 1 的 x=7。

代码清单 3.6　锁原则的演示

```
1.  public class VisibilityDemo5 {
2.      static int x = 0;
3.      static Object lock = new Object();
4.
5.      public static void main(String[] args) throws InterruptedException {
6.          Thread thread1 = new Thread(() -> {
7.              synchronized (lock) {
8.                  x = x + 3;
9.                  System.out.println("thread1 x = " + x);
10.             }
11.         });
12.         Thread thread2 = new Thread(() -> {
13.             synchronized (lock) {
14.                 x = x + 4;
15.                 System.out.println("thread2 x = " + x);
16.             }
17.         });
18.         thread1.start();
19.         thread2.start();
20.     }
21. }
```

3.4.3 volatile 原则

volatile 原则是指对某个 volatile 变量的写操作和写操作之前的所有操作都先发生于对这个 volatile 变量的读操作和读操作之后的所有操作。比如在代码清单 3.7 中，x 是 volatile 变量，而 y 为非 volatile 变量。线程 1 对 y 和 x 进行写操作，那么线程 2 在对 x 和 y 读操作时就能看到 x 和 y 的最新值，即输出 "thread2 x,y = 4,2"。这里 x 写操作前的所有操作都对 x 读操作后的所有操作可见。

代码清单 3.7　volatile 原则的演示

```java
public class VisibilityDemo6 {

    static volatile int x = 0;
    static int y = 0;

    public static void main(String[] args) throws InterruptedException {
        Thread thread1 = new Thread(() -> {
            y = 2;
            x = x + 4;
        });
        Thread thread2 = new Thread(() -> {
            System.out.println("thread2 x,y = " + x + "," + y);
        });
        thread1.start();
        Thread.sleep(1000);
        thread2.start();
    }
}
```

3.4.4 线程 start 原则

线程 start 原则是指某个线程在调用另外一个线程的 start 方法前，所有操作都先发生于刚被启动的线程中的所有操作。在代码清单 3.8 中，主线程中先执行 x=3，然后创建 thread1 并调用它的 start 方法。运行后输出为 "thread1 x = 6"，调用 start 方法前的 x=3 在线程 1 中是可见的。

代码清单 3.8　线程 start 原则的演示

```java
public class VisibilityDemo7 {

    static int x = 0;

    public static void main(String[] args) {
        x = 3;
```

```
7.      Thread thread1 = new Thread(() -> {
8.          x = x * 2;
9.          System.out.println("thread1 x = " + x);
10.     });
11.     thread1.start();
12. }
13.
14. }
```

3.4.5 线程 join 原则

线程 join 原则是指如果线程 A 调用了线程 B 的 join 方法，那么线程 B 的所有操作都先发生于线程 A 中 join 方法后面的所有操作。在代码清单 3.9 中，主线程启动线程 1 后调用 join 方法并等待线程 1 执行完才返回，那么线程 1 中的 x=x+2 操作对主线程中 join 后面的操作可见，所以输出结果为 "main-thread x = 2"。

代码清单 3.9　线程 join 原则

```
1.  public class VisibilityDemo8 {
2.
3.      static int x = 0;
4.
5.      public static void main(String[] args) throws InterruptedException {
6.          Thread thread1 = new Thread(() -> {
7.              x = x + 2;
8.          });
9.          thread1.start();
10.         thread1.join();
11.         System.out.println("main-thread x = " + x);
12.     }
13.
14. }
```

3.4.6 线程 interrupt 原则

线程 interrupt 原则是指线程 1 调用了线程 2 的 interrupt 方法，那么线程 1 在调用 interrupt 方法之前的所有操作都先发生于线程 1 中被 interrupt 后的所有操作。在代码清单 3.10 中，线程 1 启动后开始睡眠，主线程执行 x=x+2 后睡眠 2s，然后调用线程 1 的 interrupt 方法，线程 1 被中断后输出 "thread1 x = 2"。

代码清单 3.10　线程 interrupt 原则

```
1.  public class VisibilityDemo9 {
2.
3.      static int x = 0;
```

```
4.
5.      public static void main(String[] args) throws InterruptedException {
6.          Thread thread1 = new Thread(() -> {
7.              try {
8.                  Thread.sleep(20000);
9.              } catch (InterruptedException e) {
10.             }
11.             System.out.println("thread1 x = " + x);
12.         });
13.         thread1.start();
14.         x = x + 2;
15.         Thread.sleep(2000);
16.         thread1.interrupt();
17.     }
18.
19. }
```

3.4.7 finalize 原则

finalize 原则是指对象的所有操作都先发生于该对象的 finalize 方法。在代码清单 3.11 中，创建 VisibilityDemo10 对象后对其属性进行修改，然后销毁该对象。调用 System.gc()能使该对象被垃圾回收器回收，回收前会调用 finalize 方法，该方法输出"finalize 方法 x = 4"。

代码清单 3.11　finalize 原则的演示

```
1.  public class VisibilityDemo10 {
2.
3.      int x = 0;
4.
5.      private void updateX(int newX) {
6.          this.x = newX;
7.      }
8.
9.      protected void finalize() throws Throwable {
10.         System.out.println("finalize 方法 x = " + x);
11.     }
12.
13.     public static void main(String[] args) {
14.         VisibilityDemo10 demo = new VisibilityDemo10();
15.         demo.updateX(4);
16.         demo = null;
17.         System.gc();
18.     }
19.
20. }
```

3.4.8 传递原则

传递原则是指如果线程 A 先发生于线程 B，且线程 B 先发生于线程 C，那么线程 A 就先发生于线程 C。在代码清单 3.12 中，主线程中的 x=3 操作先发生于线程 1 中的所有操作，线程 1 中的 x=x*2 操作先发生于线程 2 中的所有操作，最终输出为"thread2 x = 12"。由传递原则可以知道，主线程中的 thread1.start()之前的操作先发生于线程 2 中的所有操作。

代码清单 3.12 传递原则的演示

```java
1.   public class VisibilityDemo11 {
2.
3.       static int x = 0;
4.
5.       public static void main(String[] args) {
6.           x = 3;
7.           Thread thread1 = new Thread(() -> {
8.               x = x * 2;
9.               Thread thread2 = new Thread(() -> {
10.                  x = x * 2;
11.                  System.out.println("thread2 x = " + x);
12.              });
13.              thread2.start();
14.          });
15.          thread1.start();
16.      }
17.
18.  }
```

happens-before 原则可以帮助我们从 Java 语言层面去理解可见性。常见的 8 种 happens-before 原则包括单线程原则、锁原则、volatile 原则、线程 start 原则、线程 join 原则、线程 interrupt 原则、finalize 原则以及传递原则，每个原则都给出了例子以帮助大家更好地理解。

3.5 Java 指令重排

"你看到的不一定就是你以为的！"，用这句话来描述指令重排非常贴切。我们在语言层面编写代码时，是按照思维和习惯去编写的，但编译器和 CPU 执行时的顺序却可能与代码顺序不一样。因为每个层面都有各自需要关注和考虑的事情，编译器和 CPU 可能会对我们编写的代码先优化再执行，以提高执行效率。所以指令重排其实是为了提升机器执行效率而提出的一种措施。比如在图 3.13 中，左边的是原来的代码，通过编译器编译或被 CPU 优化后的代码可能会变成右边那样的顺序。当然这里只是举个例子，并不是说一定会这样重排。

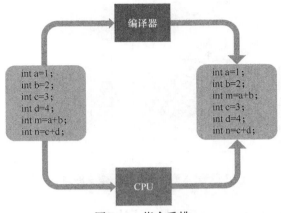

图 3.13 指令重排

为了更好地理解指令重排的作用，我们先来了解流水线相关的技术。在没有使用流水线之前，所有指令是一条条执行的，只有当一条指令执行完后才会执行下一条指令。在图 3.14 中，一共有 4 条指令，假设每条指令都需要 3 个单位的 CPU 时间，那么 4 条指令总共耗时 12 个单位的 CPU 时间。

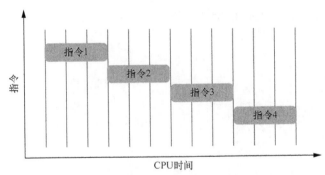

图 3.14　4 条指令的 CPU 时间耗时

那么在引入流水线技术后指令是如何执行的呢？每个指令将被拆分成若干个部分，在同一个单位的 CPU 时间内可以执行多个指令的不同部分，这样就能提升执行效率。在图 3.15 中，每个指令都被分成 3 部分，指令 1 的第 2 部分和指令 2 的第 1 部分可以同时执行，最终 4 个指令执行完后一共花费了 6 个单位的 CPU 时间。可见执行耗时优化了很多。那么流水线技术为什么能这样执行呢？主要是因为一条指令在执行时，不同阶段会涉及不同的硬件部分，比如取指阶段使用指令通路和指令寄存器，译码阶段使用指令译码器，执行阶段使用执行单元和数据通路。因此，可以将指令分成多个部分，并且不同指令的不同阶段能够同时执行。总体来说，流水线的本质就是充分利用 CPU 的各个硬件部分，由不同硬件部分执行不同指令的不同部分，看起来就像是并行的效果。因此从多个指令的整体效果来看，执行性能大大提升。

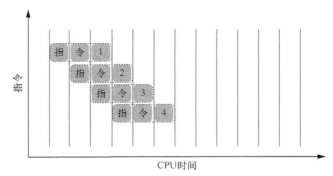

图 3.15　引入流水线技术后指令的执行耗时

接着看为什么要指令重排。指令重排是为了让 CPU 能更快地执行完所有指令，也就是说指令重排能让流水线的效果更好。代码在生成指令时，前后指令可能是相关的，比如后面一条指令依赖于前面指令执行的结果，那么在流水线的执行过程中就可能导致空等待而白白浪费 CPU 时间。我们来看图 3.16 和图 3.17，假设原来是按照指令 1、2、3、4 的顺序执行的，而且指令 2 依赖于指令 1 的结果。那么在不重排指令的情况下，需要等到指令 1 执行完后才能执行指令 2。但是，如果将指令 2 当成最后一条指令，则流水线能够充分执行，因为指令 3 和指令 4 不需要指令 1 的结果，而当要执行指令 2 时，指令 1 已经执行完并得到结果了。

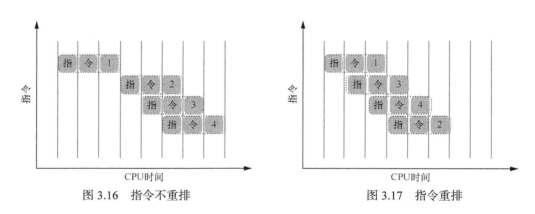

图 3.16　指令不重排　　　　　　　　图 3.17　指令重排

指令重排也不是随意重排，它需要遵守一定的原则。指令重排的原则就是不能影响到程序在单线程下的准确性，就是说，不管怎么重排都要保证其与重排前在单线程中执行的结果相同。比如在图 3.18 中，尽管指令被重排了，但最终 m 和 n 的结果与重排前相同。

但对于多线程来说，指令重排却可能导致程序执行出现错误的结果，这也是指令重排的弊端。虽然它能提升执行效率，但同样也会引入多线程问题。比如在代码清单 3.13 中，假如存在线程 1 调用 method1 方法且线程 2 调用 method2 方法的情况，如果不进行指令重排，则不会输出 a=2；如果 method1 被重排，则可能 flg=true 在 a=1 前面，程序可能会输出 a=2。

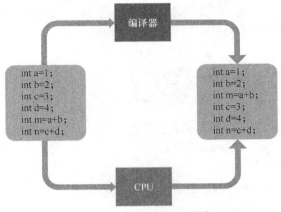

图 3.18　指令重排的原则

代码清单 3.13　指令重排引入的多线程问题

```
1.    public class InstructionReorderingDemo2 {
2.
3.        static int a = 2;
4.        static boolean flg = false;
5.
6.        public static void method1() {
7.            a = 1;
8.            flg = true;
9.        }
10.
11.       public static void method2() {
12.           if (flg && a == 2) {
13.               System.out.println("a = " + a);
14.           }
15.       }
16.
17.   }
```

　　指令重排主要发生在编译和运行时两个阶段,这两个阶段所对应的主角分别为编译器和 CPU。

　　我们先看编译阶段重排,总体来说就是源代码经过编译器编译后成为机器指令,而机器指令可能被重排。对于 Java 来说,就是 java 源文件被 javac 编译后成为字节码指令,字节码则可能被重排(见图 3.19)。

　　运行时阶段重排,指的是机器指令在被 CPU 执行时可能会先被 CPU 重排然后才执行。对于 Java 来说,就是字节码被 Java 执行器执行时可能会先重排后执行(见图 3.20)。

　　从上面的分析可以知道,指令重排的阶段和原因可能是多种多样的,特别是对 Java 来说更复杂,因为 Java 层有不同的编译器和运行时环境。那么 Java 要如何解决指令重排的问题呢?简单地说,解决指令重排的方法就是通过前面讲解的 happens-before 原则,根据这些原则就能够在某些节点上控制多线程执行的顺序。如图 3.21 所示,我们的源代码对应的指令是指令 1、2、3、4，为了防止指令 1、2 与指令 3、4 重排，我们可以根据实际情况在中间增加 happens-before 原则以阻止它们重排。

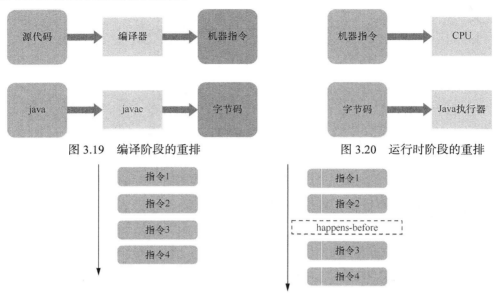

图 3.19　编译阶段的重排　　　　　　　图 3.20　运行时阶段的重排

图 3.21　增加 happens-before 原则阻止重排

对 Java 的 happens-before 原则深入研究后可知，它在 JVM 中的实现使用了内存屏障。所以，指令的重排实际上是通过内存屏障来实现的。通过内存屏障可以强制对内存进行顺序约束，而且它作用于所有线程。内存屏障通过在代码中插入内存屏障代码，使得指令无法越过屏障进行重排。在图 3.22 中，两个屏障分别将 6 条指令分割开，从而阻止它们进行重排。在编译或运行时可以将内存屏障指令插入到指定位置来阻止重排，不同厂商的 CPU 或操作系统可能提供了不同的内存屏障指令，所以在 JVM 中需要对不同的 CPU 或操作系统分别处理。

图 3.22　使用内存屏障阻止重排

第 4 章
并发知识

4.1 synchronized 互斥锁

synchronized 关键词是 Java 语言为开发人员提供的同步工具,可以将其看成一种"语法糖"。synchronized 要解决的问题是多线程并发执行过程中数据同步的问题。不像其他编程语言(比如 C++),在处理同步问题时都需要自己进行锁处理,从这点上看,Java 提供的这个关键词确实很方便。

Java 通过 synchronized 指定同步块,从而能在该指定块中避免数据竞争问题。对方法进行声明实际上也有一个对应的同步块范围,而且会指定一个对应的锁对象。同一时刻只有一个线程能进入锁中,其他线程必须等待锁里面的线程出来后才能依次进入,如图 4.1 所示。

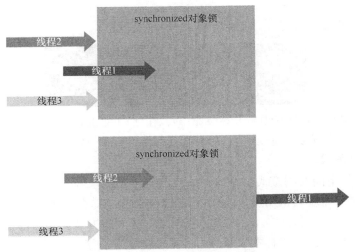

图 4.1 通过 synchronized 来指定同步块

synchronized 实现的同步语义是互斥锁,即对于某个锁来说,任意时刻只能有一个线程获得该锁。可以通过代码清单 4.1 来理解互斥,其中使用 synchronized 定义了一个方法,然后在主线程中创建两个线程并发地调用该方法。

代码清单 4.1　互斥

```
1.  public class SynchronizedDemo3 {
2.
3.      public synchronized void method(String name) {
4.          System.out.println(name + " gets the lock.");
5.          try {
6.              Thread.sleep(3000);
7.          } catch (InterruptedException e) {
8.          }
9.          System.out.println(name + " releases the lock after 3s.");
10.     }
11.
12.     public static void main(String[] args) throws InterruptedException {
13.         SynchronizedDemo3 demo = new SynchronizedDemo3();
14.
15.         new Thread(() -> {
16.             demo.method("thread1");
17.         }).start();
18.
19.         new Thread(() -> {
20.             demo.method("thread2");
21.         }).start();
22.
23.     }
24. }
```

某次的执行输出如下所示。

```
thread1 gets the lock.
thread1 releases the lock after 3s.
thread2 gets the lock.
thread2 releases the lock after 3s.
```

由输出可知，thread1 获取锁后开始休眠 3s，在此期间 thread2 没有办法执行 method 方法里面的逻辑，因为它没有办法获取锁。只有在 3s 后 thread1 释放锁后，thread2 才能获取锁，并开始执行里面的逻辑。

synchronized 的不同使用方式会导致不一样的作用范围。可通过修饰不同的对象来实现锁的范围，在代码上则体现为作用块。synchronized 实际上大致分为 4 类不同的作用方式，分别作用在对象方法上、类静态方法上、对象方法里面、类静态方法里面。不管是作用在哪里，我们只要从锁对象的角度去理解它就很容易了。下面依次看看这 4 种情况。

4.1.1　作用在对象方法上

synchronized 作用在对象方法上，即用 synchronized 来描述对象的方法，表示该对象的方法具有同步性。我们知道，在面向对象编程中，定义了类之后会通过 new 来实例化对象，而

synchronized 描述的是对象的方法,所以这个作用范围就是在对象上,对象充当了锁。需要注意的是,类可以实例化多个对象,这时每个对象都是一个锁,即每个锁的范围都是相对于自己的对象来说的。

下面看代码清单 4.2,里面定义了 add 和 add2 两个方法,其中一个方法声明为 synchronized。这两个方法都是让各自的变量完成自加操作。注意,这里的++操作实际上是非原子的。在主线程中分别启动 10 个线程,并循环 10 000 次去调用 add 和 add2 方法,然后让主线程睡眠 3s (这是为了让所有线程都执行完毕)。最终某次执行的输出结果为 count = 75608,count2 = 100000。这里我们不关心 count 的具体值,需要关注的是 count 的值肯定小于 100 000,也就是说 add 方法缺乏同步而导致线程安全问题。而 count2=100000 则是正确的结果,因为 add2 方法具有同步性。

代码清单 4.2　将 synchronized 作用在对象方法上

```
1.  public class SynchronizedDemo2 {
2.
3.      int count = 0;
4.      int count2 = 0;
5.
6.      public void add() {
7.          count++;
8.      }
9.      public synchronized void add2() {
10.         count2++;
11.     }
12.
13.     public static void main(String[] args) throws InterruptedException {
14.         SynchronizedDemo2 demo = new SynchronizedDemo2();
15.
16.         for (int i = 0; i < 10; i++)
17.             new Thread(() -> {
18.                 for (int j = 0; j < 10000; j++)
19.                     demo.add();
20.             }).start();
21.
22.         for (int i = 0; i < 10; i++)
23.             new Thread(() -> {
24.                 for (int j = 0; j < 10000; j++)
25.                     demo.add2();
26.             }).start();
27.
28.         Thread.sleep(3000);
29.         System.out.println("count = " + demo.count);
30.         System.out.println("count2 = " + demo.count2);
31.     }
32. }
```

如果用 synchronized 来声明两个不同的对象方法，情况会是怎样的呢？代码清单 4.3 分别把 method 和 method2 两个方法声明为 synchronized。

代码清单 4.3　把方法 method 和 method2 声明为 synchronized

```
1.   public class SynchronizedDemo8 {
2.
3.       public synchronized void method(String name) {
4.           System.out.println(name + " gets the lock.");
5.           sleep(3000);
6.           System.out.println(name + " releases the lock after 3s.");
7.       }
8.
9.       public synchronized void method2(String name) {
10.          System.out.println(name + " gets the lock.");
11.          sleep(3000);
12.          System.out.println(name + " releases the lock after 3s.");
13.      }
14.
15.      public static void sleep(int s) {
16.          try {
17.              Thread.sleep(s);
18.          } catch (InterruptedException e) {
19.          }
20.      }
21.
22.      public static void main(String[] args) throws InterruptedException {
23.          SynchronizedDemo8 demo = new SynchronizedDemo8();
24.
25.          new Thread(() -> {
26.              demo.method("thread1");
27.          }).start();
28.
29.          new Thread(() -> {
30.              demo.method2("thread2");
31.          }).start();
32.
33.      }
34.  }
```

此时执行的结果如下所示。同样是两个线程会产生互斥效果。声明的两个不同对象方法都以当前对象为锁，自然就会产生互斥效果。

```
thread1 gets the lock.
thread1 releases the lock after 3s.
thread2 gets the lock.
thread2 releases the lock after 3s.
```

4.1.2 作用在类静态方法上

synchronized 作用在类静态方法上，即用 synchronized 来描述类的静态方法，表示该方法具有同步性。作用在对象方法上即是以对象作为锁，而作用在类静态方法上则是以类（class）作为锁。我们要知道某个类本身也是一个对象，JVM 就是使用这个对象作为模板去生成该类的对象的，比如 xxx.class 对应的 class 对象。

类似地，在代码清单 4.4 中，add 是一个静态方法，且被定义为具有同步性。所以该方法具有同步性，不存在线程安全问题，最终的输出结果为 count = 100000。

代码清单 4.4 将 synchronized 作用在类静态方法上

```
1.   public class SynchronizedDemo4 {
2.
3.       static int count = 0;
4.
5.       public synchronized static void add() {
6.           count++;
7.       }
8.
9.       public static void main(String[] args) throws InterruptedException {
10.
11.          for (int i = 0; i < 10; i++)
12.              new Thread(() -> {
13.                  for (int j = 0; j < 10000; j++)
14.                      add();
15.              }).start();
16.
17.          Thread.sleep(3000);
18.          System.out.println("count = " + count);
19.      }
20.  }
```

作用在对象方法上与作用在类静态方法上的锁是不相同的。在代码清单 4.5 中，method 和 method2 两个方法分别是非静态方法和静态方法，这时输出结果如下。

```
thread1 gets the lock.
thread2 gets the lock.
thread1 releases the lock after 3s.
thread2 releases the lock after 3s.
```

两个线程不产生互斥效果，这是因为 method 方法以当前对象作为锁，而 method2 方法则以 SynchronizedDemo5.class 对象作为锁。锁不同自然就不存在互斥效果。

代码清单 4.5　将 synchronized 作用在对象方法上和作用在类静态方法的区别

```
1.   public class SynchronizedDemo5 {
2.
3.       public synchronized void method(String name) {
4.           System.out.println(name + " gets the lock.");
5.           sleep(3000);
6.           System.out.println(name + " releases the lock after 3s.");
7.       }
8.
9.       public static synchronized void method2(String name) {
10.          System.out.println(name + " gets the lock.");
11.          sleep(3000);
12.          System.out.println(name + " releases the lock after 3s.");
13.      }
14.
15.      public static void sleep(int s) {
16.          try {
17.              Thread.sleep(s);
18.          } catch (InterruptedException e) {
19.          }
20.      }
21.
22.      public static void main(String[] args) throws InterruptedException {
23.          SynchronizedDemo5 demo = new SynchronizedDemo5();
24.
25.          new Thread(() -> {
26.              demo.method("thread1");
27.          }).start();
28.
29.          new Thread(() -> {
30.              method2("thread2");
31.          }).start();
32.
33.      }
34.  }
```

4.1.3　作用在对象方法里面

synchronized 作用在对象方法里面，即用 synchronized 来描述方法内部的某块逻辑，表示该块逻辑具有同步性。这时需要我们指定锁对象，比如常见的 synchronized(this)就是将当前对象作为锁；也可以自己创建一个对象来作为锁。

我们看代码清单 4.6，不像前面那样在方法上声明 synchronized，而是在 add 方法内部通过 synchronized(this){xxx}的形式来声明同步块。这里的同步块包括 count++操作，所以该操作具有同步性，也就能够避免线程安全问题。在实际使用中，同步块并不要求包含整个方法的所有代码,可以是方法内的任意代码块。在主线程上分别用 10 个线程循环调用 10 000 次 add 方法和 add2 方法,最终运行的结果是 count = 100000,count2 = 100000。也就是说,两个方法都达到了同步的效果,而且 add 方法和 add2 方法都使用了当前对象作为锁,所以这两个方法其实是共用一个锁。

代码清单 4.6　将 synchronized 作用在对象方法里面

```
1.   public class SynchronizedDemo6 {
2.
3.       int count = 0;
4.       int count2 = 0;
5.
6.       public void add() {
7.           synchronized (this) {
8.               count++;
9.           }
10.      }
11.
12.      public synchronized void add2() {
13.          count2++;
14.      }
15.
16.      public static void main(String[] args) throws InterruptedException {
17.          SynchronizedDemo6 demo = new SynchronizedDemo6();
18.
19.          for (int i = 0; i < 10; i++)
20.              new Thread(() -> {
21.                  for (int j = 0; j < 10000; j++)
22.                      demo.add();
23.              }).start();
24.
25.          for (int i = 0; i < 10; i++)
26.              new Thread(() -> {
27.                  for (int j = 0; j < 10000; j++)
28.                      demo.add2();
29.              }).start();
30.
31.          Thread.sleep(3000);
32.          System.out.println("count = " + demo.count);
33.          System.out.println("count2 = " + demo.count2);
34.      }
35.  }
```

4.1.4　作用在类静态方法里面

synchronized 作用在类静态方法里面,即用 synchronized 来描述静态方法内部的某块逻辑,表示该块逻辑具有同步性。这时需要我们指定锁对象,比如 synchronized(xxx.class) {xxx}则是以该 class 对象作为锁对象；也可以自己创建一个对象来作为锁。

在代码清单 4.7 中，在 method 方法里面通过 synchronized (SynchronizedDemo7.class) {} 来声明同步块，而 method2 则直接在类静态方法上声明。

代码清单 4.7　将 synchronized 作用在类静态方法里面

```
1.  public class SynchronizedDemo7 {
2.
3.      public static void method(String name) {
4.          synchronized (SynchronizedDemo7.class) {
5.              System.out.println(name + " gets the lock.");
6.              sleep(3000);
7.              System.out.println(name + " releases the lock after 3s.");
8.          }
9.      }
10.
11.     public static synchronized void method2(String name) {
12.         System.out.println(name + " gets the lock.");
13.         sleep(3000);
14.         System.out.println(name + " releases the lock after 3s.");
15.     }
16.
17.     public static void sleep(int s) {
18.         try {
19.             Thread.sleep(s);
20.         } catch (InterruptedException e) {
21.         }
22.     }
23.
24.     public static void main(String[] args) throws InterruptedException {
25.         new Thread(() -> {
26.             method("thread1");
27.         }).start();
28.
29.         new Thread(() -> {
30.             method2("thread2");
31.         }).start();
32.
33.     }
34. }
```

运行结果如下。这两种情况其实使用了同一个对象作为锁，即都是用 SynchronizedDemo7.class 作为锁。所以 thread1 线程和 thread2 线程在任一时刻只能有一个线程持有该锁，从而达到了互斥的效果。

```
thread1 gets the lock.
thread1 releases the lock after 3s.
thread2 gets the lock.
thread2 releases the lock after 3s.
```

为了深入理解 synchronized 的实现原理，我们来看看 synchronized 被编译后生成什么指令。代码清单 4.8 是 add 方法编译前后的情况，其中黑屏部分是编译后生成的字节码指令。可以看到，通过 synchronized 声明的方法会比普通方法多了 monitorenter 和 monitorexit 两种指令，也就是由这两种指令来指定某个同步块，在这个同步块中，任意时刻只能有一个线程在里面。多出的一个 monitorexit 指令用于应对异常情况，即发生异常时要通过该命令来释放锁。

代码清单 4.8　通过 synchronized 声明的 add 方法在编译前后的情况

```
1.  public void add() {
2.      synchronized (this) {
3.          count++;
4.      }
5.  }
```

```
public void add();
  descriptor: ()V
  flags: (0x0001) ACC_PUBLIC
  Code:
    stack=3, locals=2, args_size=1
       0: aload_0
       1: dup
       2: astore_1
       3: monitorenter
       4: aload_0
       5: dup
       6: getfield      #13       // Field count:I
       9: iconst_1
      10: iadd
      11: putfield      #13       // Field count:I
      14: aload_1
      15: monitorexit
      16: goto          22
      19: aload_1
      20: monitorexit
      21: athrow
      22: return
    Exception table:
       from    to  target type
           4   16     19   any
          19   21     19   any
```

再看一下对方法声明 synchronized 的情况。对于 add2 方法（见代码清单 4.9），编译后生成的字节码指令如下面的黑屏部分所示。这时不生成 monitorenter 与 monitorexit，取而代之的是通过 ACC_SYNCHRONIZED 标识来表示同步。JVM 在执行时就会根据该标识来实现互斥锁，任意时刻都只能有一个线程执行该方法。

代码清单 4.9　声明 synchronized 的 add2 方法

```
1.  public synchronized void add2() {
2.      count2++;
3.  }
```

```
public synchronized void add2();
  descriptor: ()V
  flags: (0x0021) ACC_PUBLIC, ACC_SYNCHRONIZED
  Code:
    stack=3, locals=1, args_size=1
       0: aload_0
       1: dup
       2: getfield      #15         // Field count2:I
       5: iconst_1
       6: iadd
       7: putfield      #15         // Field count2:I
      10: return
    LineNumberTable:
      line 15: 0
      line 16: 10
    LocalVariableTable:
      Start  Length  Slot  Name  Signature
          0      11     0  this  Lcom/seaboat/thread/SynchronizedDemo6;
```

下面看一下 monitorenter 和 monitorexit 是如何协同工作的。在 JVM 执行指令的过程中，当遇到 monitorenter 指令时要去获取互斥锁，而当遇到 monitorexit 指令时则要释放互斥锁，这样就能在 Java 层面实现同步机制。此外，如果获取锁失败，则将线程放到阻塞队列中，当其他线程释放锁时，会通知阻塞队列的线程去竞争以获取锁。

我们根据图 4.2 来理解这个过程。现有 3 个线程，它们同时执行 monitorenter 指令且都使用相同的锁对象。3 个线程竞争锁，线程 2 成功获取锁，于是线程 1 和线程 3 被放到阻塞队列中。线程 2 执行 monitorexit 指令后就离开了同步区域，此时它会通知阻塞队列中的线程。线程 1 被唤醒后，获取锁并往下执行。类似地，当线程 1 执行 monitorexit 指令后，也同样会通知并唤醒线程 3 去获取锁，然后往下执行。

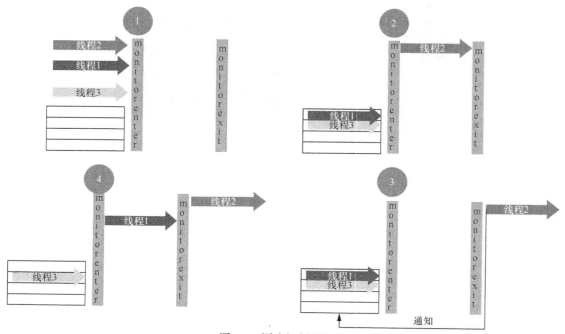

图 4.2　同步机制的实现

JVM 一般通过 Monitor（监控器）来实现 monitorenter 指令和 monitorexit 指令。Monitor 一般包括一个阻塞队列和一个等待队列。其中阻塞队列用来保存锁竞争失败的线程，它们处于阻塞状态，而等待队列则用来保存在 synchronized 块中调用 wait 方法后放置的队列。需要注意的是，调用 wait 方法后会释放锁并通知阻塞队列。

　　我们通过图 4.3 来加深对 Monitor 的理解。上面提到，Monitor 包含阻塞队列和等待队列，刚开始时 3 个线程准备进入 Monitor。其中线程 2 成功获取锁并进入 Monitor，所以线程 1 和线程 3 被加入到阻塞队列中。假如此时线程 2 调用了 wait 方法，那么该线程就会被加入到等待队列中，而且释放锁并唤醒阻塞队列中的线程 1，从而让线程 1 能够获取锁。当线程 1 执行完整个同步块后，则释放锁，并会通知阻塞队列中的线程 3。被唤醒的线程 3 将获取锁并往下执行。而等待队列中的线程 2 则需要其他线程调用 notify 方法来唤醒它，才能继续往下执行。

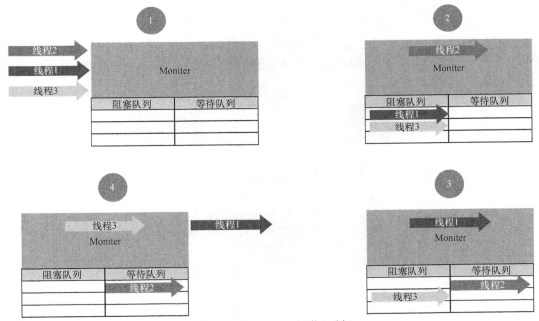

图 4.3　Monitor 的工作机制

4.2　乐观的并发策略

　　悲观者与乐观者的做事方式完全不一样。悲观者认为，我必须能够完全控制一件事情才会去做，否则在做这件事情时一定会出问题。而乐观者则相反，他们认为凡事不管最终结果如何，先做了再说，最后大不了不成功。同样，这也是悲观策略与乐观策略的区别。悲观策略会把整个对象加锁并占为己有后才去操作，而乐观策略在执行时则不必先获取锁，但会通过一定的检测手段决定是否更新数据。

4.1 节讨论的 synchronized 互斥锁属于悲观策略，它有一个明显的缺点，即不管数据是否存在竞争都加锁。实际上在数据不存在竞争的情况下，是可以允许多个线程同时访问的，比如多个线程只对数据做读取操作而不对其进行修改时，就可以让多个线程来访问该数据。有什么办法能解决这个问题呢？答案是基于冲突检测的乐观策略。这种策略没有所谓的锁概念，每个线程都先去执行操作，操作完成后再检测是否与其他线程存在共享数据冲突。如果没有则让操作成功；如果存在冲突则可以重新执行操作并检测，直到成功为止。

乐观策略的核心算法是 CAS（Compare And Swap，比较并交换），它涉及 3 个操作数：内存值、预期值、新值。当且仅当预期值和内存值相等时才将内存值修改为新值。CAS 的具体逻辑是：首先检查某个内存值是否与该线程之前读取时的值一样，如不一样则表示期间此内存值已经被别的线程更改过了，于是舍弃本次操作；否则说明期间没有其他线程更改过此内存值，于是可以用新值来更新内存值。

在图 4.4 中，两个线程并发地对某内存进行操作。线程 2 先读取某内存值作为预期值，然后执行到某个节点时决定将新值更新到内存中。如果线程 1 在此期间修改了内存值，则通过 CAS 算法就可以检测出来。假如内存值与预期值相同，则说明不存在数据冲突问题，于是线程 2 可将新值更新到内存中。

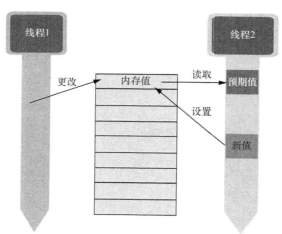

图 4.4　两个线程并发访问内存

CAS 操作具有原子性吗？是的，它的原子性由 CPU 硬件指令来保证，并通过 Java 本地接口（JNI）调用本地硬件级别指令来实现。假如我们想让 CAS 通过互斥锁来实现原子性，尽管也能实现，但用这种方式来保证原子性就没有实际意义了。

乐观策略避免了悲观策略独占对象的问题，同时也提高了并发性能。但它也存在以下缺点。

- 乐观策略只能保证一个共享变量的原子操作。如果有多个变量，CAS 就力不从心。而互斥锁却能轻易解决这个问题，不管对象数量多少及对象颗粒度大小。

- 长时间的循环操作可能导致开销较大。假如 CAS 长时间操作不成功，它就会一直进行循环操作，这会给 CPU 带来很大的开销。
- ABA 问题。CAS 的核心思想是通过比对内存值与预期值是否一样来判断内存值是否被改过，但这个判断逻辑不严谨。比如内存值原来是 A，后来被一个线程改为 B，最后又被改成了 A，则 CAS 认为此内存值并没有发生改变，但实际上它被其他线程改过。这种情况对依赖过程值的运算结果影响很大，甚至导致最终结果出现错误。该问题的解决思路是引入版本号，每次变量更新都把版本号加 1。

4.3 自旋锁

自旋锁是实现同步的一种方案。它是一种非阻塞锁，与常规锁的主要区别就在于获取锁失败后的处理方式不同。常规锁会将线程阻塞并在适当时唤醒它，而自旋锁的核心机制就在"自旋"两个字，即用自旋操作来替代阻塞操作。某一线程尝试获取某个锁时，如果该锁已经被另一个线程占用，则此线程将不断循环检查该锁是否被释放，而不是让此线程阻塞。一旦另外一个线程释放该锁，此线程就能获得该锁。自旋是一种忙等待状态，会一直消耗 CPU 的执行时间，如图 4.5 所示。

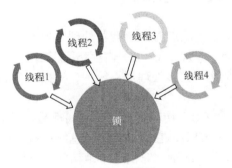

图 4.5 自旋锁

常规互斥锁有一个很大的缺点，即获取锁失败后线程会进入睡眠或阻塞状态。这个过程会涉及用户态到内核态的调度，上下文切换的开销比较大。假如某个锁的锁定时间很短，如果直接让它睡眠或阻塞，则会影响性能，因为上下文切换开销比自旋的开销更大，如图 4.6 所示。一般认为常规互斥锁更适合持有锁时间长的情况，而自旋锁更适合持有锁时间短的情况。

实际上自旋锁有多种实现方案，每种方案都是为了解决存在的缺点或为了适用其他场景。这里介绍 4 种常见的实现方案，包括原始自旋锁、排队自旋锁、CLH 锁以及 MCS 锁。每种实现方式都有自己的优缺点，不过在介绍这些实现方案之前我们先看一下两种处理器架构。

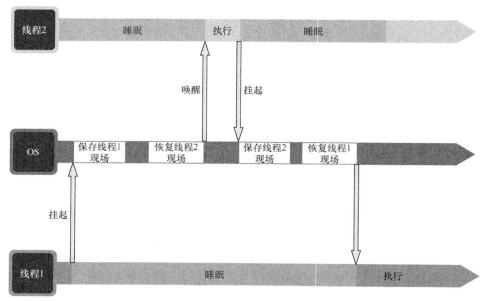

图 4.6 常规互斥锁的缺点

4.3.1 UMA 架构与 NUMA 架构

在分析 CLH 锁与 MCS 锁的缺点时，会涉及处理器架构问题，所以有必要先了解两种处理器架构：UMA（Uniform Memory Access，统一内存访问）架构和 NUMA（Non-Uniform Memory Access，非统一内存访问）架构。在多处理器系统中，根据内存的共享方式可以将处理器架构分为 UMA 和 NUMA。

就 UMA 架构来说，每个 CPU 访问主存储的时间都是一样的。下面看基于总线的 UMA 架构（见图 4.7），一共有 4 个 CPU，它们都直接与总线连接，通过总线进行通信。从这个架构中可以看到，每个 CPU 都没有区别，它们平等地访问主存储。访问主存储所需的时间都是一样的，即统一内存访问。

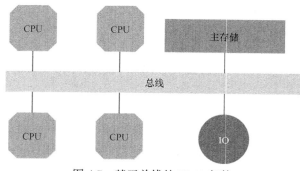

图 4.7 基于总线的 UMA 架构

当某个 CPU 想要进行读写操作时，它首先会检查总线是否空闲。只有在总线处于空闲状态时，CPU 才能与主存储进行通信，否则它将一直等待直到总线空闲。为了解决这个问题，在每个 CPU 的内部引入了缓存（Cache）。这样 CPU 的读操作就能在本地的缓存中进行，如图 4.8 所示。但这时还需要考虑 CPU 中缓存与主存的数据一致性问题，否则可能会引起脏数据问题。

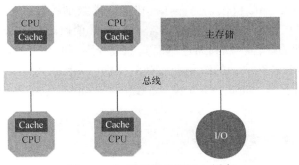

图 4.8　在 CPU 的内部引入缓存

与 UMA 架构相反，在 NUMA 架构中，并非每个 CPU 对主存储的访问时间都相同，在该架构中 CPU 能访问所有主存储。通过图 4.9 可以看到，如果 CPU 通过本地总线来访问相应的本地主存储，则访问时间较短。如果访问的是非本地主存储（远程主存），则访问时间将较长。也就是说，CPU 访问本地主存和访问远程主存的速度不相同。NUMA 架构的优点在于，它具有优秀的可扩展性，能够将数百个 CPU 组合起来。

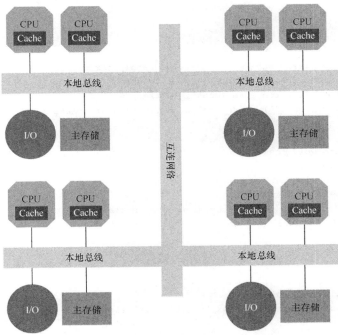

图 4.9　CPU 访问本地主存和访问远程主存的速度不同

4.3.2 原始自旋锁

原始自旋锁就是多个线程不断自旋，都不断尝试获取锁。我们看代码清单 4.10 中的 lock 和 unlock 这两个方法（Unsafe 仅仅是为操作提供了硬件级别的原子 CAS 操作）。对于 lock 方法，假如有若干线程存在竞争，能成功通过 CAS 将 value 值修改为 newV 的线程就是成功获取锁的线程。成功获取锁的线程将顺利通过，而其他线程则不断循环检测 value 值是否改回 0，将 value 改为 0 的操作就是获取锁的线程释放锁的操作。unlock 方法用于释放锁，释放后其他线程又继续对该锁进行竞争。如此一来，没获得锁的线程也不会被挂起或阻塞，而是处于不断循环检查的状态。

代码清单 4.10　原始自旋锁的演示

```
1.   public class SpinLock {
2.       private static Unsafe unsafe = null;
3.       private static final long valueOffset;
4.       private volatile int value = 0;
5.       static {
6.           try {
7.               unsafe = getUnsafeInstance();
8.               valueOffset = unsafe.objectFieldOffset(SpinLock.class.getDeclaredField("value"));
9.           } catch (Exception ex) {
10.              throw new Error(ex);
11.          }
12.      }
13.
14.      private static Unsafe getUnsafeInstance() throws Exception {
15.        Field theUnsafeInstance = Unsafe.class.getDeclaredField("theUnsafe");
16.        theUnsafeInstance.setAccessible(true);
17.        return (Unsafe) theUnsafeInstance.get(Unsafe.class);
18.      }
19.
20.      public void lock() {
21.          for (;;) {
22.              int newV = value + 1;
23.              if (newV == 1)
24.                  if (unsafe.compareAndSwapInt(this, valueOffset, 0, newV)) {
25.                      return;
26.                  }
27.          }
28.      }
29.
30.      public void unlock() {
31.          unsafe.compareAndSwapInt(this, valueOffset, 1, 0);
32.      }
33.  }
```

我们通过图 4.10 来加深对自旋锁的理解。现在有 5 个线程，它们都轮询 value 变量。t1 率先成功修改 value 的值，即成功获取锁。t1 将 value 设置为 1 后，其他线程都无法得到锁，只能继续循环检测。当 t1 释放锁后，将 value 置为 0，此时剩下的线程继续竞争锁；以此类推。这样就能保证某个区域的线程安全性。

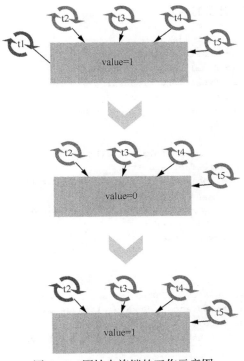

图 4.10　原始自旋锁的工作示意图

原始自旋锁存在以下缺点。
- 它不具有公平性，不能保证先到的线程先获取锁，所有线程都得一起竞争锁。
- 它需要保证各个 CPU 的缓存与主存之间的数据一致性，因此通信开销很大。如果在多处理器系统上，这个问题则更严重。

4.3.3　排队自旋锁

为了解决原始自旋锁的公平性问题，于是引入了一种排队机制，这就是排队自旋锁。所有线程在尝试获取锁之前得先拿到一个排队号，然后再不断轮询是否已经轮到自己了。判断的依据就是当前处理号是否等于自己的排队号，如果两者相等，则表示已经轮到自己了，于是得到锁并往下执行。

我们看代码清单 4.11 中的 lock 和 unlock 这两个方法（Unsafe 仅仅是为操作提供了硬件

级别的原子 CAS 操作）。对于 lock 方法，首先通过不断循环去尝试拿到一个排队号，一旦成功拿到排队号，就开始通过 while(processingNum != nowNum)轮询是否已经轮到自己了。而 unlock 方法则是直接修改当前处理号，并直接加 1，表示自己已经不需要锁了，可以让给下一位了。

代码清单 4.11　排队自旋锁的演示

```
1.   public class TicketLock {
2.       private static Unsafe unsafe = null;
3.       private static final long ticketNumOffset;
4.       private static final long processingNumOffset;
5.       private volatile int ticketNum = 0;
6.       private volatile int processingNum = 0;
7.       static {
8.           try {
9.               unsafe = getUnsafeInstance();
10.              ticketNumOffset = unsafe
11.                      .objectFieldOffset(TicketLock.class.getDeclaredField("ticketNum"));
12.              processingNumOffset = unsafe
13.                      .objectFieldOffset(TicketLock.class.getDeclaredField("processingNum"));
14.          } catch (Exception ex) {
15.              throw new Error(ex);
16.          }
17.      }
18.
19.      private static Unsafe getUnsafeInstance() throws Exception {
20.        Field theUnsafeInstance = Unsafe.class.getDeclaredField("theUnsafe");
21.        theUnsafeInstance.setAccessible(true);
22.        return (Unsafe) theUnsafeInstance.get(Unsafe.class);
23.      }
24.
25.     public int lock() {
26.         int nowNum;
27.         for (;;) {
28.             nowNum = ticketNum;
29.             if (unsafe.compareAndSwapInt(this, ticketNumOffset, ticketNum, ticketNum + 1)) {
30.                 break;
31.             }
32.         }
33.         while (processingNum != nowNum) {
34.         }
35.
36.         return nowNum;
37.     }
38.
39.     public void unlock(int ticket) {
40.         int next = ticket + 1;
41.         unsafe.compareAndSwapInt(this, processingNumOffset, ticket, next);
42.     }
```

```
43.
44. }
```

在图 4.11 中，每个线程一到达就会先去拿一个排队号，然后观察当前处理号是否等于自己所持有的排队号。如果排队号等于当前处理号，则成功获取锁，往下执行。

虽然排队自旋锁解决了公平性问题，但是 CPU 的缓存与主存之间的数据一致性的问题还是没有解决。因为每个线程都对同一个变量操作，这将导致大量的同步操作，从而影响整体性能。

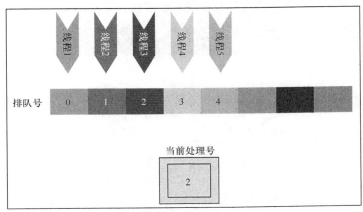

图 4.11　排队自旋锁的工作示意图

4.3.4　CLH 锁

为了解决同步操作带来的花销问题，Craig、Landin、Hagersten 三人发明了 CLH 锁。CLH 锁的核心思想是：通过一定手段将所有线程对某一共享变量的轮询竞争转化为一个线程队列，且队列中的线程各自轮询自己的本地变量。

这个转化过程有下面两个要点。

- 应该构建怎样的队列以及如何构建队列？
  为了保证公平性，我们构建的是一个 FIFO（先进先出）队列。构建时主要通过移动尾部节点 tail 来实现队列的排队，每个想获取锁的线程创建一个新节点并通过 CAS 原子操作将新节点赋给 tail，然后让当前线程轮询前一节点的某个状态位。在图 4.12 中可以清晰看到队列结构及自旋操作，这样就成功构建了线程排队队列。
- 如何释放队列？
  执行完线程后，只需将当前线程所对应的节点状态位设置为解锁状态即可。由于下一节点一直在轮询，所以可获取到锁。

CLH 锁将众多线程长时间对某资源的竞争，通过有序化这些线程将其转化为只需对本地变量检测。而唯一存在竞争的地方就是在入队列之前对尾部节点 tail 的竞争，但此时竞争的线程数量已经少了很多，比起所有线程直接对某资源竞争的轮询次数也减少了很多。这大大节省了 CPU 缓存同步的消耗，从而显著提升了系统性能。

下面来看一个简单的 CLH 锁实现代码（见代码清单 4.12），以便更好地理解 CLH 锁的原理。其中 lock 与 unlock 这两个方法提供加锁和解锁操作，每次加锁、解锁时必须将一个 CLHNode 对象作为参数传入。lock 方法的 for 循环是通过 CAS 操作将新节点插入队列，而 while 循环则是检测前驱节点的锁状态位。一旦前驱节点的锁状态位允许，就结束检测，让线程往下执行。解锁操作先判断当前节点是否为尾部节点，如果是则直接将尾节点设置为空，此时说明仅仅只有一个线程在执行，否则将当前节点的锁状态位设置为解锁状态。

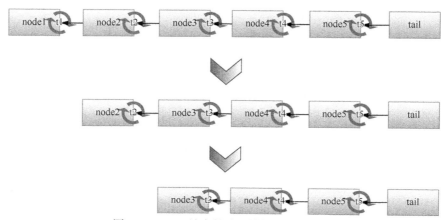

图 4.12 CLH 锁中的队列结构和自旋操作

代码清单 4.12 简单的 CLH 锁实现代码

```
1.   public class CLHLock {
2.       private static Unsafe unsafe = null;
3.       private static final long valueOffset;
4.       private volatile CLHNode tail;
5.
6.       public class CLHNode {
7.           private volatile boolean isLocked = true;
8.       }
9.
10.      static {
11.          try {
12.              unsafe = getUnsafeInstance();
13.              valueOffset = unsafe.objectFieldOffset(CLHLock.class.getDeclaredField("tail"));
14.          } catch (Exception ex) {
15.              throw new Error(ex);
16.          }
17.      }
18.
19.      public void lock(CLHNode currentThreadNode) {
20.          CLHNode preNode = null;
```

```
21.        for (;;) {
22.            preNode = tail;
23.            if (unsafe.compareAndSwapObject(this, valueOffset, preNode, currentThreadNode))
24.                break;
25.        }
26.        if (preNode != null)
27.            while (preNode.isLocked) {
28.            }
29.    }
30.
31.    public void unlock(CLHNode currentThreadNode) {
32.        if (!unsafe.compareAndSwapObject(this, valueOffset, currentThreadNode, null))
33.            currentThreadNode.isLocked = false;
34.    }
35.
36.    private static Unsafe getUnsafeInstance() throws Exception {
37.        Field theUnsafeInstance = Unsafe.class.getDeclaredField("theUnsafe");
38.        theUnsafeInstance.setAccessible(true);
39.        return (Unsafe) theUnsafeInstance.get(Unsafe.class);
40.    }
41. }
```

虽然 CLH 锁解决了大量线程同时操作同一个变量时所带来的开销问题,但它的自旋的对象是前驱节点,这在 NUMA 架构下可能会存在性能问题。因为如果前驱节点和当前节点不在同一个本地主存储中,则访问时间会很长,由此引发性能问题。

4.3.5 MCS 锁

MCS 锁由 John Mellor-Crummey 和 Michael Scott 两人发明,它旨在解决 CLH 锁存在的问题。它也是基于 FIFO 队列,与 CLH 不同的地方在于轮询的对象不同。MCS 锁中的线程只对本地变量自旋,而前驱节点则负责通知 MCS 锁中的线程结束自旋操作。这就减少了 CPU 缓存与主存储之间不必要的同步操作,也减少了同步带来的性能损耗。

在图 4.13 中,每个线程都对应着队列中的一个节点。节点内有一个 spin 变量,表示是否需要旋转。一旦前驱节点使用完锁,就修改后继节点的 spin 变量,通知其不必继续进行自旋操作,已成功获取锁。

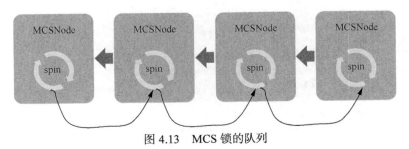

图 4.13 MCS 锁的队列

下面来看一个简单的 MCS 锁实现代码（见代码清单 4.13），以便更好地理解 MCS 锁的原理。其中 lock 与 unlock 这两个方法提供加锁和解锁操作，每次加锁、解锁时必须将一个 MCSNode 对象作为参数传入。lock 方法的 for 循环通过 CAS 操作将新节点赋给队列尾部节点 tail。如果存在前驱节点，新节点就开始自旋操作，等待前驱节点解锁时通知自己。一旦前驱节点执行解锁，则会将本节点的 spin 变量修改为 false，本节点则获取锁并停止自旋，让线程往下执行。解锁操作先判断当前节点是否为尾部节点，如果是，则什么都不用处理，此时说明仅仅只有一个线程在执行。否则将后继节点的 spin 变量设置为 false，此时要考虑特殊情况，即如果不存在后继节点则将尾部节点 tail 设为 null。在此期间可能又有线程进来，这时 tail 的 CAS 修改会失败，所以只能自旋，等后继节点不为空再往下执行。

代码清单 4.13　简单的 MCS 锁实现代码

```java
1.  public class MCSLock {
2.      private static Unsafe unsafe = null;
3.      volatile MCSNode tail;
4.      private static final long valueOffset;
5.
6.      public static class MCSNode {
7.          MCSNode next;
8.          volatile boolean spin = true;
9.      }
10.
11.     static {
12.         try {
13.             unsafe = getUnsafeInstance();
14.             valueOffset = unsafe.objectFieldOffset(MCSLock.class.getDeclaredField("tail"));
15.         } catch (Exception ex) {
16.             throw new Error(ex);
17.         }
18.     }
19.
20.     private static Unsafe getUnsafeInstance() throws Exception {
21.         Field theUnsafeInstance = Unsafe.class.getDeclaredField("theUnsafe");
22.         theUnsafeInstance.setAccessible(true);
23.         return (Unsafe) theUnsafeInstance.get(Unsafe.class);
24.     }
25.
26.     public void lock(MCSNode currentThreadMcsNode) {
27.         MCSNode predecessor = null;
28.         for (;;) {
29.             predecessor = tail;
30.             if (unsafe.compareAndSwapObject(this, valueOffset, tail, currentThreadMcsNode))
31.                 break;
32.         }
33.         if (predecessor != null) {
34.             predecessor.next = currentThreadMcsNode;
35.             while (currentThreadMcsNode.spin) {
```

```
36.            }
37.        }
38.    }
39.
40.    public void unlock(MCSNode currentThreadMcsNode) {
41.        if (tail != currentThreadMcsNode) {
42.            if (currentThreadMcsNode.next == null) {
43.                if (unsafe.compareAndSwapObject(this, valueOffset,
currentThreadMcsNode, null)) {
44.                    return;
45.                } else {
46.                    while (currentThreadMcsNode.next == null) {
47.                    }
48.                }
49.            }
50.            currentThreadMcsNode.next.spin = false;
51.        }
52.    }
53. }
```

4.4 线程饥饿

线程饥饿是一种因为长期无法获取共享资源或 CPU 而导致线程无法执行的现象，在并发过程中更多的是指线程分配不到 CPU 而无法执行，因为其他贪婪的线程把 CPU 都占用了。比如 Java 在使用 synchronized 对资源加锁时，如果不断有大量线程去竞争获取锁，那么就有可能会产生线程饥饿。这就好比源源不断的球通过仅有的两个通道（见图 4.14），由于通道有限，所以会导致入口拥挤，使得箭头所指处的球长期无法进入通道。这也就是所谓的饥饿现象。

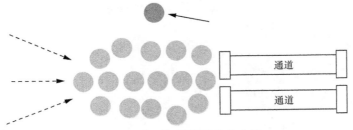

图 4.14　线程饥饿的形象化表示

4.4.1　synchronized 饥饿

造成线程饥饿的情况有多种。我们先看 synchronized 的例子。synchronized 其实就是加锁，因此多个线程到达该节点后就会去竞争锁，而 synchronized 没有要求公平性，于是便可能导致有些线程一直得不到锁而无法执行。为了让大家更加形象地看到效果，代码清单 4.14 使用了图形进度条（分别使用 10 个进度条代表线程被执行的次数）。每个线程里面都会不断循环去获取锁，然后将自己的进度条加 1，这样就能够通过进度条来看线程获得 CPU 执行的情况了。

代码清单 4.14　synchronized 饥饿的演示代码

```java
1.  public class StarvationDemo {
2.
3.      private static Object lock = new Object();
4.
5.      public static void main(String[] args) {
6.
7.          JFrame frame = new JFrame("线程饥饿—synchronized");
8.          frame.setDefaultCloseOperation(JFrame.EXIT_ON_CLOSE);
9.          frame.setLayout(new FlowLayout(FlowLayout.LEFT));
10.         frame.setSize(new Dimension(350, 200));
11.         for (int i = 0; i < 10; i++) {
12.             JProgressBar progressBar = new JProgressBar();
13.             progressBar.setStringPainted(true);
14.             progressBar.setMinimum(0);
15.             progressBar.setMaximum(1000);
16.             frame.add(progressBar);
17.             new Thread(() -> {
18.                 progressBar.setString(Thread.currentThread().getName());
19.                 int c = 0;
20.                 while (true) {
21.                     synchronized (lock) {
22.                         if (c >= 1000)
23.                             break;
24.                         progressBar.setValue(++c);
25.                         try {
26.                             Thread.sleep(1);
27.                         } catch (InterruptedException e) {
28.                         }
29.                     }
30.                 }
31.             }).start();
32.         }
33.         frame.setVisible(true);
34.     }
35.
36.  }
```

某个时刻的运行情况如图 4.15 所示，可以看到 Thread-4 在这段时间内几乎没有得到 CPU 的执行。当竞争的线程数达到很大的量级时，就有可能会导致某个或某些线程在很长时间内都得不到执行，从而产生线程饥饿。

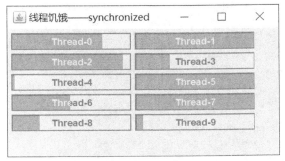

图 4.15　synchronized 饥饿现象

4.4.2　优先级饥饿

我们继续看线程优先级造成线程饥饿的例子。Java 中的每个线程都有自己的优先级，一般情况下使用的是默认优先级。如果线程优先级不同，就可能会引起线程饥饿。比如在代码清单 4.15 中，我们还是以 10 个线程为观察对象，图形进度条表示对应线程获得 CPU 执行的次数。其中 Thread-0 线程的优先级被设置为 1，其他线程优先级为 10。

代码清单 4.15　优先级饥饿的演示代码

```java
1.  public class StarvationDemo2 {
2.
3.      static List<Thread> threads = new ArrayList<Thread>();
4.
5.      public static void main(String[] args) {
6.
7.          JFrame frame = new JFrame("线程优先级—线程饥饿");
8.          frame.setDefaultCloseOperation(JFrame.EXIT_ON_CLOSE);
9.          frame.setLayout(new FlowLayout(FlowLayout.LEFT));
10.         frame.setSize(new Dimension(350, 200));
11.         for (int i = 0; i < 10; i++) {
12.             JProgressBar progressBar = new JProgressBar();
13.             progressBar.setStringPainted(true);
14.             progressBar.setMinimum(0);
15.             progressBar.setMaximum(1000);
16.             frame.add(progressBar);
17.             Thread t = new Thread(() -> {
18.                 progressBar.setString(Thread.currentThread().getName());
19.                 int c = 0;
20.                 while (true) {
21.                     if (c >= 1000)
22.                         break;
23.                     progressBar.setValue(++c);
24.                     int a = 0;
25.                     for (long l = 0; l < 10000000; l++)
26.                         a++;
```

```
27.                });
28.            });
29.            if (i == 0)
30.                t.setPriority(1);
31.            else
32.                t.setPriority(10);
33.            threads.add(t);
34.        }
35.        frame.setVisible(true);
36.        for (Thread t : threads)
37.            t.start();
38.    }
39.
40. }
```

某个时刻的运行情况如图 4.16 所示，可以看到 Thread-0 只得到很少的 CPU 执行，而其他优先级高的线程则得到了大量的 CPU 执行。由此可以看出，当竞争很激烈时，低优先级的线程可能会长时间获取不到 CPU，从而产生线程饥饿。

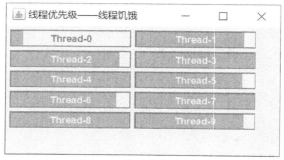

图 4.16　优先级饥饿现象

4.4.3　线程自旋饥饿

第三种线程饥饿的例子与自旋相关。在并发中经常会使用自旋锁，它的实现核心就是自旋操作，自旋同样会导致线程饥饿现象。在代码清单 4.16 中同样是 10 个线程，每个线程对应一个图形进度条，通过进度条可以反映出线程饥饿。

代码清单 4.16　线程自旋解饿的演示代码

```
1.   public class StarvationDemo3 {
2.
3.       static SpinLock spinLock = new SpinLock();
4.
5.       public static void main(String[] args) {
6.
7.           JFrame frame = new JFrame("线程自旋饥饿");
```

```
8.              frame.setDefaultCloseOperation(JFrame.EXIT_ON_CLOSE);
9.              frame.setLayout(new FlowLayout(FlowLayout.LEFT));
10.             frame.setSize(new Dimension(350, 300));
11.             for (int i = 0; i < 10; i++) {
12.                 JProgressBar progressBar = new JProgressBar();
13.                 progressBar.setStringPainted(true);
14.                 progressBar.setMinimum(0);
15.                 progressBar.setMaximum(10);
16.                 frame.add(progressBar);
17.                 new Thread(() -> {
18.                     progressBar.setString(Thread.currentThread().getName());
19.                     int c = 0;
20.                     while (true) {
21.                         if (c >= 10)
22.                             break;
23.                         spinLock.lock();
24.                         progressBar.setValue(++c);
25.                         int a = 0;
26.                         for (long l = 0; l < 100000000; l++)
27.                             a++;
28.                         spinLock.unlock();
29.                     }
30.                 }).start();
31.             }
32.             frame.setVisible(true);
33.         }
34.
35. }
```

某个时刻的运行情况如图 4.17 所示,可以看到 Thread-9 获得很少的 CPU 执行的机会,其他线程则在自旋中得到更多的 CPU 执行。这种情况下如果存在大量自旋线程,则可能导致某个线程长期得不到 CPU 执行,从而产生线程饥饿。

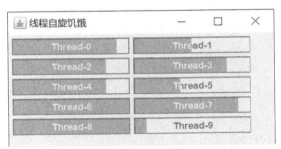

图 4.17 线程自旋饥饿现象

4.4.4 等待唤醒饥饿

就即将介绍的第四种线程饥饿的情况,我们主要理解它形成的原因。不一定能模拟出这种情况,因为不同的 JVM 对 wait 和 notify 的实现可能不同。前面章节讲过 wait/notify 模式的实现机制,一个线程在调用 Object 的 wait 方法时会被放到一个等待集中,而当另外一个线程调用该 Object 的 notify 方法时则会从等待集中唤醒一个线程。需要注意的是,JVM 规范中并没有规定等待集(也称为等待队列)的数据结构,也没有规定先进入等待集的一定会先唤醒,也可能是随机唤醒其中一个线程。这就导致在某些 JVM 中会产生线程饥饿问题。假如等待集包含了很多线程,而且不断有其他线程进入等待集,而其他线程在调用 notify 方法时可能会随机唤醒某个线程,这时就可能导致有些线程长期无法被唤醒。代码清单 4.17 是模拟该种情况的代码,但我们可能看不到饥饿现象,因为常用的 JVM HotSpot 的 wait/notify 的等待集是一种先入先出结构,即先进入等待集的线程会被先唤醒。

代码清单 4.17　等待唤醒饥饿的演示代码

```
1.   public class StarvationDemo4 {
2.       private static Object lock = new Object();
3.
4.       public static void main(String[] args) {
5.           for (int i = 0; i < 10; i++) {
6.               new Thread(() -> {
7.                   while (true) {
8.                       synchronized (lock) {
9.                           try {
10.                              lock.wait();
11.                          } catch (InterruptedException e) {
12.                          }
13.                      }
14.                  }
15.              }).start();
16.          }
17.          new Thread(() -> {
18.              while (true) {
19.                  synchronized (lock) {
20.                      lock.notify();
21.                  }
22.              }
23.          }).start();
24.      }
25.  }
```

4.4.5　公平性解决饥饿

从前文中可以看到,由于很多线程存在竞争而导致某些线程无法被 CPU 执行,或者是因为线程优先级的原因导致某些线程无法被 CPU 执行,所以解决饥饿问题的方法就是引入公平机制。为了给线程竞争添加公平性,可以引入队列。在图 4.18 中,5 个线程都在竞争一个锁,假如线程 1 获得了锁,那么就可以把剩下的 4 个线程放到一个队列中,然后按顺序去获得锁,这样就提供了公平机制，避免了线程饥饿。

JDK 提供了一些具备公平机制的锁，可以直接拿来用。锁的公平机制的实现会在第 5 章进行深入分析。下面看看如何用 ReentrantLock（重入锁）来实现公平的锁，从而避免线程饥饿，如代码清单 4.18 所示。ReentrantLock 可以很方便地通过一个 boolean 来声明要使用公平锁还是非公平锁，这里只需传入 true。

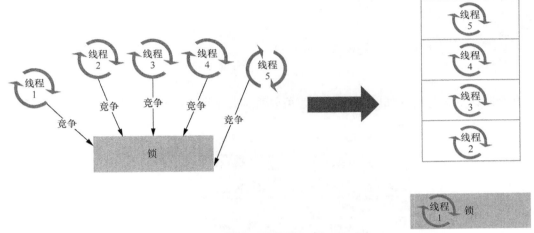

图 4.18　添加公平性来解决饥饿问题

代码清单 4.18　通过公平机制锁解决线程饥饿的演示代码

```java
1.  public class StarvationDemo5 {
2.      static boolean isFair = true;
3.      private static ReentrantLock lock = new ReentrantLock(isFair);
4.
5.      public static void main(String[] args) {
6.
7.          JFrame frame = new JFrame("公平机制锁解决线程饥饿");
8.          frame.setDefaultCloseOperation(JFrame.EXIT_ON_CLOSE);
9.          frame.setLayout(new FlowLayout(FlowLayout.LEFT));
10.         frame.setSize(new Dimension(350, 200));
11.         for (int i = 0; i < 10; i++) {
12.             JProgressBar progressBar = new JProgressBar();
13.             progressBar.setStringPainted(true);
14.             progressBar.setMinimum(0);
15.             progressBar.setMaximum(1000);
16.             frame.add(progressBar);
17.             new Thread(() -> {
18.                 progressBar.setString(Thread.currentThread().getName());
19.                 int c = 0;
20.                 while (true) {
21.                     if (c >= 1000)
```

```
22.                    break;
23.                lock.lock();
24.                progressBar.setValue(++c);
25.                try {
26.                    Thread.sleep(1);
27.                } catch (InterruptedException e) {
28.                }
29.                lock.unlock();
30.            }
31.        }).start();
32.    }
33.    frame.setVisible(true);
34.    }
35. }
```

某个时刻的运行情况如图 4.19 所示，可以看到几乎所有线程都公平地被 CPU 执行。

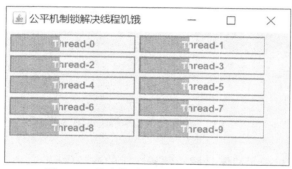

图 4.19　通过公平机制锁解决线程饥饿

4.5　数据竞争

所谓数据竞争，是指存在至少两个线程去读写某个共享内存，其中至少有一个线程对该共享内存进行写操作。在图 4.20 中，线程 1 和线程 2 在并发执行的过程中都对某个共享内存进行读写操作，这种情况下如果没有其他措施来保证，可能就会导致执行结果出现错误。简单来说就是多个线程在同时对一个内存进行写操作时，在写的过程中其他线程读取到的该内存的数值并非是预期的。

代码清单 4.19 所示为一个数据竞争的示例。为了更容易产生数据竞争现象，这里首先定义一个 Memory 类表示共享内存，然后定义 update 方法来表示对共享内存的写操作，它包含了对 Memory 中 a、b 两个变量的更新。接着在主线程中创建并启动一个新线程去调用 update 方法，可以看到该方法并非是原子的。最后在主线程中打印出 a、b 两个变量，结果可能是 "0,0" "0,1" 或 "1,1"。实际上，如果把 update 作为一个操作的话，它要么就是两个变量都加 1，要么就是都不加 1，但却出现了 "0,1" 的情况，也就是 update 在尚未执行完时主线程读了该内存，由此造成了数据竞争。

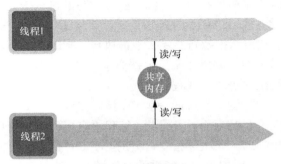

图 4.20 数据竞争

代码清单 4.19 数据竞争的示例

```
1.   public class DataRaceDemo {
2.
3.       Memory mem = new Memory();
4.
5.       public void update() {
6.           mem.b++;
7.           mem.a++;
8.       }
9.
10.      public void print_result() {
11.          System.out.println(mem);
12.      }
13.
14.      public static void main(String[] args) throws InterruptedException {
15.          DataRaceDemo demo = new DataRaceDemo();
16.          Thread thread1 = new Thread(() -> {
17.              demo.update();
18.          });
19.          thread1.start();
20.          for (int i = 0; i < 5000; i++)
21.              ;
22.          demo.print_result();
23.      }
24.
25.      static class Memory {
26.          public int a = 0;
27.          public int b = 0;
28.
29.          public String toString() {
30.              return (a + "," + b);
31.          }
32.      }
33.  }
```

产生数据竞争的根本原因是一个 CPU 在任意时刻都只能执行一条机器指令,但对某个内存的写操作可能会用到若干条机器指令,这就可能导致写的过程中还未完全修改完内存,其他线程就进行读取,从而导致执行结果不可预知,也就产生了数据竞争。从 CPU 底层来看,一条机器指令是原子性的,它要么被执行,要么不被执行。所以,如果某个操作只需一条机器指令,则该操作天生具备原子性。如果要避免数据竞争,则需要保证写内存的操作是原子性的,这样就能避免在修改未结束时被其他线程读取到。

在图 4.21 中,其中一个线程执行 count++ 操作,该操作分成 4 步,期间另外一个线程对 count 进行读写操作,这就产生了数据竞争。实际上,编程语言层面上很简单的 count=1 赋值操作在某些硬件平台上并非是单独的一条机器指令,即使是对某个变量进行赋值操作都可能导致数据竞争。此外,一些编程语言(比如 Java)提供的并发模型中的每个线程栈都会保存一个共享变量的副本,而某个线程在修改完该副本后可能需要一定的时间才能刷新到主存,并且其他线程读取到最新值可能也需要一些时间,这种情况也会造成数据竞争。

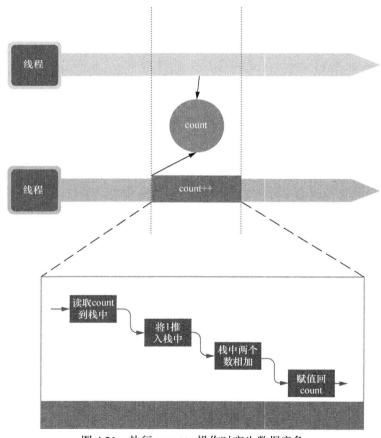

图 4.21　执行 count++ 操作时产生数据竞争

解决数据竞争的方法其实很简单,就是将对共享内存的更新操作原子化,同时保证内存的可见性。在代码清单 4.20 中,我们对原来的代码进行改动,引入 AtomicReference 类,使得对 Memory 对象的更新具备原子性,而且还将变量 a、b 声明为 volatile,保证其可见性。可以看到,针对 update 方法主要是通过自旋+CAS 来达到原子更新,此时再运行就不会产生"0,1"这种中间状态了,即 update 方法是一个原子方法,a、b 只能同时为 0 或 1。

代码清单 4.20 通过引入 AtomicReference 类来添加原子性

```
1.  public class DataRaceDemo2 {
2.
3.      Memory mem = new Memory();
4.      AtomicReference<Memory> aMem = new AtomicReference<Memory>(mem);
5.
6.      public void update() {
7.          for (;;) {
8.              Memory newMem = new Memory();
9.              newMem.a = aMem.get().a + 1;
10.             newMem.b = aMem.get().b + 1;
11.             if (aMem.compareAndSet(aMem.get(), newMem))
12.                 break;
13.         }
14.     }
15.
16.     public void print_result() {
17.         System.out.println(aMem.get());
18.     }
19.
20.     public static void main(String[] args) throws InterruptedException {
21.         DataRaceDemo demo = new DataRaceDemo();
22.         Thread thread1 = new Thread(() -> {
23.             demo.update();
24.         });
25.         thread1.start();
26.         for (int i = 0; i < 50000; i++)
27.             ;
28.         demo.print_result();
29.     }
30.
31.     static class Memory {
32.         public volatile int a = 0;
33.         public volatile int b = 0;
34.
35.         public String toString() {
36.             return (a + "," + b);
37.         }
38.     }
39. }
```

4.6　竞争条件

在讲解竞争条件之前，需要先了解一下什么是临界区。从代码角度来看，临界区就是一块代码区域，在该区域中多个线程不同的执行顺序以及线程的并发交叉执行都可能导致执行结果与预期的不一样。以代码清单 4.21 为例，临界区其实就是 incrementValue 方法的代码块。由于 a++ 其实对应好几条机器指令，所以当多个线程并发执行时，会导致执行结果并非是预期的结果。

代码清单 4.21　临界区的演示代码

```
1.  public class CriticalSectionDemo {
2.
3.      static int a = 0;
4.
5.      public static int incrementValue() {
6.          return a++;
7.      }
8.
9.      public static void main(String[] args) {
10.         for (int i = 0; i < 10; i++)
11.             new Thread(() -> {
12.                 for (int j = 0; j < 1000; j++)
13.                     incrementValue();
14.             }).start();
15.         try {
16.             Thread.sleep(1000);
17.         } catch (InterruptedException e) {
18.             e.printStackTrace();
19.         }
20.         System.out.println(a);
21.     }
22. }
```

多线程在执行临界区的代码时就会产生竞争条件。如果只有一个线程，那么执行临界区的代码后不会产生任何问题，但在多线程下却会产生不可预期的结果。用一个形象点的比喻就是：临界区好比一条赛道，若干线程进入该赛道后开始竞争，它们的竞争结果将直接影响临界区中代码的执行结果。在图 4.22 中，3 个线程并发地进入临界区，临界区可以产生的竞争对象包括内存、数据库和文件等。

代码清单 4.22 是一个竞争条件的简单示例，与数据竞争示例不同的地方在于 a、b 这两个变量都已经具有原子性了，update 方法就是一个临界区，临界区内涉及 a 和 b 的原子增加运算，其中的 for 循环是为了能更容易产生竞争条件。主线程启动了 10 个线程去执行 update 方法，因为单个线程执行 update 方法时会将 a 加 1 并且 b 为加上 a 加 1 后的值，所以 10 个线程执行完后，预期值应该是"10,55"。但实际情况可能是"10,70"（b 可能是 55~100 之间的值）。

4.6 竞争条件

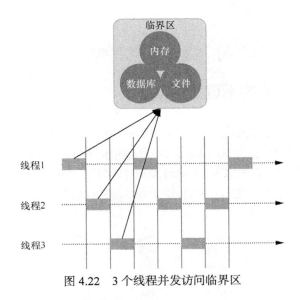

图 4.22 3 个线程并发访问临界区

代码清单 4.22 竞争条件的简单示例

```
1.  public class ConditionRaceDemo {
2.
3.      AtomicInteger a = new AtomicInteger(0);
4.      AtomicInteger b = new AtomicInteger(0);
5.      int delta;
6.
7.      public void update() {
8.          delta = a.incrementAndGet();
9.          for (int i = 0; i < 10000; i++)
10.             ;
11.         b.addAndGet(delta);
12.     }
13.
14.     public void print_result() {
15.         System.out.println(a + "," + b);
16.     }
17.
18.     public static void main(String[] args) throws InterruptedException {
19.         ConditionRaceDemo demo = new ConditionRaceDemo();
20.         for (int i = 0; i < 10; i++) {
21.             Thread thread1 = new Thread(() -> {
22.                 demo.update();
23.             });
24.             thread1.start();
25.         }
26.         Thread.sleep(2000);
27.         demo.print_result();
28.     }
```

```
29.
30. }
```

造成竞争条件的根本原因主要有两个：线程执行顺序的不确定性和并发机制。

4.6.1 线程执行顺序的不确定性

因为现代的操作系统多数是抢占式任务型的，所有的任务调度都由操作系统完全控制（见图 4.23），这就导致每个线程在启动后被执行的顺序并非是编码顺序，而是由操作系统的调度算法来决定。比如在代码中先编写 thread1.start()再编写 thread2.start()，并不意味着 thread1 比 thread2 先执行。这便是不确定性，执行结果也无法与预期相同。

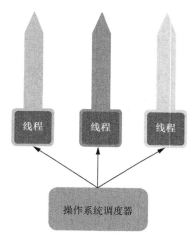

图 4.23　操作系统负责调度任务

比如在代码清单 4.23 中，虽然从编码上来看 thread1 比 thread2 更早调用，但实际上却不一定先执行 thread1，程序的最终结果可能是 8 或 10。

代码清单 4.23　线程执行顺序的不确定性

```
1.  public class ConditionRaceDemo2 {
2.
3.      static volatile int a = 3;
4.
5.      public synchronized static void calc() {
6.          a = a + 2;
7.      }
8.
9.      public synchronized static void calc2() {
10.         a = a * 2;
11.     }
12.
```

```
13.     public static void main(String[] args) throws InterruptedException {
14.         Thread thread1 = new Thread(() -> calc());
15.         Thread thread2 = new Thread(() -> calc2());
16.         thread1.start();
17.         thread2.start();
18.         Thread.sleep(100);
19.         System.out.println(a);
20.     }
21.
22. }
```

我们先看结果为 8 的情况。线程 2 读取 a，此时为 3，然后执行 3*2=6 并将结果写回变量 a。接着切换到线程 1，线程 1 读取 a，此时为 6，然后执行 6+2=8 并将结果写回变量 a。最终变量 a 的值为 8（见图 4.24）。

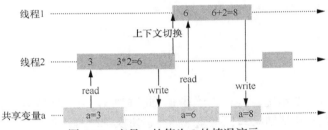

图 4.24　变量 a 的值为 8 的情况演示

继续看结果为 10 的情况。线程 1 读取 a，此时为 3，然后执行 3+2=5 并将结果写回变量 a，此时变量 a 为 5。接着切换到线程 2，它先读取变量 a。然后执行 5*2=10 并将结果写回变量 a。最终变量 a 的值为 10（见图 4.25）。

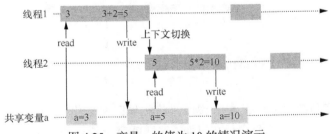

图 4.25　变量 a 的值为 10 的情况演示

4.6.2　并发机制

在并发过程中，多个线程会进行上下文切换，并交叉着执行。为了更好地理解这个问题，我们来看一下代码清单 4.24。其中 calc 方法对变量 a 加 2 并赋值，主线程中启动两个线程分别执行这个方法。实际上我们想要的结果是 7，因为 3+2+2=7，但是多次运行的结果还可能是 5。这便是并发机制自身造成的。

代码清单 4.24　并发机制的演示代码

```
1.  public class ConditionRaceDemo4 {
2.
3.      static volatile int a = 3;
4.
5.      public static void calc() {
6.          a = a + 2;
7.      }
8.
9.      public static void main(String[] args) throws InterruptedException {
10.         Thread thread1 = new Thread(() -> calc());
11.         Thread thread2 = new Thread(() -> calc());
12.         thread2.start();
13.         thread1.start();
14.         Thread.sleep(100);
15.         System.out.println(a);
16.     }
17.
18. }
```

我们分析结果为 5 的情况。线程 1 读取 a，此时为 3。然后切换到线程 2 执行，也读取 a，此时为 3，执行 3+2=5 并将结果写回变量 a。接着切换到线程 1 继续执行 3+2=5，并将结果写回变量 a。最终变量 a 的结果为 5（见图 4.26）。

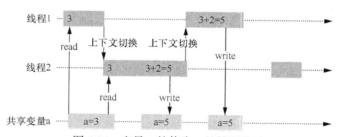

图 4.26　变量 a 的值为 5 的情况演示

那么如何解决竞争条件呢？简单来说就是将临界区原子化，一旦让临界区具有原子性，就能够保证临界区同时只能有一个线程在里面，这样就避免了竞争。Java 中最简单的实现方式就是使用语言层面提供的 synchronized 关键词。我们对 calc 方法进行 synchronized 声明，使得该方法具有原子性，这里的原子性是由互斥锁机制实现的。两个线程不管谁先执行，都能原子地对变量 a 加 2，确保最终结果为 7，代码如代码清单 4.25 所示。

代码清单 4.25　通过 synchronized 声明来添加原子性

```
1.  public class ConditionRaceDemo5 {
2.
3.      static volatile int a = 3;
4.
```

```
 5.         public synchronized static void calc() {
 6.             a = a + 2;
 7.         }
 8.
 9.         public static void main(String[] args) throws InterruptedException {
10.             Thread thread1 = new Thread(() -> calc());
11.             Thread thread2 = new Thread(() -> calc());
12.             thread2.start();
13.             thread1.start();
14.             Thread.sleep(100);
15.             System.out.println(a);
16.         }
17.
18.     }
```

4.7 死锁

死锁是一种由两个或两个以上的线程或进程构成一个无限互相等待的环状状态。以两个线程为例，线程 1 持有 A 锁的同时在等待 B 锁，而线程 2 持有 B 锁的同时在等待 A 锁，从而导致两个线程互相等待，无法运行。现实生活中一个经典的死锁情形就是 4 辆汽车通过没有红绿灯的十字路口（见图 4.27）。假如 4 辆汽车同时到达中心，那么它们将形成一个死锁状态。每辆车拥有自己所在车道的使用权，但同时也在等另外一辆汽车让出它的车道的使用权。

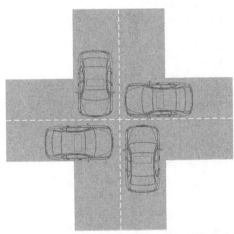

图 4.27 4 辆汽车同时出现在没有红绿灯的十字路口

下面看一下代码清单 4.26 所示的死锁示例。该示例中一共有 lock1 和 lock2 两个锁。线程 1 启动后先尝试获取 lock1 锁，在成功获取 lock1 后再继续尝试获取 lock2 锁。而线程 2 则是先尝试获取 lock2 锁，成功获取 lock2 锁后再继续尝试获取 lock1 锁。

代码清单 4.26　死锁示例代码

```java
1.  public class DeadLockDemo {
2.
3.      static String lock1 = "lock1";
4.      static String lock2 = "lock2";
5.
6.      public static void main(String[] args) {
7.
8.          new Thread(() -> {
9.              System.out.println("thread1 trying to get lock1");
10.             synchronized (lock1) {
11.                 System.out.println("thread1 gets lock1");
12.                 try {
13.                     Thread.sleep(100);
14.                 } catch (Exception e) {
15.                 }
16.                 System.out.println("thread1 trying to get lock2");
17.                 synchronized (lock2) {
18.                     System.out.println("thread1 gets lock2");
19.                 }
20.             }
21.         }).start();
22.
23.         new Thread(() -> {
24.             System.out.println("thread2 trying to get lock2");
25.             synchronized (lock2) {
26.                 System.out.println("thread2 gets lock2");
27.                 try {
28.                     Thread.sleep(100);
29.                 } catch (Exception e) {
30.                 }
31.                 System.out.println("thread2 trying to get lock1");
32.                 synchronized (lock1) {
33.                     System.out.println("thread2 gets lock1");
34.                 }
35.             }
36.         }).start();
37.
38.     }
39.
40.  }
```

在某次启动程序后，可能的输出情况如下所示。这说明进入了死锁状态。但并非每次都一定会进入死锁状态，让每个线程睡眠 100ms 是为了增加死锁的可能。最终两个线程处于无限互相等待的状态——获得 lock1 锁的线程 1 在等 lock2 锁，而获得 lock2 锁的线程 2 却在等 lock1 锁。

```
thread1 trying to get lock1
thread1 gets lock1
thread2 trying to get lock2
thread2 gets lock2
thread2 trying to get lock1
thread1 trying to get lock2
```

死锁的形成需要满足以下几个条件。

- **互斥条件**：资源具有排它性。某个线程或进程获取该资源后，其他线程或进程便无法再获得该资源的使用权，除非占有资源的线程或进程释放该资源的使用权，在释放之前其他线程或进程只能处于等待状态。
- **阻塞不释放条件**：线程或进程因请求某个资源而进入阻塞状态时，不释放已获取的资源。
- **占有并等待条件**：某个线程或进程应该占有至少一个资源，然后等待获取另外一个资源，且该资源由其他线程或进程所占有。
- **非抢占条件**：资源一旦被某个线程或进程所获取，其他线程或进程不能对该资源进行抢占，只能由持有资源的线程或进程自愿释放。
- **环形条件**：死锁时所有等待的线程与等待的资源形成一个环形，比如最简单的环形就是线程 P1 持有资源 R1 而等待资源 R2，线程 P2 持有资源 R2 而等待资源 R3，线程 P3 持有资源 R3 而等待资源 R1（见图 4.28）。

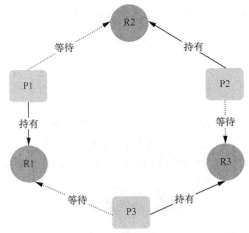

图 4.28 等待的线程与资源形成一个环形

由于死锁的检测涉及很多复杂的场景，而且它在运行时才会产生，所以语言编译器一般不会提供死锁的检测功能。这其实就是鸵鸟算法，即如果我们对于某件事没有很好的处理方法，那么就学鸵鸟把头埋入沙中假装什么都看不见。于是死锁的处理就交给了程序开发人员，开发人员需要自己去避免死锁的发生，或者制定某些措施来处理发生的死锁。

常见的死锁处理方式大致分为两类。

- **事前的预防措施**：包括锁的顺序化、资源合并、避免锁嵌套等。

- **事后的处理措施**：包括锁超时机制、抢占资源机制、撤销线程机制等。

下面详细看看每种措施的情况。

4.7.1 锁的顺序化

针对死锁形成条件中的环形条件，可以采用破坏这个条件的方式来避免死锁的发生。具体来说就是将锁的获取顺序化，所有线程和进程对锁的获取都按指定的顺序进行。在图 4.29 中，P1、P2、P3 这 3 个线程都先尝试持有 R1 锁，再尝试持有 R2 锁，最后尝试持有 R3 锁。当然也可以看成要获取 R3 锁就必须先获取 R2 锁，而要获取 R2 锁就必须先获取 R1 锁。这样就能破坏环形条件，从而避免死锁。

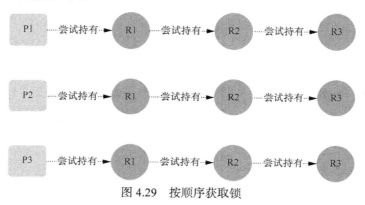

图 4.29 按顺序获取锁

4.7.2 资源合并

资源合并即将多个资源进行合并，将其当成一个资源来看待，这样就能将对多个资源的获取变成只对一个资源的获取，从而避免了死锁的发生。在图 4.30 中，将资源 R1、资源 R2 和资源 R3 合并成一个资源 R，然后 3 个线程对其进行获取操作，就可以避免死锁。

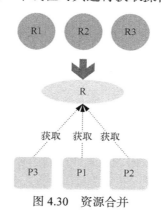

图 4.30 资源合并

4.7.3 避免锁嵌套

锁获取操作的嵌套行为可能导致死锁发生,所以可以通过去除锁嵌套来避免死锁。每个线程在使用完某个资源后就释放,然后才能再获取另外一个资源,而且使用完后又进行释放,这就是去除锁嵌套。在图 4.31 中,线程 P1 持有 R1 锁后释放,然后持有 R2 锁后释放,最后持有 R3 锁并释放,其他线程也进行类似操作,这也就避免了死锁。

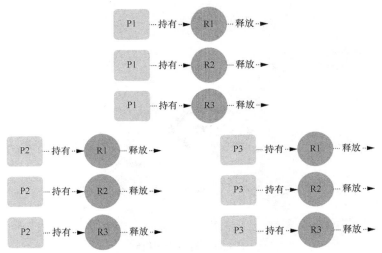

图 4.31 避免锁嵌套

4.7.4 锁超时机制

事后处理的第一种措施是锁超时机制,核心就是对锁的等待并非是永久的,而是具有超时机制。某个线程对某个锁的等待时间如果超过了指定的时间则进行超时处理,直接结束掉该线程。比如在图 4.32 中,3 个线程已经进入死锁状态,假如线程 P1 等待 R2 锁的时间超时,此时 P1 将结束并且释放对 R1 锁的占有权。这时线程 P3 则能够获取 R1 锁,于是能够解除死锁状态。

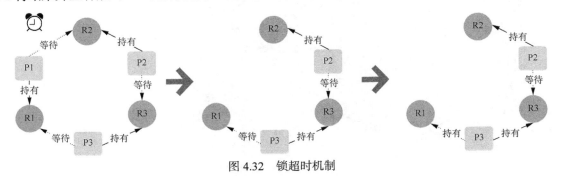

图 4.32 锁超时机制

4.7.5 抢占资源机制

事后处理的第二种措施是抢占资源机制，它的处理核心是在发生死锁时对资源进行抢占。某个线程在发生死锁后，它可以去抢占别的线程已经拥有的资源，从而让自己能够往下执行。比如在图 4.33 中，3 个线程已经进入死锁状态，这时线程 P3 就会去抢占线程 P1 持有的资源 R1，从而使得 P3 拥有 R1 和 R3 两个资源，线程 P3 在使用完后释放这两个资源给其他线程使用。

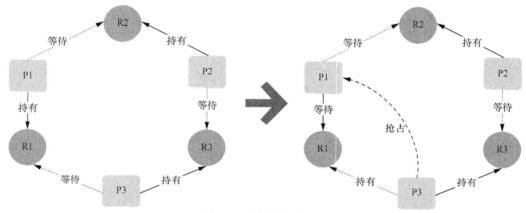

图 4.33 抢占资源机制

4.7.6 撤销线程机制

事后处理的第三种措施是撤销线程机制，它是在发生死锁时对线程进行撤销。比如在图 4.34 中，3 个线程进入死锁状态，这时线程 P1 被撤销，从而使得线程 P3 能持有资源 R1。

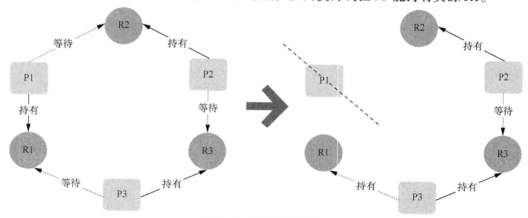

图 4.34 撤销线程机制

4.7.7 死锁的检测

事后处理的所有措施都需要建立在死锁检测的基础上。如果无法检测出死锁，就谈不上对其进行处理。那么如何才能检测出死锁呢？思路就是将线程和资源转换成有向图，P→R 表示线程请求资源，P←R 表示线程持有资源。这样原来死锁的图就可以转化成图 4.35 中右边的图，线程与资源之间的关系就能够用有向图来表示。而产生死锁时会构成一个环，对线程与资源关系的图进行有向化处理就能检测出是否存在环，如果存在则说明产生了死锁。一般可以使用广度优先算法或深度优先算法进行死锁检测。

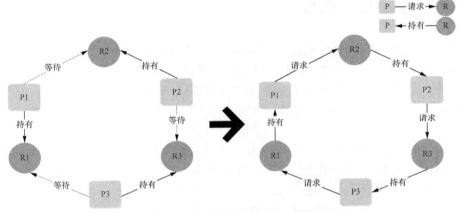

图 4.35 将线程和死锁转换为有向图

前文介绍了这么多死锁的内容，那么是否有活锁这个概念呢？实际上是有的，活锁与死锁的现象是一样的，即线程无法往前执行。死锁是因为都在等待资源而进入阻塞状态，而活锁则是因为某些逻辑导致一直在做无用功使得线程无法向前执行。同样是以十字路口的场景为例，假如汽车行驶的逻辑被设置为"如果左边有汽车开来，则先礼让"，那么图 4.36 中的这 4 辆汽车同时到达路口后，因为司机都在一直互相礼让，最终大家都无法通行。活锁在实际中很少出现，我们只要知道有这回事便可以，这里不再深入。

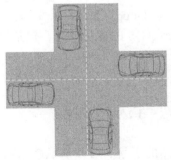

图 4.36 汽车行驶到十字路口的场景

第 5 章

AQS 同步器

5.1 什么是 AQS 同步器

同步器是专门为多线程并发而设计的同步机制，在这种机制下，多线程并发执行时线程之间通过某种共享状态来实现同步，只有当状态满足某种条件时线程才能往下执行。在不同的应用场景中，对同步器的需求也不同。JDK 将各种同步器中相同的部分抽象成一个统一的基础同步器，然后以基础同步器为模板通过继承的方式来实现不同的同步器，这个基础同步器就是 AQS 同步器。

1995 年，SUN 公司发布了第一个 Java 语言版本，可以说从 JDK 1.1 到 JDK 1.4 期间，Java 主要用于开发移动应用和中小型企业应用，基本不会涉及大型并发场景。随着互联网的发展，SUN 公司开始注重并发模块，于是在 J2SE 5.0（开发代号为 Tiger）的版本中增加了更加强大的并发工具包——java.util.concurrent，即 JUC。此后，由于 Java 在高并发场景中表现优异，很多大型互联网公司开始使用 Java 作为主要开发语言，例如阿里巴巴、eBay 等。这些公司的系统能够处理世界级的大型并发场景，这也反映了 Java 在大型并发场景是可行、可靠的。

JDK 的并发包提供了各种同步工具，其中大多数同步工具的实现基于 AbstractQueuedSynchronizer 类，即 AQS 同步器。它为不同场景提供了实现锁及同步机制的基本框架，为同步状态的原子性管理、线程的阻塞、线程的阻塞解除及排队管理提供了一种通用的机制。

JDK 并发工具包 JUC 的作者是 Doug Lea，但实现思想却结合了多位大师的智慧。如果想深入理解相关理论，可以阅读 Doug Lea 的论文 *The java.util.concurrent Synchronizer Framework*。从该论文中可以找到 AQS 的理论基础，包括框架的基本原理、需求、设计、实现思路、用法及性能等。

5.2 AQS 的等待队列与状态转换

要对 AQS 同步器有一个整体的了解，可以从 AQS 内部结构图（见图 5.1）着手。实际上，AQS 同步器有 5 个核心要素：同步状态、等待队列、独占模式、共享模式和条件

（Condition）队列。同步状态用于实现锁机制，等待队列用于存放等待锁的线程，独占模式和共享模式分别用于实现独占锁和共享锁，条件队列提供了可替代 wait/notify 机制的条件队列模式。

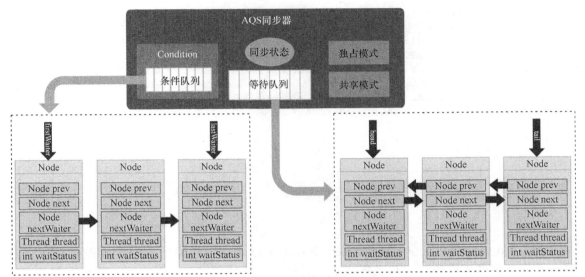

图 5.1 AQS 内部结构

AQS 同步器将线程封装到一个 Node 里面，并维护一个 CHL Node FIFO 队列，这是一个非阻塞的 FIFO 队列，意味着在并发条件下往此队列进行插入或移除操作时不会阻塞。它通过自旋+CAS 来保证节点插入和移除的原子性，从而实现快速插入。

下面分析 AbstractQueuedSynchronizer 类的实现原理（见代码清单 5.1），这里的代码并非严格与 JDK 一致。它是一个继承了 AbstractOwnableSynchronizer 类的抽象类，将同步器的通用特性和方法都抽象到该类中。主要类变量 head 和 tail 是等待队列的结构，而 state 则是同步状态，它的类型为 32 位整型，在更新 state 时必须保证是原子性的。另外 3 个 VarHandle 变量分别对应 head、tail 和 state 的句柄，用于对它们执行 CAS 操作。JDK 的 CAS 操作主要通过 Unsafe 类来实现，但从 JDK 9 版本后开始使用 VarHandle 来替代 Unsafe。而 SPIN_FOR_TIMEOUT_THRESHOLD 是一个阈值，它是决定使用自旋方式消耗时间还是使用系统阻塞方式消耗时间的分割线。也就是说，竞争时如果耗时小于 1 000ns 则自旋，如果超过 1 000ns 则调用系统阻塞。

代码清单 5.1 AbstractQueuedSynchronizer 类的实现原理

```
1.  public abstract class AbstractQueuedSynchronizer extends AbstractOwnableSynchronizer {
2.
3.      private transient volatile Node head;
4.      private transient volatile Node tail;
5.      private volatile int state;
```

```
 6.      private static final VarHandle STATE;
 7.      private static final VarHandle HEAD;
 8.      private static final VarHandle TAIL;
 9.      static final long SPIN_FOR_TIMEOUT_THRESHOLD = 1000L;
10.
11.      static {
12.          try {
13.              MethodHandles.Lookup l = MethodHandles.lookup();
14.              STATE = l.findVarHandle(AbstractQueuedSynchronizer.class, "state", int.class);
15.              HEAD = l.findVarHandle(AbstractQueuedSynchronizer.class, "head", Node.class);
16.              TAIL = l.findVarHandle(AbstractQueuedSynchronizer.class, "tail", Node.class);
17.          } catch (ReflectiveOperationException e) {
18.              throw new Error(e);
19.          }
20.      }
21.
22.      protected AbstractQueuedSynchronizer() {
23.      }
24.
25.      ...
26. }
```

父类 AbstractOwnableSynchronizer 主要用于记录在独占模式下哪个线程拥有该同步器，其实现也非常简单，仅仅是提供了设置和获取同步器拥有者的方法，如代码清单 5.2 所示。AQS 中如果某个线程以独占模式得到锁，那么就可以通过父类提供的方法来维护锁拥有者。

代码清单 5.2　父类 AbstractOwnableSynchronizer 的实现

```
 1. public abstract class AbstractOwnableSynchronizer {
 2.
 3.     protected AbstractOwnableSynchronizer() {
 4.     }
 5.
 6.     private transient Thread exclusiveOwnerThread;
 7.
 8.     protected final void setExclusiveOwnerThread(Thread thread) {
 9.         exclusiveOwnerThread = thread;
10.     }
11.
12.     protected final Thread getExclusiveOwnerThread() {
13.         return exclusiveOwnerThread;
14.     }
15.
16. }
```

AQS 同步器的等待队列由它内部的 Node 来实现（见图 5.2），Node 类属于 AbstractQueuedSynchronizer 类的内部类，该类的主要属性如代码清单 5.3 所示。

5.2 AQS 的等待队列与状态转换

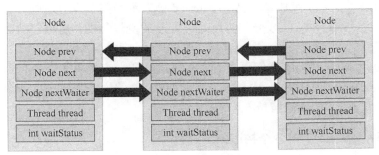

图 5.2 Node 的构成

代码清单 5.3 Node 类的主要属性

```
1.   static final class Node {
2.       static final Node SHARED = new Node();
3.       static final Node EXCLUSIVE = null;
4.       static final int CANCELLED =  1;
5.       static final int SIGNAL    = -1;
6.       static final int CONDITION = -2;
7.       static final int PROPAGATE = -3;
8.       volatile int waitStatus;
9.       volatile Node prev;
10.      volatile Node next;
11.      Node nextWaiter;
12.      volatile Thread thread;
13.
14.      private static final VarHandle NEXT;
15.      private static final VarHandle PREV;
16.      private static final VarHandle THREAD;
17.      private static final VarHandle WAITSTATUS;
18.      static {
19.          try {
20.              MethodHandles.Lookup l = MethodHandles.lookup();
21.              NEXT = l.findVarHandle(Node.class, "next", Node.class);
22.              PREV = l.findVarHandle(Node.class, "prev", Node.class);
23.              THREAD = l.findVarHandle(Node.class, "thread", Thread.class);
24.              WAITSTATUS = l.findVarHandle(Node.class, "waitStatus", int.class);
25.          } catch (ReflectiveOperationException e) {
26.              throw new Error(e);
27.          }
28.      }
29.      ...
30.  }
```

SHARED 和 EXCLUSIVE 用于标识该节点是独占模式还是共享模式。waitStatus 标识该节点的等待状态，可能值为 0、CANCELLED、SIGNAL、CONDITION、PROPAGATE。prev 和 next 用于组成双向链表，即等待队列的结构。nextWaiter 表示节点是独占模式还是共享模式，但对于条件队列来说则用于组成单向链表。thread 则表示当前节点对应的线程。此外，NEXT、PREV、THREAD、WAITSTATUS 分别为所对应变量的变量句柄，通过句柄能方便地对变量进行访问和修改。

了解了 Node 类的属性后，我们再来看它提供的方法。它提供了 3 个构造方法：无参构造方法、入参为 nextWaiter 的构造方法和入参为 waitStatus 的构造方法（这两个参数分别表示节点的模式和节点状态），如代码清单 5.4 所示。

代码清单 5.4　Node 类的 3 个构造方法

```
1.  static final class Node {
2.
3.      ...
4.
5.      Node() {
6.      }
7.
8.      Node(Node nextWaiter) {
9.          this.nextWaiter = nextWaiter;
10.         THREAD.set(this, Thread.currentThread());
11.     }
12.
13.     Node(int waitStatus) {
14.         WAITSTATUS.set(this, waitStatus);
15.         THREAD.set(this, Thread.currentThread());
16.     }
17.
18.     final boolean isShared() {
19.         return nextWaiter == SHARED;
20.     }
21.
22.     final Node predecessor() {
23.         Node p = prev;
24.         if (p == null)
25.             throw new NullPointerException();
26.         else
27.             return p;
28.     }
29.
30.     final boolean compareAndSetWaitStatus(int expect, int update) {
31.         return WAITSTATUS.compareAndSet(this, expect, update);
32.     }
33.
34.     final boolean compareAndSetNext(Node expect, Node update) {
35.         return NEXT.compareAndSet(this, expect, update);
36.     }
37.
38.     final void setPrevRelaxed(Node p) {
39.         PREV.set(this, p);
```

```
40.     }
41.
42. }
```

在构造方法中可直接将参数值赋值给对应变量,也可通过变量句柄进行赋值。isShared 方法用于判断等待队列是否为共享模式,predecessor 方法用于获取该节点对应的前驱节点,如果为空则抛空指针异常。compareAndSetWaitStatus 方法和 compareAndSetNext 方法用于以 CAS 方式修改对应的变量,setPrevRelaxed 方法用于设置当前节点的前驱节点。

等待队列由节点组成,每个节点都有自己的状态,它的值可能为 0、CANCELLED、SIGNAL、CONDITION 或 PROPAGATE。为了搞明白各种可能的状态转换,可参考图 5.3 所示的节点状态转换图。新建节点的状态默认值为 0,即无状态,它可能转为 SIGNAL、PROPAGATE 或 CANCELLED,而 SIGNAL、CONDITION 和 PROPAGATE 都可能转为无状态。PROPAGATE 还可能转为 SIGNAL,SIGNAL 和 CONDITION 也可以转为 CANCELLED。

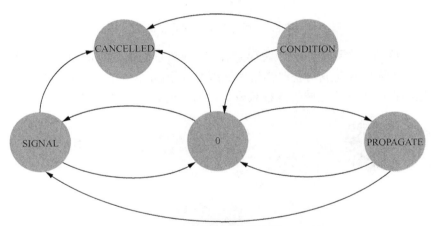

图 5.3　节点状态转换图

- SIGNAL:值为-1,表示后续节点中的线程通过 park 进入等待状态,当前节点在释放锁或取消时,要通过 unpark 解除后继节点的等待。
- CANCELLED:值为 1,表示当前节点的线程因为超时或中断被取消。
- CONDITION:值为-2,表示当前节点在条件队列中。
- PROPAGATE:值为-3,共享模式的头节点可能处于此状态,表示无条件往下传播。
- 0:除了以上 4 种状态的第五种状态,表示节点为无状态。

5.3　AQS 的独占锁与共享锁

多个线程在并发执行时,经常会导致数据竞争和竞争条件问题,那么该如何解决呢?答案就是引入同步机制，通过同步机制来控制共享数据和临界区的访问。同步机制可以通过锁来实现，所以 AQS 同步器也抽象出了锁的获取操作和释放操作（见图 5.4）。而且它还提供了独占锁和共享锁两种模式，便于上层各种同步工具的实现。

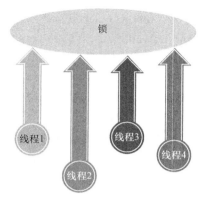

图 5.4　通过锁来实现同步机制

独占锁是指该锁一次只能由一个线程持有，其他线程则无法获得，除非已持有锁的线程释放了该锁。一个线程只有在成功获取锁后才能继续往下执行，当离开竞争区域时则释放锁，释放的锁供其他即将进入竞争区域的线程获取。独占锁的释放和获取过程如图 5.5 所示。

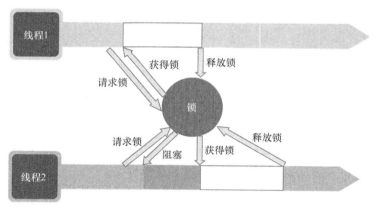

图 5.5　独占锁的释放和获取过程

获取独占锁和释放独占锁分别对应 acquire 方法和 release 方法。获取独占锁的主要逻辑为：先尝试获取锁，成功则往下执行，否则把线程放到等待队列中并可能将线程挂起。释放独占锁的主要逻辑为：唤醒等待队列中一个或多个线程去尝试获取锁。这里先了解大概的逻辑，后面会有详细的代码分析。

共享锁是指该锁可以由多个线程所持有，多个线程都能同时获得该锁，而不必等到持有锁的线程释放该锁。比如一般我们所说的读锁就是共享锁，一个共享数据是可以被多个线程读取，只要不改变共享数据就不会有问题。共享锁的释放和获取过程如图 5.6 所示。

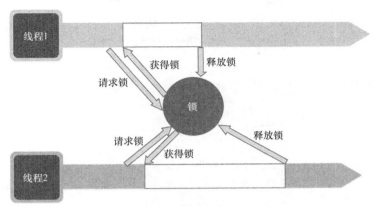

图 5.6 共享锁的释放和获取过程

获取共享锁和释放共享锁分别对应 acquireShared 方法和 releaseShared 方法。获取共享锁的主要逻辑为：先尝试获取锁，成功则往下执行，否则把线程放到等待队列中并可能将线程挂起。释放共享锁的主要逻辑为：唤醒等待队列中一个或多个线程去尝试获取锁。

5.4 AQS 独占锁获取与释放

在分析核心方法之前，我们先看一些辅助方法（见代码清单 5.5），因为核心方法会调用到它们。getState 方法和 setState 方法用于读写同步状态 state，需要注意这里额外提供了 compareAndSetState 方法，它用于通过 CAS 方式修改 state。compareAndSetTail 方法提供了 CAS 方式来修改队列尾 tail，setHead 方法用于设置队列头。parkAndCheckInterrupt 方法使当前线程进入等待状态，当它被唤醒后会判断当前线程是否被中断，并在中断时清理中断标识。

代码清单 5.5　辅助方法

```
1.  public abstract class AbstractQueuedSynchronizer extends AbstractOwnableSynchronizer {
2.
3.      ...
4.
5.      protected final int getState() {
6.          return state;
7.      }
8.
9.      protected final void setState(int newState) {
10.         state = newState;
11.     }
12.
13.     protected final boolean compareAndSetState(int expect, int update) {
14.         return STATE.compareAndSet(this, expect, update);
15.     }
```

```
16.
17.     private final boolean compareAndSetTail(Node expect, Node update) {
18.         return TAIL.compareAndSet(this, expect, update);
19.     }
20.
21.     private void setHead(Node node) {
22.         head = node;
23.         node.thread = null;
24.         node.prev = null;
25.     }
26.
27.     private final boolean parkAndCheckInterrupt() {
28.         LockSupport.park(this);
29.         return Thread.interrupted();
30.     }
31.
32.     ...
33.
34. }
```

shouldParkAfterFailedAcquire 方法用于判断当线程获取锁失败后是否要让该线程进入等待状态，传入的参数是前驱节点和当前节点，如代码清单 5.6 所示。如果前驱节点的状态为 SIGNAL，则返回 true，表示可以让线程进入等待状态，因为前驱节点会负责唤醒该线程。如果前驱节点的状态值大于 0，则表示前驱节点已经被取消，于是通过一个 do-while 循环往前清除所有已经被取消的节点。否则尝试通过 CAS 方式将前驱节点的状态改为 SIGNAL，并返回 false，表示不让线程进入等待状态。

代码清单 5.6　shouldParkAfterFailedAcquire 方法

```
1.  private static boolean shouldParkAfterFailedAcquire(Node pred, Node node) {
2.      int ws = pred.waitStatus;
3.      if (ws == Node.SIGNAL)
4.          return true;
5.      if (ws > 0) {
6.          do {
7.              node.prev = pred = pred.prev;
8.          } while (pred.waitStatus > 0);
9.          pred.next = node;
10.     } else {
11.         pred.compareAndSetWaitStatus(ws, Node.SIGNAL);
12.     }
13.     return false;
14. }
```

5.4.1　获取独占锁的逻辑

AQS 同步器提供了独占模式，该模式对应的方法为 acquire。在代码清单 5.7 中可以看到，该方法包含了 tryAcquire、addWaiter 和 acquireQueued 这 3 个核心方法。其中 tryAcquire 方法预留给子类实现，用于实现尝试获取锁的操作。addWaiter 方法实现对当前线程的入队操作，它会新建一个节点并添加到队列尾部，这里通过 Node.EXCLUSIVE 指定创建独占模式的节点。acquireQueued 方法实现节点入队后的操作，比如队列的头节点才有资格去尝试获取锁，当然尝试若干次后也可能进入等待状态，而非头节点则直接进入等待状态。

代码清单 5.7　acquire 方法

```
1.  public final void acquire(int arg) {
2.      if (!tryAcquire(arg) && acquireQueued(addWaiter(Node.EXCLUSIVE), arg))
3.          Thread.currentThread().interrupt();
4.  }
5.
6.  protected boolean tryAcquire(int arg) {
7.      throw new UnsupportedOperationException();
8.  }
```

总之，整个独占锁的获取分为 3 个步骤。
- 首先，在进入等待队列之前进行一次尝试获取锁操作（这相当于闯入策略）。
- 其次，若闯入失败，则将当前线程加入到等待队列中。
- 最后，在加入等待队列后还会执行一些操作，从而保证队列中头节点有资格获取锁而其他节点则直接进入等待状态。

我们以图 5.7 为例进行介绍，以增强对这 3 个步骤的理解。

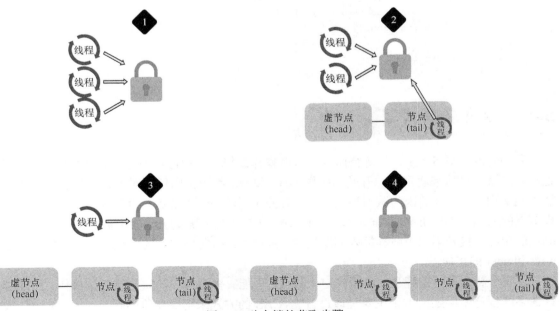

图 5.7　独占锁的获取步骤

1. 开始时独占锁已经被其他线程占有，此时有 3 个线程想要获取锁，它们都尝试闯入，但是都失败。
2. 其中一个线程先加入到等待队列中，此时队列为空，于是先创建一个虚节点（head）作为头节点。然后将该节点添加到头节点之后，并让该节点尝试获取锁。
3. 第二个线程继续加入到等待队列中，因为它的前驱节点不是 head 节点，所以现在还没有资格获取锁。它只需要让自己进入等待状态，前驱节点会在适当的时候唤醒它。此时第一个线程也已经进入等待状态。
4. 同样，第三个线程也加入到等待队列中，并使自己进入等待状态。

5.4.2　尝试获取独占锁

尝试获取独占锁的操作由 AQS 提供给子类定义，对应的是 tryAcquire 方法，子类只需在该方法中实现锁的获取逻辑即可。简单的实现如代码清单 5.8 所示，这是一个独占模式锁的获取操作，直接通过 CAS 尝试修改同步状态 state。如果修改成功则表示成功获取独占锁，此时返回 true，于是 AQS 的 acquire 方法也成功获取锁。否则返回 false，那么 AQS 的 acquire 方法会继续将其加入到等待队列中。

代码清单 5.8　tryAcquire 方法的简单实现

```
1.    public boolean tryAcquire(int acquires) {
2.        if (compareAndSetState(0, 1)) {
3.            return true;
4.        }
5.        return false;
6.    }
```

5.4.3　入队操作逻辑

我们先通过图 5.8 来理解等待队列的入队操作逻辑。前面已经说过，队列是由 Node 节点组成双向链表结构来实现的。开始的时候 head 和 tail 都不指向任何对象，假如有一个线程要插入到队列中，刚开始因为队列为空，所以需要先创建一个虚节点，然后将该线程节点插入到虚节点的后面。其中 head 指向虚节点，而 tail 则指向该线程节点，同时维护节点之间 prev 和 next 的指向。接着第二个线程插入到队列中，此时 tail 指向新加入的节点，同时维护节点之间 prev 和 next 的指向。

接着分析入队时的竞争。对于同一个队列，可能有多个线程同时执行入队操作，所以必然存在多线程竞争问题。这里主要是通过自旋+CAS 的方式来解决竞争问题。在图 5.9 中，假如队列中已经存在一个非 head 节点，然后一共有 3 个线程同时执行入队操作，它们就会通过自旋的方式来入队。假如其中一个线程成功入队，则队列中一共有两个非 head 节点，此时只剩下两个线程在自旋。然后一个线程接着一个线程进入队列，最终所有竞争线程都成功入队。

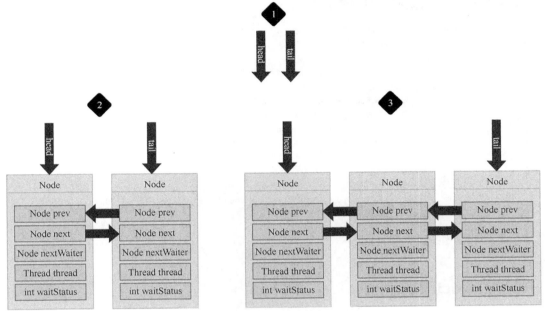

图 5.8 等待队列的入队操作逻辑

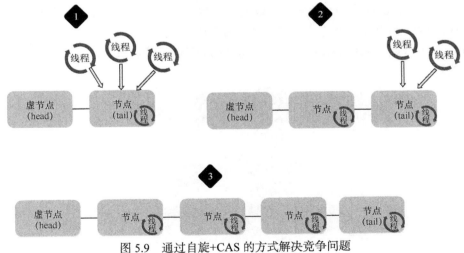

图 5.9 通过自旋+CAS 的方式解决竞争问题

最后我们对实现代码进行分析（见代码清单 5.9）。AQS 的入队操作由 addWaiter 方法实现，即在该方法中通过自旋+CAS 来实现多线程下的安全入队操作，其主要逻辑是先创建 Node 节点，当前线程会被保存在该节点中。然后通过 for(;;) 实现自旋，在自旋中按照双向链表结构来组织新加入的节点。我们看一下当队列为空时需要处理的特殊情况。如果队列中还不存在任何节点（即 tail 为空），则需要先创建一个虚节点并让 head 指向它，而 tail 则指向当前线程节点，最后维护 prev 和 next 的指向，从而完成入队操作。

代码清单 5.9　入队操作逻辑的实现

```
1.    private Node addWaiter(Node mode) {
2.        Node node = new Node(mode);
3.        for (;;) {
4.            Node oldTail = tail;
5.            if (oldTail != null) {
6.                node.setPrevRelaxed(oldTail);
7.                if (compareAndSetTail(oldTail, node)) {
8.                    oldTail.next = node;
9.                    return node;
10.               }
11.           } else {
12.               Node h;
13.               if (HEAD.compareAndSet(this, null, (h = new Node())))
14.                   tail = h;
15.           }
16.       }
17.   }
```

5.4.4　入队后的操作

当线程被加入到等待队列后，仍然需要做一些额外的操作。如果该节点的前驱节点是 head 节点，则有资格去尝试获取锁，尝试一次以上且失败后才进入等待状态。假如该节点的前驱节点不是 head 节点，则直接让其进入等待状态。图 5.10 的上半部分是有资格尝试获取锁的情况，下半部分是 3 个节点入队后的情况，第一个非 head 节点已经尝试获取锁，而后面两个节点则没有资格尝试。

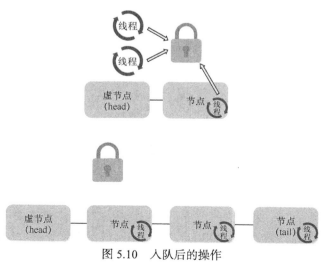

图 5.10　入队后的操作

acquireQueued 方法传入的参数为线程节点和锁值,如代码清单 5.10 所示,代码核心逻辑主要是 for(;;)循环中的内容。首先获取节点的前驱节点,如果前驱节点为 head 节点,则调用 tryAcquire 方法尝试获取锁,如果成功获取锁,则将该节点设为 head 节点。如果不成功则调用 shouldParkAfterFailedAcquire 方法判断是否应该让线程进入等待状态。如果条件成立,则调用 parkAndCheckInterrupt 方法使当前线程进入等待状态,等待过程中如果被中断则要进行标识。注意,这里即使被中断,它仍然会继续获取锁。也就是说,中断后仍然在循环中继续下一次的获取锁尝试或进入等待状态,所以实际上对中断并不敏感。整个过程中如果发生异常,则调用 cancelAcquire 方法取消当前节点。如果已被中断,则要设置线程的中断标识。

代码清单 5.10　acquireQueued 方法

```
1.  final boolean acquireQueued(final Node node, int arg) {
2.      boolean interrupted = false;
3.      try {
4.          for (;;) {
5.              final Node p = node.predecessor();
6.              if (p == head && tryAcquire(arg)) {
7.                  setHead(node);
8.                  p.next = null;
9.                  return interrupted;
10.             }
11.             if (shouldParkAfterFailedAcquire(p, node))
12.                 interrupted |= parkAndCheckInterrupt();
13.         }
14.     } catch (Throwable t) {
15.         cancelAcquire(node);
16.         if (interrupted)
17.             Thread.currentThread().interrupt();
18.         throw t;
19.     }
20. }
```

5.4.5　虚节点可能消失

实际上,虚节点是在队列初始化时用来充当 head 节点,它并非一直保持着,可能会被后面入队的节点所替代。在图 5.11 中,3 个线程竞争锁,其中一个线程入队,开始时因为队列为空所以创建了虚节点,该线程的节点则作为虚节点的后继。假如此时节点 A 尝试获取锁成功,节点 A 则会被设置为 head 节点,同时节点 A 对应的线程也被清除。此后以节点 A 为 head 节点,其他线程加入到它的后面。

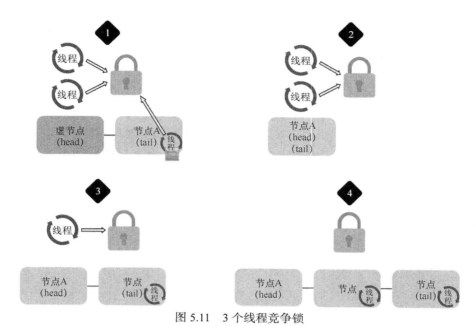

图 5.11　3 个线程竞争锁

5.4.6　取消锁获取操作

入队后的节点可能在某些情况下需要取消锁获取操作，比如发生异常或超时时，就需要将对应节点取消。取消时需要考虑多种情况（见图 5.12）。

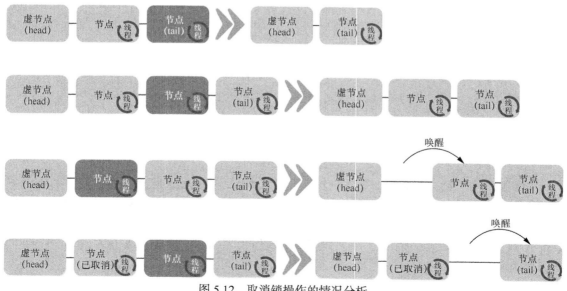

图 5.12　取消锁操作的情况分析

- 当节点为尾节点时，直接将该节点删除。
- 当节点为非尾节点且前驱节点不为头节点时，直接将该节点删除。
- 当节点为非尾节点且前驱节点为头节点时，除了删除该节点外还需要唤醒后继节点，以免永远无法唤醒后面的节点。
- 当节点为非尾节点且前驱节点的状态为已取消时，除了删除该节点外还需要唤醒后继节点，以免永远无法唤醒后面的节点。

取消锁获取操作由 cancelAcquire 方法实现，传入的参数为待取消节点，如代码清单 5.11 所示。该方法的主要逻辑为：首先将节点包含的线程设置为空并清除，然后通过一个 while 循环向前清除已经被取消的前驱节点，将当前节点的状态设置为 CANCELLED。接着分两种情况处理，如果该节点是尾节点 tail，则将前驱节点设为尾节点，并将前驱节点的 next 引用设为空，从而使得节点被垃圾回收器回收。如果该节点不是尾节点，则继续分为两种情况：一种是该节点的前驱节点不为头节点且前驱节点未被取消，则直接将该节点从队列中删除；另一种是该节点的前驱节点为头节点或前驱节点已被取消，则除了删除节点，还需要唤醒该节点的后继节点，以保证后面的节点不会进入永久等待状态。

代码清单 5.11　cancelAcquire 方法

```
1.  private void cancelAcquire(Node node) {
2.      if (node == null)
3.          return;
4.      node.thread = null;
5.      Node pred = node.prev;
6.      while (pred.waitStatus > 0)
7.          node.prev = pred = pred.prev;
8.      Node predNext = pred.next;
9.      node.waitStatus = Node.CANCELLED;
10.
11.     if (node == tail && compareAndSetTail(node, pred)) {
12.         pred.compareAndSetNext(predNext, null);
13.     } else {
14.         int ws;
15.         if (pred != head && ((ws = pred.waitStatus) == Node.SIGNAL || (ws <= 0 && pred.compareAndSetWaitStatus(ws, Node.SIGNAL))) && pred.thread != null) {
16.             Node next = node.next;
17.             if (next != null && next.waitStatus <= 0)
18.                 pred.compareAndSetNext(predNext, next);
19.         } else {
20.             unparkSuccessor(node);
21.         }
22.         node.next = node;
23.     }
24. }
```

5.4.7 唤醒后继节点

等待队列中的节点在进入等待状态后，需要有某个线程将其唤醒，这是十分重要的一项工作，若考虑得不周全则会导致线程永久阻塞等待。AQS 队列节点的唤醒工作由前驱节点负责，即每个节点负责唤醒自己的后继节点。这里分 3 种情况（见图 5.13）：
- 当节点为尾节点时不存在后继节点，无须唤醒；
- 当节点为非尾节点且后继节点没有被取消时，则直接唤醒后继节点；
- 当节点为非尾节点且后继节点已被取消时，则找到最近的未被取消的节点并将其唤醒。

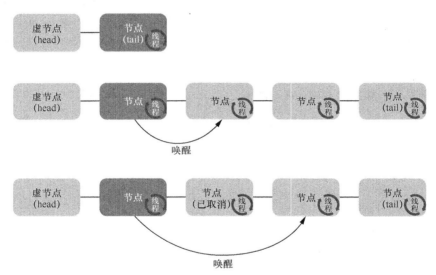

图 5.13 唤醒后继节点的情况分析

AQS 设计的等待队列中的线程是由前驱节点负责唤醒的，所以我们需要一个唤醒后继节点的操作。首先将节点的状态改为无状态（0），然后获取节点的后继节点。如果后继节点为空则代表已经到队列尾，不存在后继节点。如果后继节点处于已取消状态，那么就从队列尾部向前寻找最近的未被取消的节点来作为后继节点，之所以从尾部向前查找，是因为整个过程中尾部可能已经被插入其他节点。最后调用 LockSupport.unpark 方法唤醒后继节点对应的线程。整个过程如代码清单 5.12 所示。

代码清单 5.12　唤醒后继节点的代码实现

```
1.    private void unparkSuccessor(Node node) {
2.        int ws = node.waitStatus;
3.        if (ws < 0)
```

```
4.            node.compareAndSetWaitStatus(ws, 0);
5.        Node s = node.next;
6.        if (s == null || s.waitStatus > 0) {
7.            s = null;
8.            for (Node p = tail; p != node && p != null; p = p.prev)
9.                if (p.waitStatus <= 0)
10.                   s = p;
11.       }
12.       if (s != null)
13.           LockSupport.unpark(s.thread);
14.    }
```

5.4.8 释放独占锁的逻辑

　　一个线程持有锁并在使用完后需要将锁释放，而释放锁时需要考虑到等待队列中的线程，因为队列中的线程可能已经处于等待状态，需要将其唤醒。在图 5.14 中，首先，某个线程已经持有锁，而其他线程则处于等待队列中。然后，持有锁的线程释放锁，并唤醒头节点的下一个节点的线程。此时将该节点从队列中删除，而被唤醒的线程则成功获取到锁。最后，持有锁的线程又释放锁并唤醒下一个节点。

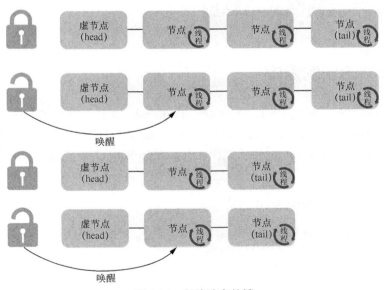

图 5.14　释放独占的锁

　　ASQ 同步器对独占锁的释放由 release 方法实现，如代码清单 5.13 所示。该方法的逻辑非常简单，即先尝试释放锁再唤醒后继节点。尝试释放锁的对应方法为 tryRelease，它预留给子类实现。如果成功释放锁则调用 unparkSuccessor 方法唤醒后继节点，传入的参数为头节点。最简单的 tryRelease 实现就是直接将同步状态改为 0。

代码清单 5.13　release 方法

```
1.   public final boolean release(int arg) {
2.       if (tryRelease(arg)) {
3.           Node h = head;
4.           if (h != null && h.waitStatus != 0)
5.               unparkSuccessor(h);
6.           return true;
7.       }
8.       return false;
9.   }
10.
11.  protected boolean tryRelease(int arg) {
12.      throw new UnsupportedOperationException();
13.  }
14.  protected boolean tryRelease(int releases) {
15.      setState(0);
16.      return true;
17.  }
```

5.5　AQS 共享锁获取与释放

5.5.1　获取共享锁的逻辑

AQS 提供的共享模式所对应的方法为 acquireShared，如代码清单 5.14 所示。从中可以看到该方法包含了 tryAcquireShared 方法和 doAcquireShared 方法。其中 tryAcquireShared 方法预留给子类实现，用于实现尝试释放锁的操作。doAcquireShared 方法则包括入队操作和入队后的操作。

代码清单 5.14　acquireShared 方法

```
1.   public final void acquireShared(int arg) {
2.       if (tryAcquireShared(arg) < 0)
3.           doAcquireShared(arg);
4.   }
5.
6.   protected int tryAcquireShared(int arg) {
7.       throw new UnsupportedOperationException();
8.   }
```

与独占锁的获取步骤类似，可以将共享锁的获取分为 3 步：首先，在进入等待队列之前尝试获取锁，因为共享锁可由多个线程获取，所以可以通过自旋锁和 CAS 来解决锁竞争问题。

其次，当所有锁都被获取完后则将当前线程加入到等待队列中。最后，在加入等待队列后还会执行一些操作，从而保证队列中头节点才有资格获取锁，其他节点则直接进入等待状态。可以通过图 5.15 来理解共享锁的获取过程。

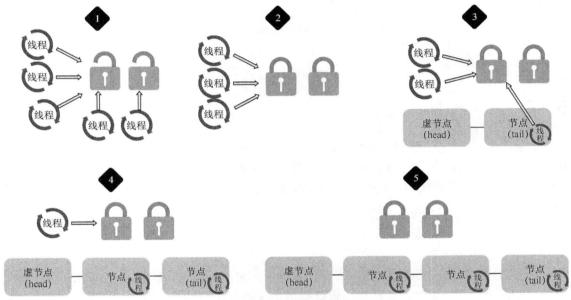

图 5.15　共享锁的获取过程

1. 开始时有两个未被获取的锁，一共有 5 个线程去竞争这两个锁。
2. 通过自旋和 CAS，其中两个线程成功获取锁，而剩下的 3 个线程则获取锁失败。
3. 其中一个线程先加入到等待队列中，此时队列为空，于是先创建一个虚节点作为头节点。然后将该节点添加到头节点之后，并让该节点尝试获取锁。
4. 第二个线程继续加入到等待队列中，因为它不是第一个非 head 节点，所以它现在还没有资格获取锁。它只需要让自己进入等待状态，前驱节点会在适当的时候唤醒它。此时第一个线程也已经进入等待状态。
5. 同样，第三个线程也加入到等待队列中，并使自己进入等待状态。

尝试获取共享锁的操作由 AQS 的子类来定义，对应 tryAcquireShared 方法（见代码清单 5.15），子类在该方法中实现共享锁的获取逻辑。与独占锁不同的是，共享锁存在多个锁供不同线程获取，同步状态 state 的值即表示剩下多少个锁可以获取，当其值等于 0 时则表示没有可用的锁。我们来看尝试获取锁的一种实现，它通过 for(;;) 实现自旋操作去竞争共享锁，每成功获取一个锁则将同步状态值减 1，当同步状态值为 0 时则返回小于 0 的值。返回小于 0 的值时，AQS 的 acquireShared 方法会将当前线程加入到等待队列中。

代码清单 5.15　tryAcquireShared 方法

```
1.    public int tryAcquireShared(int interval) {
2.        for (;;) {
3.            int current = getState();
4.            int newCount = current - 1;
5.            if (newCount < 0 || compareAndSetState(current, newCount))
6.                return newCount;
7.        }
8.    }
```

5.5.2　入队操作

实际上共享锁的入队操作与独占锁的入队操作一样，它们都使用 addWaiter 方法进行入队操作，具体的逻辑请参考前面的独占锁。这里需要说明的是，在调用 addWaiter 执行入队操作时，需要指定是独占模式节点还是共享模式节点。比如下面两行代码分别表示向等待队列中增加共享模式节点和独占模式节点。

```
1.    addWaiter(Node.SHARED);
2.    addWaiter(Node.EXCLUSIVE);
```

EXCLUSIVE 和 SHARED 被声明为静态变量，它们仅仅充当标识。addWaiter 方法中创建节点所使用的是下面的构造函数，也就是借用了条件队列的 nextWaiter 来保存标识。当它为 null 时表示独占模式节点，而当它为 SHARED 对象时表示共享模式节点，如代码清单 5.16 所示。

代码清单 5.16　使用 nextWaiter 来保存标识

```
1.    static final Node SHARED = new Node();
2.    static final Node EXCLUSIVE = null;
3.
4.    Node(Node nextWaiter) {
5.        this.nextWaiter = nextWaiter;
6.        THREAD.set(this, Thread.currentThread());
7.    }
```

5.5.3　入队后的操作

与独占锁的情况类似，对于共享模式，当线程加入到等待队列时，只有第一个非 head 节点才有资格去尝试获取共享锁，尝试一次以上且失败后才会进入等待状态。而如果该节点不是第一个非 head 节点，则让其直接进入等待状态。图 5.16 的上方为第一个非 head 节点在尝试获取锁，下方是节点直接进入等待状态。

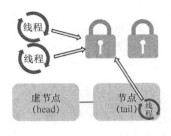

图 5.16　入队后的操作

　　doAcquireShared 方法的第一行代码是入队操作，剩余的代码为入队后的操作，如代码清单 5.17 所示。该方法的核心逻辑为 for(;;) 循环块的内容。首先获取节点的前驱节点，如果前驱节点为 head 节点，则调用 tryAcquireShared 方法尝试获取锁。如果返回的值大于等于 0，则表示成功获取锁，此时调用 setHeadAndPropagate 方法将该节点设为 head 节点，并在必要的情况下唤醒后继节点。接着调用 shouldParkAfterFailedAcquire 方法判断是否应该让线程进入等待状态，如果条件成立则调用 parkAndCheckInterrupt 方法使当前线程进入等待状态。整个过程中如果发生异常，则调用 cancelAcquire 方法取消当前节点。如果已被中断，则需要设置线程的中断标识。

代码清单 5.17　doAcquireShared 方法

```
1.    private void doAcquireShared(int arg) {
2.        final Node node = addWaiter(Node.SHARED);
3.        boolean interrupted = false;
4.        try {
5.            for (;;) {
6.                final Node p = node.predecessor();
7.                if (p == head) {
8.                    int r = tryAcquireShared(arg);
9.                    if (r >= 0) {
10.                       setHeadAndPropagate(node, r);
11.                       p.next = null;
12.                       return;
13.                   }
14.               }
15.               if (shouldParkAfterFailedAcquire(p, node))
16.                   interrupted |= parkAndCheckInterrupt();
17.           }
18.       } catch (Throwable t) {
19.           cancelAcquire(node);
```

```
20.            throw t;
21.        } finally {
22.            if (interrupted)
23.                Thread.currentThread().interrupt();
24.        }
25.    }
```

setHeadAndPropagate 方法（见代码清单 5.18）除了设置头节点外，还会在某些条件下唤醒后继节点。比如 propagate 大于 0 时，表示 AQS 同步器的同步状态值大于 0，其他线程还能够继续获取锁，所以通过 doReleaseShared 方法来唤醒后继节点。

代码清单 5.18　setHeadAndPropagate 方法

```
1.    private void setHeadAndPropagate(Node node, int propagate) {
2.        Node h = head;
3.        setHead(node);
4.        if (propagate > 0 || h == null || h.waitStatus < 0 || (h = head) == null
5.            || h.waitStatus < 0) {
6.            Node s = node.next;
7.            if (s == null || s.isShared())
8.                doReleaseShared();
9.        }
10.   }
```

独占模式的队列节点在获取到锁后不需要唤醒后继节点，但为什么共享模式下要让队列的第一个非 head 节点唤醒后继节点呢？这主要是因为共享模式存在多个锁可用，如果仅仅靠锁释放时去唤醒后继节点，则可能导致锁存在但没法获取的情况。比如在图 5.17 所示的情况中，

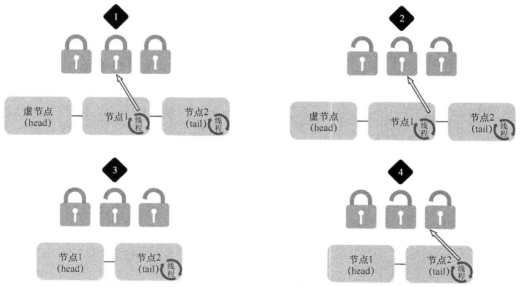

图 5.17　共享模式下存在后继节点唤醒后无法获取锁的情况

刚开始 3 个锁已经被 3 个线程占有，并且队列中有两个非 head 节点。接着假设 3 个线程都释放了锁，与此同时节点 1 的线程正在尝试获取锁。然后节点 1 成功获取锁，将节点 1 设置为 head 节点，并且唤醒节点 1 的后继节点，最后节点 2 继续去获取锁。可以看到，如果没有节点 1 的唤醒操作，就会导致另外两个锁空闲，而节点 2 又在排队等待。

5.5.4 引入 PROPAGATE 状态

我们知道 PROPAGATE 状态是为共享模式而引入的，为什么要这么做呢？在回答这个问题之前，我们先看独占模式的唤醒过程。因为独占模式下只有一个锁，所以在唤醒后继节点时不存在任何竞争问题。

我们通过图 5.18 来理解。开始时有一个线程和等待队列中的第一个非 head 节点竞争锁，最终由不在队列的线程成功获取锁，而且第一个非 head 节点也进入等待状态。线程在使用完锁后释放锁，同时它负责唤醒 head 节点的后继节点，最后队列中的节点继续去获取锁。可以看到释放锁时唤醒队列的节点是不存在竞争的。如果第 2 步中是第一个非 head 节点成功获取锁，也没关系，只要它释放锁的时候也唤醒后继节点即可。

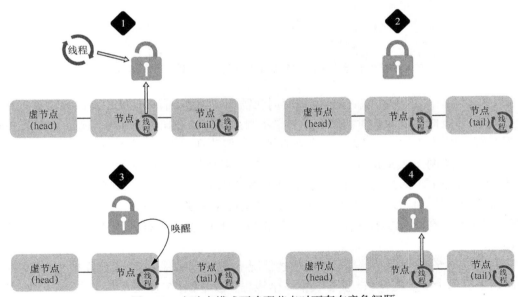

图 5.18　在独占模式下唤醒节点时不存在竞争问题

对于共享模式，如果仅仅在释放锁时才唤醒后继节点，会产生什么特殊情况呢？这可能导致有锁可用却无线程去获取。我们通过图 5.19 来理解这种情况。开始时有 3 个可用的锁，队列的第一个非 head 节点去获取锁，当它成功获取锁后将节点 1 的线程清除并成为 head 节点。如果节点 1 在获取锁的同时没有唤醒它的后继节点，那么实际上还有两个锁可用，但队列中的节点 2 却无法获取，必须一直等到锁释放时去唤醒节点 2，此时节点 2 才能够去获取锁。这种情况明显存在问题，所以必须在队列节点获取锁操作中加入唤醒后继节点的操作。

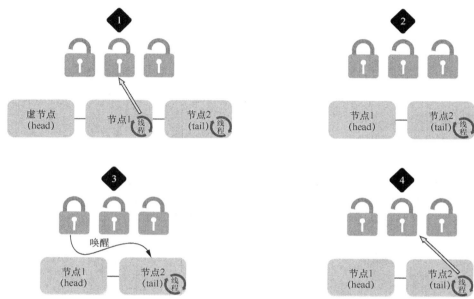

图 5.19　在共享模式下释放锁时才唤醒后继节点导致有锁可用却无线程获取的情况

从中可以看出，共享模式的唤醒有两个时机需要考虑：持有锁的线程释放锁时和队列中第一个非 head 节点成功获取锁时。那么，如果不引入额外的 PROPAGATE 状态，能否完成这个任务呢？答案是不完全能，因为可能存在有锁可用但队列线程却等待的情况。

必须理清这两个唤醒时机的触发条件。线程释放锁时的触发条件是队列中 head 节点的状态为 SIGNAL，即只有 head 节点的状态为 SIGNAL 时才会唤醒后继节点。而队列中第一个非 head 节点成功获取锁时的触发条件是可用锁的数量大于 0，即有锁可用时才会唤醒后继节点。在具体实现时，在多并发的情况下可能导致有锁可用但队列线程却等待的情况。

我们通过图 5.20 所示的过程来理解上述情况。在某个时刻，3 个锁已经被 3 个线程占有，等待队列中有两个节点在排队。接着其中一个线程释放了锁，它唤醒了头节点的后继节点并且将头节点的状态改为 0。接下去因为 3 个锁都已经被占有（即可用锁为 0），所以节点 1 不会唤醒它的后继节点。假如这时又有另外一个线程释放了锁，而且此时节点 1 还未被设置为 head 节点，那么该线程读到的头节点状态还是为 0，因此它不会唤醒后继节点。最终的状态就是存在一个可用锁，但等待队列的线程却不会去获取锁。

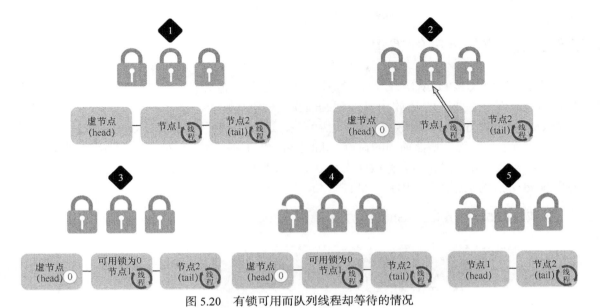

图 5.20 有锁可用而队列线程却等待的情况

这时就要引入一个额外的状态来解决上述问题,这个额外的状态定义为 PROPAGATE,值为 −3。在图 5.21 中,前面 3 步与图 5.20 中的一样,从第 4 步开始不相同。假如这时有一个线程释放锁,它发现头节点的状态为 0,就会把状态改为 −3,另外规定第一个非 head 节点发现头节点为 −3 时要唤醒它的后继节点,所以节点 1 尽管发现可用锁为 0,但它仍然会唤醒后继节点。最终的情况就是节点 1 成为头节点,而节点 2 继续去获取锁。问题得到解决。

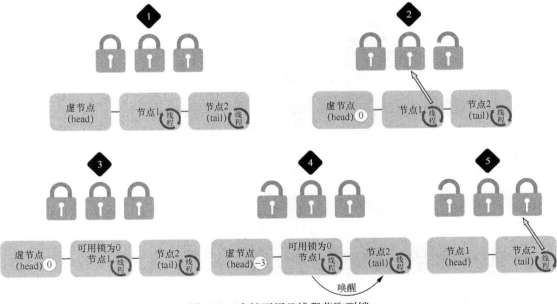

图 5.21 有锁可用且线程获取到锁

5.5.5 释放共享锁的逻辑

经过讨论引入 PROPAGATE 状态的原因。我们已经清楚共享锁释放时需要做哪些工作了。ASQ 同步器对共享锁的释放由 releaseShared 实现，如代码清单 5.19 所示。它先调用 tryReleaseShared 方法尝试释放锁（该方法由子类实现），然后再调用 doReleaseShared 方法来唤醒后继节点。注意，按照前面的讨论，doReleaseShared 方法需要处理两种情况：当 head 节点的状态为 SIGNAL 时，需要修改状态为 0 并唤醒后继节点；当 head 节点的状态为 0 时，需要将 head 节点的状态改为 PROPAGATE。

代码清单 5.19　releaseShared 方法

```
1.   public final boolean releaseShared(int arg) {
2.       if (tryReleaseShared(arg)) {
3.           doReleaseShared();
4.           return true;
5.       }
6.       return false;
7.   }
8.
9.   protected boolean tryReleaseShared(int arg) {
10.      throw new UnsupportedOperationException();
11.  }
12.
13.  private void doReleaseShared() {
14.      for (;;) {
15.          Node h = head;
16.          if (h != null && h != tail) {
17.              int ws = h.waitStatus;
18.              if (ws == Node.SIGNAL) {
19.                  if (!h.compareAndSetWaitStatus(Node.SIGNAL, 0))
20.                      continue;
21.                  unparkSuccessor(h);
22.              } else if (ws == 0 && !h.compareAndSetWaitStatus(0, Node.PROPAGATE))
23.                  continue;
24.          }
25.          if (h == head)
26.              break;
27.      }
28.  }
```

5.6　AQS 的阻塞与唤醒

AQS 同步器的阻塞与唤醒使用的是 LockSupport 类的 park 与 unpark 方法，该类分别调用

5.6 AQS 的阻塞与唤醒

的是 Unsafe 类的 park 与 unpark 本地方法，这两个本地方法依赖于不同操作系统的实现。AQS 同步器的阻塞与唤醒操作会在获取锁的操作中使用，即如果获取不到锁的线程进入排队队列后则需要阻塞，阻塞使用的是 LockSupport.park() 方法。相应地，排队队列中前驱节点负责唤醒后继节点包含的线程，唤醒使用的是 LockSupport.unpark(thread) 方法。

完整的逻辑为：一个线程参与锁竞争，首先它会尝试获取锁；如果失败就创建节点并插入到队列尾部；然后再次尝试获取锁，如成功则直接返回，否则设置节点状态为待运行状态；最后使用 LockSupport 的 park 阻塞当前线程。前驱节点运行完后将尝试唤醒后继节点，使用的是 LockSupport 的 unpark 唤醒。

LockSupport 类为构建锁和同步器提供了基本的线程阻塞与唤醒原语，AQS 同步器使用它来控制线程的阻塞和唤醒，当然其他的同步器或锁也会使用它。我们熟悉的阻塞与唤醒操作是 wait/notify 方式，它主要是从 Object 的角度来设计的。而 LockSupport 提供的 park/unpark 则是从线程的角度来设计的，真正解耦了线程之间的同步。为了更好地理解 JDK 的这些并发工具，需要具体分析该类的实现。该类主要包含阻塞操作和唤醒操作，如图 5.22 所示。

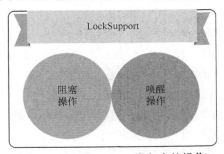

图 5.22　LockSupport 类包含的操作

下面通过图 5.23 所示的思维导图来理解 LockSupport 类的几个核心方法。这些方法总体可以分为 park 开头的方法和 unpark 方法。park 开头的方法用于执行阻塞操作，它又分为两类：参数包含阻塞对象的方法和参数不包含阻塞对象的方法。

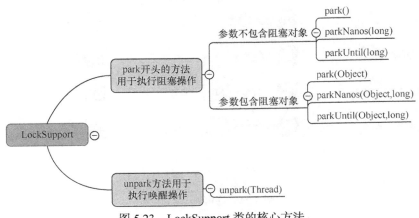

图 5.23　LockSupport 类的核心方法

下面对每个方法进行说明。

- park()方法：对当前线程执行阻塞操作，直到获取到可用许可后才解除阻塞（有关许可的概念，请见 5.6.1 节）。
- parkNanos(long)方法：对当前线程执行阻塞操作，等待获取到可用许可后才解除阻塞；最大的等待时间由传入的参数来指定，一旦超过最大时间，也会解除阻塞。
- parkUntil(long)方法：对当前线程执行阻塞操作，等待获取到可用许可后才解除阻塞；最大的等待时间为参数所指定的最后期限时间。
- park(Object)方法：与 park()方法同义，但传入的参数多为阻塞对象。
- parkNanos(Object,long)方法：与 parkNanos(long)同义，但指定了阻塞对象。
- parkUntil(Object,long)方法：与 parkUntil(long)同义，但指定了阻塞对象。
- unpark(Thread)方法：将指定线程的许可置为可用，也就相当于唤醒了该线程。

5.6.1　许可机制

在介绍 LockSupport 核心方法时出现了一个高频的词——许可。什么是许可呢？其实可以将 LockSupport 使用的许可看成一种二元信号变量，该信号变量分为两种状态，分别用 0 和 1 来表示，其中 0 表示无许可，1 表示有许可。每个线程都对应一个信号变量，当线程调用 park 时其实就是去获取许可，如果成功获取到许可则能够往下执行，否则将阻塞，直到成功获取许可为止。而当线程调用 unpark 时则是释放许可，供线程去获取。如图 5.24 所示，开始时有 4 个线程调用 park 方法，由于无许可，因此都处于阻塞状态。接着假设线程 1 和线程 4 的信号变量变为有许可，则它们都能继续往下执行。最后线程 1 和线程 4 获取许可后，其信号又变为无许可。

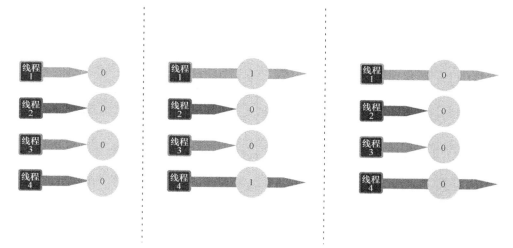

图 5.24　许可机制的应用

5.6.2 LockSupport 示例

下面看一个通过 LockSupport 进行阻塞和唤醒的示例，如代码清单 5.20 所示。其中，主线程分别创建 thread1 和 thread2，然后先启动 thread1，它会调用 LockSupport.park()方法进入阻塞状态。接着主线程睡眠 3s 后启动 thread2，thread2 会调用 LockSupport.unpark(thread1)方法让 thread1 得到许可从而唤醒 thread1。最终 thread1 输出"thread2 wakes up thread1"。

代码清单 5.20 通过 LockSupport 进行阻塞和唤醒

```java
1.  public class LockSupportDemo2 {
2.
3.      private static String message;
4.
5.      public static void main(String[] args) throws InterruptedException {
6.          Thread thread1 = new Thread(() -> {
7.              LockSupport.park();
8.              System.out.println(message);
9.          });
10.
11.         Thread thread2 = new Thread(() -> {
12.             message = "thread2 wakes up thread1";
13.             LockSupport.unpark(thread1);
14.         });
15.
16.         thread1.start();
17.         Thread.sleep(3000);
18.         thread2.start();
19.     }
20. }
```

需要注意的是，在调用 LockSupport 的 park 方法时，一般会使用 while(condition)循环体，如代码清单 5.21 所示。这样能保证在线程被唤醒后再一次判断条件是否符合。

代码清单 5.21 在调用 park 方法时使用 while(condition)循环体

```java
1.  public class LockSupportDemo3 {
2.
3.      private static String message;
4.
5.      public static void main(String[] args) throws InterruptedException {
6.          Thread thread1 = new Thread(() -> {
7.              while (message == null) {
8.                  LockSupport.park();
9.              }
10.             System.out.println(message);
11.         });
12.
```

```
13.        Thread thread2 = new Thread(() -> {
14.            message = "thread2 wakes up thread1";
15.            LockSupport.unpark(thread1);
16.        });
17.
18.        thread1.start();
19.        Thread.sleep(3000);
20.        thread2.start();
21.    }
22. }
```

5.6.3　park 与 unpark 的顺序

下面来探讨 LockSupport 的 park 和 unpark 的执行顺序的问题。一般先执行 park，线程阻塞，然后再执行 unpark 来唤醒该线程。如果先执行 unpark，再执行 park，park 是否会永远没办法被唤醒了呢？在往下分析之前，我们先来看看 wait/notify 方式的执行顺序示例，如代码清单 5.22 所示。主线程分别创建了 thread1 和 thread2，然后启动它们。由于 thread1 会睡眠 3s，所以先由 thead2 执行了 notify 去唤醒阻塞在锁对象上的线程，而在 3s 后 thread1 才会执行 wait 方法，此时它将一直阻塞在那里。所以 wait/notify 方式的执行顺序会对唤醒产生影响。

代码清单 5.22　wait/notify 方式的执行顺序

```
1.  public class LockSupportDemo {
2.      private static String message;
3.
4.      public static void main(String[] args) {
5.          Object lock = new Object();
6.          Thread thread1 = new Thread(() -> {
7.              try {
8.                  Thread.sleep(3000);
9.              } catch (InterruptedException e) {
10.                 e.printStackTrace();
11.             }
12.             synchronized (lock) {
13.                 try {
14.                     lock.wait();
15.                 } catch (InterruptedException e) {
16.                 }
17.             }
18.             System.out.println(message);
19.         });
20.         Thread thread2 = new Thread(() -> {
21.             synchronized (lock) {
22.                 message = "thread2 wakes up thread1";
23.                 lock.notify();
24.             }
25.         });
26.
```

```
27.        thread1.start();
28.        thread2.start();
29.
30.    }
31. }
```

park 和 unpark 的执行顺序如代码清单 5.23 所示。代码清单 5.23 与代码清单 5.22 的逻辑几乎相同,唯一不同的是阻塞和唤醒操作改为 park 和 unpark。实际上代码清单 5.23 能正确运行,且最终输出了 "thread2 wakes up thread1"。可以看出 park/unpark 方式的执行顺序不影响唤醒,这是因为 park/unpark 使用了许可机制。如果先调用 unpark 去释放许可,那么调用 park 时就能直接获取到许可而不必等待。

代码清单 5.23 park/unpark 方式的执行顺序

```
1.  public class LockSupportDemo1 {
2.      private static String message;
3.
4.      public static void main(String[] args) {
5.
6.          Thread thread1 = new Thread(() -> {
7.              try {
8.                  Thread.sleep(3000);
9.              } catch (InterruptedException e) {
10.                 e.printStackTrace();
11.             }
12.             LockSupport.park();
13.             System.out.println(message);
14.         });
15.
16.         Thread thread2 = new Thread(() -> {
17.             message = "thread2 wakes up thread1";
18.             LockSupport.unpark(thread1);
19.         });
20.
21.         thread1.start();
22.         thread2.start();
23.
24.     }
25. }
```

5.6.4 park 对中断的响应

park 方法支持中断。一个线程在调用 park 方法进入阻塞后,如果该线程被中断,则能够解除阻塞并立即返回。但需要注意的是,它不会抛出中断异常,所以我们不必去捕获 InterruptedException。代码清单 5.24 是一个中断示例,thread1 启动后调用 park 方法进入阻塞状态,然后主线程睡眠 1s 后中断 thread1,此时 thread1 将解除阻塞状态并输出 null。接着主线程睡眠 3s 后启动 thread2，thread2 将调用 unpark，但 thread1 已经因中断而解除阻塞了。

代码清单 5.24　中断示例

```
1.  public class LockSupportDemo4 {
2.
3.      private static String message;
4.
5.      public static void main(String[] args) throws InterruptedException {
6.          Thread thread1 = new Thread(() -> {
7.              LockSupport.park();
8.              System.out.println(message);
9.          });
10.
11.         Thread thread2 = new Thread(() -> {
12.             message = "thread2 wakes up thread1";
13.             LockSupport.unpark(thread1);
14.         });
15.
16.         thread1.start();
17.         Thread.sleep(1000);
18.         thread1.interrupt();
19.         Thread.sleep(3000);
20.         thread2.start();
21.     }
22. }
```

5.6.5　park 是否会释放锁

我们再思考这样一个问题，当线程调用 LockSupport 的 park 方法时，是否会释放该线程所持有的锁资源呢？答案是不会。我们通过代码清单 5.25 来理解。thread1 和 thread2 这两个线程都通过 synchronized(lock) 来获取锁。thread1 启动后获得锁，而且它调用 park 方法使得 thread1 进入阻塞状态，但是 thread1 却不会释放锁 lock。thread2 因为一直无法获取锁而无法进入同步块，也就没办法执行 unpark 操作。最终的结果造成了死锁，thread1 在等 thread2 的 unpark 操作，而 thread2 却在等 thread1 释放锁。

代码清单 5.25　调用 park 方法时不释放线程持有的锁资源

```
1.  public class LockSupportDemo5 {
2.
3.      private static String message;
4.
5.      public static void main(String[] args) throws InterruptedException {
6.          Object lock = new Object();
7.          Thread thread1 = new Thread(() -> {
8.              synchronized (lock) {
9.                  LockSupport.park();
```

```
10.                System.out.println(message);
11.            }
12.        });
13.
14.        Thread thread2 = new Thread(() -> {
15.            synchronized (lock) {
16.                message = "thread2 wakes up thread1";
17.                LockSupport.unpark(thread1);
18.            }
19.        });
20.
21.        thread1.start();
22.        Thread.sleep(3000);
23.        thread2.start();
24.    }
25. }
```

前面在介绍 park 相关的方法时将其分为两大类：不带阻塞对象和带阻塞对象。阻塞对象（Blocker）就是线程调用 park 方法时所在的对象，它的主要作用是供监视诊断工具使用，通过 getBlocker 方法能获取阻塞对象并对其进行分析。在代码清单 5.26 中，定义了一个 Blocker 类，其中 doPark 方法调用 LockSupport 的 park 方法，并传入 this 作为阻塞对象。thread1 启动后调用 Blocker 的 doPark 方法，thread2 则能够通过 LockSupport 的 getBlocker 方法来获取阻塞对象，然后可以对阻塞对象进行分析，这里会输出"我是 blocker"。

代码清单 5.26　获取阻塞对象并进行分析

```
1.  public class LockSupportDemo6 {
2.
3.      public static void main(String[] args) throws InterruptedException {
4.
5.          Blocker blocker = new Blocker();
6.
7.          Thread thread1 = new Thread(() -> {
8.              blocker.doPark();
9.          });
10.
11.         Thread thread2 = new Thread(() -> {
12.             Object o = LockSupport.getBlocker(thread1);
13.             System.out.println(o);
14.         });
15.
16.         thread1.start();
17.         Thread.sleep(3000);
18.         thread2.start();
19.     }
20.
21.     static class Blocker {
22.         public void doPark() {
```

```
23.            LockSupport.park(this);
24.        }
25.
26.    public String toString() {
27.        return "我是blocker";
28.    }
29.   }
30. }
```

5.6.6 LockSupport 的实现

实际上，LockSupport 的实现很简单，它间接调用了 Unsafe 的方法。Unsafe 类提供了很多底层的和不安全的操作，主要使用了 park、unpark、putObject 和 getObjectVolatile 等方法。park 和 unpark 方法对底层线程进行操作，而 putObject 方法是将对象设置为线程对象中的 parkBlocker 属性，getObjectVolatile 方法则是获取线程对象中的 parkBlocker 属性值，如图 5.25 所示。

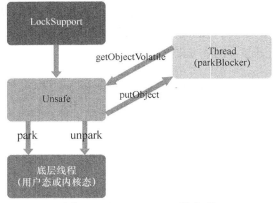

图 5.25　LockSupport 的实现

下面来分析 LockSupport 类的实现，如代码清单 5.27 所示。该代码并非与 JDK 完全相同，但核心实现是一样的。该类中主要包含一个 Unsafe 对象和 long 变量，由于无法直接通过 new 来实例化 Unsafe 对象，所以在 static 块中通过反射来获取 Unsafe 对象，而 long 变量表示 Thread 对象的 parkBlocker 属性的偏移，有了偏移就可以操作该属性。park、parkNanos 和 parkUntil 这 3 个方法都是间接调用了 Unsafe 对象的 park 方法。

代码清单 5.27　LockSupport 类的实现

```
1.  public class LockSupport {
2.
3.      private static final Unsafe U;
```

5.6 AQS 的阻塞与唤醒

```
4.      private static final long PARKBLOCKER;
5.
6.      static {
7.          try {
8.              Field theUnsafeInstance = Unsafe.class.getDeclaredField("theUnsafe");
9.              theUnsafeInstance.setAccessible(true);
10.             U = (Unsafe) theUnsafeInstance.get(Unsafe.class);
11.             PARKBLOCKER = U.objectFieldOffset(Thread.class.getDeclaredField("parkBlocker"));
12.         } catch (ReflectiveOperationException e) {
13.             throw new Error(e);
14.         }
15.     }
16.
17.     private LockSupport() {
18.     }
19.
20.     public static void park() {
21.         U.park(false, 0L);
22.     }
23.
24.     public static void parkNanos(long nanos) {
25.         if (nanos > 0)
26.             U.park(false, nanos);
27.     }
28.
29.     public static void parkUntil(long deadline) {
30.         U.park(true, deadline);
31.     }
32.
33.         ...
34.
35. }
```

传入阻塞对象时，会涉及 park、parkNonos 和 parkUntil 3 个方法（见代码清单 5.28）。这 3 个方法都会先通过 Thread.currentThread() 获取当前线程对象，然后调用 setBlocker 将阻塞对象设置为 Thread 对象中的 parkBlocker 属性，接着调用 Unsafe 对象的 park 方法，最终调用 setBlocker 将 Thread 对象中的 parkBlocker 属性清空。注意，这里的 setBlocker 方法间接调用了 Unsafe 对象的 putObject 方法，该方法能够通过属性偏移来修改属性值，而 getBlocker 方法则是通过间接调用 Unsafe 对象的 getObjectVolatile 方法来获取 Thread 对象的 parkBlocker 属性值。

代码清单 5.28　park、parkNonos 和 parkUntil 方法

```
1.  public static void park(Object blocker) {
2.      Thread t = Thread.currentThread();
3.      setBlocker(t, blocker);
4.      U.park(false, 0L);
5.      setBlocker(t, null);
```

```
6.      }
7.
8.      public static void parkNanos(Object blocker, long nanos) {
9.          if (nanos > 0) {
10.             Thread t = Thread.currentThread();
11.             setBlocker(t, blocker);
12.             U.park(false, nanos);
13.             setBlocker(t, null);
14.         }
15.     }
16.
17.     public static void parkUntil(Object blocker, long deadline) {
18.         Thread t = Thread.currentThread();
19.         setBlocker(t, blocker);
20.         U.park(true, deadline);
21.         setBlocker(t, null);
22.     }
23.
24.     public static Object getBlocker(Thread t) {
25.         if (t == null)
26.             throw new NullPointerException();
27.         return U.getObjectVolatile(t, PARKBLOCKER);
28.     }
29.
30.     private static void setBlocker(Thread t, Object arg) {
31.         U.putObject(t, PARKBLOCKER, arg);
32.     }
```

unpark 方法则是间接调用 Unsafe 对象的 unpark 方法，如代码清单 5.29 所示。

代码清单 5.29　间接调用 unpark 方法

```
1.  public static void unpark(Thread thread) {
2.      if (thread != null)
3.          U.unpark(thread);
4.  }
```

5.7　AQS 的中断机制

AQS 同步器提供了中断机制，也就是说，某些操作是支持中断的。实际上，AQS 同步器的中断机制主要用于控制获取锁时的等待策略。比如，在不支持中断的锁获取方式中，线程必须一直等待锁，而如果加了中断支持，则可以随时中断该线程的等待。在图 5.26 中，对于不支持中断的锁获取方式，每个节点只能无限等待前驱节点使用完锁后 unpark。而支持中断的获取方式则可以中断某个线程节点，相当于直接取消了该节点获取锁的资格。

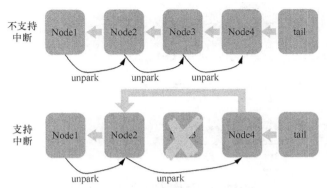

图 5.26 支持中断和不支持中断的锁获取方式

5.7.1 synchronized 不支持中断

Java 从语言层面提供了 synchronized 关键词来实现锁机制，虽然它使用起来非常便捷，但也存在某些不足的地方。中断机制便是其中之一，线程一旦进入锁等待，就失去了对这个线程的控制，唯一能做的就是等待，直到成功为止。在图 5.27 中，3 个线程中的线程 1 成功获取了对象锁，并且线程 1 持有锁的时间非常长。此时线程 2 和线程 3 将一直等待下去。假如想让线程 2 取消等待，则没有任何办法。即使对线程 2 调用中断方法去中断它，也是于事无补，最终这两个线程仍然将无限期等待下去。

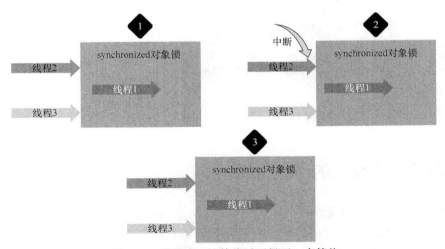

图 5.27 线程进入锁等待后不得不一直等待

在代码清单 5.30 中，先让线程 1 获取锁，输出"Thread1 gets the lock"后进入无限睡眠，它会永远占有锁。然后线程 2 尝试获取锁，它将一直等待下去。最后通过 interrupt() 方法对线程 2 进行中断，但实际上没有任何效果，主线程输出"Thread2 can't be interrupted"后，线程 2 将继续无限等待下去。

代码清单 5.30　线程一直等待的实现代码

```
1.  public class SynchronizedInterruptDemo {
2.
3.      public static void main(String[] args) {
4.          Object lock = new Object();
5.
6.          Thread thread1 = new Thread(() -> {
7.              synchronized (lock) {
8.                  System.out.println("Thread1 gets the lock");
9.                  try {
10.                     Thread.sleep(100000000000000000L);
11.                 } catch (InterruptedException e) {
12.                 }
13.             }
14.         });
15.
16.         Thread thread2 = new Thread(() -> {
17.             synchronized (lock) {
18.                 System.out.println("Thread2 can never get the lock");
19.             }
20.         });
21.
22.         try {
23.             thread1.start();
24.             Thread.sleep(1000);
25.             thread2.start();
26.             Thread.sleep(2000);
27.             thread2.interrupt();
28.             System.out.println("Thread2 can't be interrupted");
29.         } catch (InterruptedException e) {
30.         }
31.     }
32.
33. }
```

5.7.2　AQS 独占模式的中断

为了解决 synchronized 无法支持中断的问题，AQS 提供了独占锁的中断支持，对应的方法是 acquireInterruptibly，如代码清单 5.31 所示。该方法的核心流程与 acquire 方法是相同的，不同的地方是增加了抛出 InterruptedException 异常的判断。该过程涉及两个层面的中断机制：通过 if (Thread.interrupted())在 Java 语言层面进行中断标识判断；通过 LockSupport.park()在 JVM 层面实现中断。两个层面都是捕获到中断时向上抛出 InterruptedException 异常。这会涉及两个抛中断异常的位置，前一个对应 Java 层面，后一个对应 JVM 层面。

代码清单 5.31　acquireInterruptibly 方法

```
1.   public final void acquireInterruptibly(int arg) throws InterruptedException {
2.       if (Thread.interrupted())
3.           throw new InterruptedException();
4.       if (!tryAcquire(arg))
5.           doAcquireInterruptibly(arg);
6.   }
7.
8.   private void doAcquireInterruptibly(int arg) throws InterruptedException {
9.       final Node node = addWaiter(Node.EXCLUSIVE);
10.      try {
11.          for (;;) {
12.              final Node p = node.predecessor();
13.              if (p == head && tryAcquire(arg)) {
14.                  setHead(node);
15.                  p.next = null;
16.                  return;
17.              }
18.              if (shouldParkAfterFailedAcquire(p, node) && parkAndCheckInterrupt())
19.                  throw new InterruptedException();
20.          }
21.      } catch (Throwable t) {
22.          cancelAcquire(node);
23.          throw t;
24.      }
25.  }
```

parkAndCheckInterrupt 方法的代码只有如下几行，它的作用是执行 LockSupport.park(this) 后，线程进入等待状态，而如果其他线程对该线程进行中断，则会使其跳出等待状态并将线程的中断标识置为 true，然后通过 Thread.interrupted() 便能得知是否中断。

```
1.   private final boolean parkAndCheckInterrupt() {
2.       LockSupport.park(this);
3.       return Thread.interrupted();
4.   }
```

LockSupport.park 方式由 JVM 层的中断提供支持，我们通过代码清单 5.32 来理解它的中断过程。首先线程 1 输出 "thread1 is running..." 后进入等待状态，然后线程 2 对线程 1 执行中断，此时线程 1 跳出等待状态后往下执行。因为线程 1 已经被中断，所以输出 true。

代码清单 5.32　LockSupport.park 的中断过程

```
1.   public class ParkInterrupt {
2.       public static void main(String[] args) {
3.           Thread thread1 = new Thread(() -> {
4.               System.out.println("thread1 is running...");
5.               LockSupport.park();
```

```
6.         });
7.         thread1.start();
8.         try {
9.             Thread.currentThread().sleep(2000);
10.        } catch (InterruptedException e) {
11.            e.printStackTrace();
12.        }
13.        thread1.interrupt();
14.        System.out.println(thread1.isInterrupted());
15.     }
16. }
```

5.7.3　AQS 共享模式的中断

共享模式的中断实现与独占模式的中断实现相同，都使用了 Java 层和 JVM 层的中断机制，当中断标识为 true 时向上抛出 InterruptedException 异常。在代码清单 5.33 中，两处抛中断异常的位置对应了 Java 层和 JVM 层中断的判断，总体流程与独占模式类似，这里不再赘述。

代码清单 5.33　共享模式的中断

```
1.  public final void acquireSharedInterruptibly(int arg) throws InterruptedException {
2.      if (Thread.interrupted())
3.          throw new InterruptedException();
4.      if (tryAcquireShared(arg) < 0)
5.          doAcquireSharedInterruptibly(arg);
6.  }
7.
8.  private void doAcquireSharedInterruptibly(int arg) throws InterruptedException {
9.      final Node node = addWaiter(Node.SHARED);
10.     try {
11.         for (;;) {
12.             final Node p = node.predecessor();
13.             if (p == head) {
14.                 int r = tryAcquireShared(arg);
15.                 if (r >= 0) {
16.                     setHeadAndPropagate(node, r);
17.                     p.next = null;
18.                     return;
19.                 }
20.             }
21.             if (shouldParkAfterFailedAcquire(p, node) && parkAndCheckInterrupt())
22.                 throw new InterruptedException();
23.         }
24.     } catch (Throwable t) {
25.         cancelAcquire(node);
26.         throw t;
27.     }
28. }
```

5.8　AQS 的超时机制

5.8.1　synchronized 不支持超时

synchronized 的另外一个不足是它没提供锁获取超时机制。也就是说，一旦线程进入锁等待，它就会一直等待下去，不管等待多久都不会失败。在某些场景中，需要指定获取锁的超时时间，将超时的线程直接当成失败处理，这时就需要用到超时机制。在图 5.28 中，不管线程 1 持有锁的时间多长，线程 2 和线程 3 都将一直等待且永不超时。

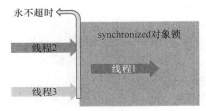

图 5.28　线程永不超时

在代码清单 5.34 中，线程 1 获得锁后输出 "Thread1 gets the lock" 并进入无限睡眠，它将永远占有锁。接着线程 2、线程 3 开始去获取锁，但它们将永远等待下去。

代码清单 5.34　线程永不超时

```
1.  public class SynchronizedTimeoutDemo {
2.      public static void main(String[] args) {
3.          Object lock = new Object();
4.          Thread thread1 = new Thread(() -> {
5.              synchronized (lock) {
6.                  System.out.println("Thread1 gets the lock");
7.                  try {
8.                      Thread.sleep(1000000000000000000L);
9.                  } catch (InterruptedException e) {
10.                     e.printStackTrace();
11.                 }
12.             }
13.         });
14.         Thread thread2 = new Thread(() -> {
15.             synchronized (lock) {
16.                 System.out.println("Thread2 can never get the lock");
17.             }
18.         });
19.         Thread thread3 = new Thread(() -> {
```

```
20.            synchronized (lock) {
21.                System.out.println("Thread3 can never get the lock");
22.            }
23.        });
24.        try {
25.            thread1.start();
26.            Thread.sleep(1000);
27.            thread2.start();
28.            thread3.start();
29.        } catch (InterruptedException e) {
30.            e.printStackTrace();
31.        }
32.    }
33. }
```

5.8.2 AQS 独占模式的超时

AQS 通过 tryAcquireNanos 方法提供了超时机制，其核心流程与 acquire 方法相同，不同的地方在于获取锁的过程中增加了时间限制，如代码清单 5.35 所示。总体来说，超时机制的思想就是先计算截止时间，然后在锁检查操作中计算是否已到截止时间，如果已到截止时间，则取消队列中的该节点并跳出循环。向 doAcquireNanos 方法传入的 nanosTimeout 参数值就是超时时间，该方法首先计算 deadline（截止时间），然后在循环中计算最新的剩余时间是否小于等于 0，如果是则直接取消队列中的该节点。如果剩余时间超过 SPIN_FOR_TIMEOUT_THRESHOLD，则使用 LockSupport.parkNanos 方法让线程进入阻塞状态，当它超时后又会进入下一轮的超时判断。

代码清单 5.35　提供了超时机制的 tryAcquireNanos 方法

```
1.  public final boolean tryAcquireNanos(int arg, long nanosTimeout) throws InterruptedException {
2.      if (Thread.interrupted())
3.          throw new InterruptedException();
4.      return tryAcquire(arg) || doAcquireNanos(arg, nanosTimeout);
5.  }
6.
7.  private boolean doAcquireNanos(int arg, long nanosTimeout) throws InterruptedException {
8.      if (nanosTimeout <= 0L)
9.          return false;
10.     final long deadline = System.nanoTime() + nanosTimeout;
11.     final Node node = addWaiter(Node.EXCLUSIVE);
12.     try {
13.         for (;;) {
14.             final Node p = node.predecessor();
15.             if (p == head && tryAcquire(arg)) {
16.                 setHead(node);
17.                 p.next = null;
```

```
18.              return true;
19.          }
20.          nanosTimeout = deadline - System.nanoTime();
21.          if (nanosTimeout <= 0L) {
22.              cancelAcquire(node);
23.              return false;
24.          }
25.          if (shouldParkAfterFailedAcquire(p, node)
26.              && nanosTimeout > SPIN_FOR_TIMEOUT_THRESHOLD)
27.              LockSupport.parkNanos(this, nanosTimeout);
28.          if (Thread.interrupted())
29.              throw new InterruptedException();
30.      }
31.  } catch (Throwable t) {
32.      cancelAcquire(node);
33.      throw t;
34.  }
35. }
```

AQS 的超时控制有两点需要注意:
- 超时时间包括竞争入队的时间,如果在竞争入队时就把超时时间消耗完,则直接当作超时处理;
- SPIN_FOR_TIMEOUT_THRESHOLD 阈值是决定使用自旋方式消耗时间还是使用阻塞方式消耗时间的分割线。

JUC 工具包的作者通过测试将默认值设置为 1 000ns,即如果在成功插入等待队列后剩余时间大于 1 000ns 则调用系统底层阻塞,否则不调用系统底层阻塞,而且仅仅让其在 Java 层不断循环消耗时间(这属于性能优化的措施)。

5.8.3 AQS 共享模式的超时

AQS 共享模式的超时与独占模式的实现相同,这里不再赘述,读者可以参考代码清单 5.35 来理解代码清单 5.36。

代码清单 5.36 共享模式的超时机制

```
1.  public final boolean tryAcquireSharedNanos(int arg, long nanosTimeout)
2.          throws InterruptedException {
3.      if (Thread.interrupted())
4.          throw new InterruptedException();
5.      return tryAcquireShared(arg) >= 0 || doAcquireSharedNanos(arg, nanosTimeout);
6.  }
7.
8.  private boolean doAcquireSharedNanos(int arg, long nanosTimeout) throws InterruptedException {
9.      if (nanosTimeout <= 0L)
10.         return false;
```

```java
11.     final long deadline = System.nanoTime() + nanosTimeout;
12.     final Node node = addWaiter(Node.SHARED);
13.     try {
14.         for (;;) {
15.             final Node p = node.predecessor();
16.             if (p == head) {
17.                 int r = tryAcquireShared(arg);
18.                 if (r >= 0) {
19.                     setHeadAndPropagate(node, r);
20.                     p.next = null;
21.                     return true;
22.                 }
23.             }
24.             nanosTimeout = deadline - System.nanoTime();
25.             if (nanosTimeout <= 0L) {
26.                 cancelAcquire(node);
27.                 return false;
28.             }
29.             if (shouldParkAfterFailedAcquire(p, node)
30.                 && nanosTimeout > SPIN_FOR_TIMEOUT_THRESHOLD)
31.                 LockSupport.parkNanos(this, nanosTimeout);
32.             if (Thread.interrupted())
33.                 throw new InterruptedException();
34.         }
35.     } catch (Throwable t) {
36.         cancelAcquire(node);
37.         throw t;
38.     }
39. }
```

5.9 AQS 的原子性如何保证

我们在研究 AQS 同步器时，发现 AbstractQueuedSynchronizer 这个类的很多地方都使用了 CAS 操作。在并发实现中，CAS 操作必须具有原子性，而且是硬件级别的原子性。Java 被隔离在硬件之上，因此硬件级别的操作明显力不从心。为了能够执行操作系统层面的操作，就必须通过用 C/C++甚至是汇编来编写本地方法。在 Java 中一般可以通过 JNI 方式让 Java 代码调用 C/C++本地代码。

Java 从一开始就被定为一个安全的编程语言，它屏蔽了指针和内存的管理，从而减少犯错的风险。但 Java 仍然为我们留下了一个后门，通过这个后门能够进行一些底层的、不安全的操作，比如内存的申请/释放/访问等操作、底层硬件的原子操作、内存屏障的设置、对象的操作等。这个后门就是 Unsafe 类，该类位于 sun.misc 包下。在新版本的 JDK 中，sun.misc 包下的 Unsafe 类会间接调用 jdk.internal.misc 包下的 Unsafe 类。也就是说，sun.misc.Unsafe 的实现将全部委托给 jdk.internal.misc.Unsafe。

有时，我们在新版本的 JDK 中会看到用 VarHandle 替代了 Unsafe 的一部分功能，实际上它们实现的本质都类似，但官方的说法是 VarHandle 更安全，更易用，且性能更高，并且官方推荐不要使用 Unsafe。不管如何，只要我们掌握了实现的原理，无论接口如何变化都能轻松应对。

在图 5.29 中，Java 语言层能够通过 Unsafe 通道来执行底层的操作，这些操作可能涉及 JVM 层的堆和方法区，当然也可能涉及操作系统层，甚至更深的硬件层。

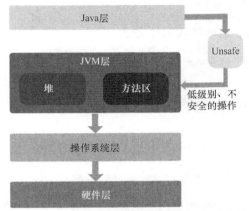

图 5.29　Java 层通过 Unsafe 通道执行底层操作

Unsafe 提供了很多与底层相关的操作，如图 5.30 所示。但对于并发和线程来说，我们主要关注与 CAS 和线程调度相关的方法。其中 CAS 包括 compareAndSwapInt、compareAndSwapLong 和 compareAndSwapObject 这 3 个方法，而线程调度包括 park 和 unpark 这两个方法。

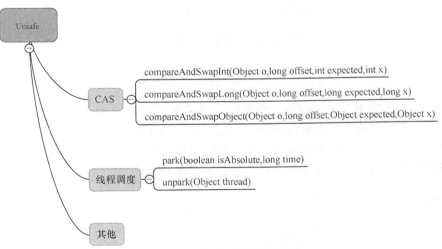

图 5.30　Unsafe 提供的底层操作

实际上，JDK 开发人员做了一些措施来避免 Unsafe 的滥用。因为 Unsafe 主要供 JDK 内部类库使用，所以在实例化时增加了使用安全校验，使得只有受信任代码才能对其进行实例化。

该校验主要通过类加载器来判断，后面会讲到详细的判断逻辑。Unsafe 提供了 getUnsafe 方法来获取 Unsafe 对象，所以第一种实例化方式就是直接调用该方法，但是这种方式对于 Java 语言层面的开发人员来说是行不通的，因为没办法通过安全校验，会抛出 SecurityException 异常。而第二种实例化方式则是通过反射机制来绕过安全检查，直接去修改 Unsafe 类中 theUnsafe 字段的访问权限，使其能被访问，然后获取该字段的值，从而成功得到 Unsafe 对象。这两种实例化方式如代码清单 5.37 所示。

代码清单 5.37　Unsafe 的两种实例化方式

```java
public class UnsafeInstanceTest {

    //实例化方式 1
    public static Unsafe getUnsafeInstance_1() {
        return Unsafe.getUnsafe();
    }

    //实例化方式 2
    public static Unsafe getUnsafeInstance_2() {
        try {
            Field theUnsafeField = Unsafe.class.getDeclaredField("theUnsafe");
            theUnsafeField.setAccessible(true);
            return (Unsafe) theUnsafeField.get(null);
        } catch (Exception e) {
            return null;
        }
    }

    public static void main(String[] args) {
        Unsafe unsafe = UnsafeInstanceTest.getUnsafeInstance_1();//失败
        Unsafe unsafe2 = UnsafeInstanceTest.getUnsafeInstance_2();//成功
    }

}
```

我们对 Unsafe 实例化的工作做一个总结：当 JDK 源码中涉及 Unsafe 时，可以直接通过 getUnsafe 方法进行实例化，如果在 Java 语言层面实例化 Unsafe，则需要通过反射的方式来实现。也就是说，本书在模拟 JDK 并发类时都会使用反射方式去获取 Unsafe 对象。

在知道如何去实例化 Unsafe 对象后，接下来了解 Unsafe 类，如代码清单 5.38 所示。

代码清单 5.38　Unsafe 类

```java
public class UnsafeTest {

    private int flag = 100;
    private static long offset;
    private static Unsafe unsafe = null;

```

```
7.    static {
8.        try {
9.            Field theUnsafeField = Unsafe.class.getDeclaredField("theUnsafe");
10.           theUnsafeField.setAccessible(true);
11.           unsafe = (Unsafe) theUnsafeField.get(null);
12.           offset = unsafe.objectFieldOffset(UnsafeTest.class.getDeclaredField("flag"));
13.       } catch (Exception e) {
14.           e.printStackTrace();
15.       }
16.   }
17.
18.   public static void main(String[] args) throws Exception {
19.       int expect = 100;
20.       int update = 101;
21.       UnsafeTest unsafeTest = new UnsafeTest();
22.       System.out.println("flag 字段的地址偏移为：" + offset);
23.       unsafe.compareAndSwapInt(unsafeTest, offset, expect, update);
24.       System.out.println("CAS 操作后 flag 的值为：" + unsafeTest.flag);
25.   }
26.
27. }
```

在代码清单 5.38 中，使用 Unsafe 对象进行硬件级别的 CAS 操作，也就是修改 UnsafeTest 对象的 flag 字段。其中 offset 表示 flag 字段的地址偏移，可由 Unsafe 的 objectFieldOffset 方法获得，这个地址偏移在调用 compareAndSwapInt 方法时将会作为其中一个参数。除了这个参数外还需要预期值和更新值。最终的输出结果如下。

```
1. flag 字段的地址偏移为：12
2. CAS 操作后 flag 的值为：101
```

下面对 Unsafe 实现源码进行分析。由于涉及 C/C++ 代码，如果读者在阅读下面的代码时有压力，则可以跳过这部分。我们只关注 CAS、线程调度和构造函数相关的几个方法。在代码清单 5.39 中，registerNatives 方法用于注册本地方法。构造函数是 private 的，所以不能通过构造函数来实例化 Unsafe 对象。而 getUnsafe 方法则是用来获取 Unsafe 对象的方法，虽然它是 public 的，但它会对调用者进行安全检查，判断调用者是否由 bootstrap 类加载器加载，否则会抛出 SecurityException 异常。其他的就是 CAS 操作和线程调度相关的方法，它们都是本地方法，后面会逐个分析。

代码清单 5.39　Unsafe 的实现源码

```
1. public final class Unsafe {
2.
3.     private static native void registerNatives();
4.
5.     static {
6.         registerNatives();
7.     }
```

```
8.
9.      private Unsafe() {
10.     }
11.
12.     private static final Unsafe theUnsafe = new Unsafe();
13.
14.     public static Unsafe getUnsafe() {
15.         Class<?> caller = Reflection.getCallerClass();
16.         if (!VM.isSystemDomainLoader(caller.getClassLoader()))
17.             throw new SecurityException("Unsafe");
18.         return theUnsafe;
19.     }
20.
21.     public final native boolean compareAndSwapInt(Object o, long offset, int expected, int x);
22.
23.     public final native boolean compareAndSwapLong(Object o, long offset, long expected, long x);
24.
25.     public final native boolean compareAndSwapObject(Object o, long offset, Object expected,
26.             Object x);
27.
28.     public native void park(boolean isAbsolute, long time);
29.
30.     public native void unpark(Object thread);
31.
32. }
```

前面说到，CompareAndSwapInt 是一个本地方法，该方法对应的本地实现位于 /openjdk/hotspot/src/share/vm/prims/unsafe.cpp 中，对应的代码如代码清单 5.40 所示。具体的 3 步逻辑为：首先获取对象的 oop，oop 是 JVM 层中的 Java 对象表示；然后获取该 oop 对象中对应的偏移地址，也就是 Java 层对象中某个字段的地址；最后通过 Atomic::cmpxchg 对指定地址执行 CPU 级别的 CAS 操作。

代码清单 5.40　CompareAndSwapInt 的实现代码

```
1.  UNSAFE_ENTRY(jboolean, Unsafe_CompareAndSwapInt(JNIEnv *env, jobject unsafe,
2.              jobject obj, jlong offset, jint e, jint x))
3.  {
4.      oop p = JNIHandles::resolve(obj);
5.      jint* addr = (jint *)index_oop_from_field_offset_long(p, offset);
6.
7.      return (jint)(Atomic::cmpxchg(x, addr, e)) == e;
8.  }UNSAFE_END
```

我们知道，不同类型的 CPU 的指令集是不同的，不同的操作系统的汇编语言也可能不同，这导致需要为不同类型的 CPU 和不同的操作系统编写不一样的汇编语言。这里以 X86 架构的

CPU 为例。在 Linux 系统下，汇编代码如代码清单 5.41 所示。我们主要关注其中的 cmpxchgl 指令，它是 CPU 级别的 CAS 操作指令。

代码清单 5.41　Linux 系统下 CAS 操作指令的汇编代码

```
1.    inline jint Atomic::cmpxchg(jint exchange_value, volatile jint *dest,
2.            jint compare_value, cmpxchg_memory_order order) {
3.        int mp = os::is_MP();
4.        __asm__ volatile (LOCK_IF_MP(%4) "cmpxchgl %1,(%3)"
5.                : "=a" (exchange_value)
6.                : "r" (exchange_value), "a" (compare_value), "r" (dest), "r" (mp)
7.                : "cc", "memory");
8.        return exchange_value;
9.    }
```

X86 架构的 CPU 在 Windows 系统下的汇编代码如代码清单 5.42 所示。其中 cmpxchg 指令是 CPU 级别的 CAS 操作指令。

代码清单 5.42　Windows 系统下 CAS 操作指令的汇编代码

```
1.    inline jint Atomic::cmpxchg(jint exchange_value, volatile jint *dest,
2.    jint compare_value, cmpxchg_memory_order order) {
3.        int mp = os::is_MP();
4.        __asm {
5.            mov edx, dest
6.            mov ecx, exchange_value
7.            mov eax, compare_value
8.            LOCK_IF_MP(mp)
9.            cmpxchg dword ptr [edx], ecx
10.       }
11.   }
```

compareAndSwapLong 对应的本地实现也是在 unsafe.cpp 中。由于 int 是 4 字节，而 long 为 8 字节，所以底层的实现与 compareAndSwapInt 有些差别，具体的实现代码如代码清单 5.43 所示。前面两行仍然通过偏移量来获取对象中某个字段的地址，接下来根据 VM_Version::supports_cx8()分两种情况处理。第一种是 CPU 指令支持 8 字节的 CAS，此时直接通过 Atomic::cmpxchg 来执行 CAS 操作。第二种是 CPU 不支持 8 字节的 CAS，此时需要 MutexLockerEx 锁的协助才能完成 CAS，步骤是先加锁，再通过 Atomic::load 获取对应地址的值。如果值与期望值不相等，则直接返回 false，表示失败，否则继续调用 Atomic::store 来修改内存值，最终自动释放锁。

代码清单 5.43　compareAndSwapLong 的实现代码

```
1.    UNSAFE_ENTRY(jboolean, Unsafe_CompareAndSwapLong(JNIEnv *env, jobject unsafe,
2.            jobject obj, jlong offset, jlong e, jlong x)) {
3.        Handle p(THREAD, JNIHandles::resolve(obj));
4.        jlong* addr = (jlong*)index_oop_from_field_offset_long(p(), offset);
```

```
5.      if (VM_Version::supports_cx8()) {
6.          return (jlong)(Atomic::cmpxchg(x, addr, e)) == e;
7.      } else {
8.          MutexLockerEx mu(UnsafeJlong_lock, Mutex::_no_safepoint_check_flag);
9.          jlong val = Atomic::load(addr);
10.         if (val != e) {
11.             return false;
12.         }
13.         Atomic::store(x, addr);
14.         return true;
15.     }
16. }UNSAFE_END
```

在 CPU 支持 8 字节的 CAS 情况下，同样需要为不同的 CPU 类型和不同的操作系统编写不同的汇编语言。以 X86 为例，在 Linux 下的汇编代码如代码清单 5.44 所示。此时会用到 cmpxchgq 指令来实现硬件级别的 CAS 操作。

代码清单 5.44　Linux 系统下的汇编代码

```
1.  inline jlong Atomic::cmpxchg (jlong exchange_value, volatile jlong* dest, jlong compare_value, cmpxchg_memory_order order) {
2.      bool mp = os::is_MP();
3.      __asm__ __volatile__ (LOCK_IF_MP(%4) "cmpxchgq %1,(%3)"
4.                            : "=a" (exchange_value)
5.                            : "r" (exchange_value), "a" (compare_value), "r" (dest), "r" (mp)
6.                            : "cc", "memory");
7.      return exchange_value;
8.  }
```

Windows 系统下的汇编代码如代码清单 5.45 所示，其中主要使用了 cmpxchg8b 指令来实现 CAS 操作。

代码清单 5.45　Windows 系统下的汇编代码

```
1.  inline jlong Atomic::cmpxchg(jlong exchange_value, volatile jlong *dest,
2.      jlong compare_value, cmpxchg_memory_order order) {
3.      int mp = os::is_MP();
4.      jint ex_lo = (jint) exchange_value;
5.      jint ex_hi = *(((jint*) &exchange_value) + 1);
6.      jint cmp_lo = (jint) compare_value;
7.      jint cmp_hi = *(((jint*) &compare_value) + 1);
8.      __asm {
9.      push ebx
10.     push edi
11.     mov eax, cmp_lo
12.     mov edx, cmp_hi
13.     mov edi, dest
14.     mov ebx, ex_lo
```

```
15.     mov ecx, ex_hi
16.     LOCK_IF_MP(mp)
17.     cmpxchg8b qword ptr [edi]
18.     pop edi
19.     pop ebx
20.   }
21. }
```

compareAndSwapObject 的实现如代码清单 5.46 所示。首先分别获取 3 个对象的 oop，分别对应更新后的对象 x、期望的对象 e 和待更新的对象 p。然后获取待更新对象 p 的地址，接着调用 oopDesc::atomic_compare_exchange_oop 对 JVM 层面的对象进行 CAS 操作。如果返回值不等于期望对象 e，则表示更新失败，直接返回 false。最后是设置内存屏障，保证执行的顺序性和对其他线程的可见性。

代码清单 5.46　compareAndSwapObject 的实现代码

```
1.  UNSAFE_ENTRY(jboolean, Unsafe_CompareAndSwapObject(JNIEnv *env, jobject unsafe,
2.        jobject obj, jlong offset, jobject e_h, jobject x_h)) {
3.      oop x = JNIHandles::resolve(x_h);
4.      oop e = JNIHandles::resolve(e_h);
5.      oop p = JNIHandles::resolve(obj);
6.      HeapWord* addr = (HeapWord *)index_oop_from_field_offset_long(p, offset);
7.      oop res = oopDesc::atomic_compare_exchange_oop(x, addr, e, true);
8.      if (res != e) {
9.          return false;
10.     }
11.     update_barrier_set((void*)addr, x);
12.     return true;
13. }UNSAFE_END
```

oopDesc::atomic_compare_exchange_oop 核心方法如代码清单 5.47 所示。该方法根据 UseCompressedOops 分两种情况处理。UseCompressedOops 表示 JVM 中对象的指针是否使用了压缩指针（压缩指针是为了节约内存）。对于压缩指针的情况，首先将值都转换成 narrowOop 类型（该类型其实是无符号整型），然后再调用 Atomic::cmpxchg 进行 CPU 级别的 CAS 操作，最后再转成未压缩指针。而对于非压缩指针的情况，则调用 Atomic::cmpxchg_ptr 进行 CPU 级别的 CAS 操作。此时的指针大小为 64 位，可转成 long 型再执行 CPU 级别的 64 位的 CAS 操作。

代码清单 5.47　oopDesc::atomic_compare_exchange_oop 核心方法

```
1.  oop oopDesc::atomic_compare_exchange_oop(oop exchange_value,
2.                                   volatile HeapWord *dest,
3.                                   oop compare_value,
4.                                   bool prebarrier) {
5.      if (UseCompressedOops) {
6.        if (prebarrier) {
7.          update_barrier_set_pre((narrowOop*)dest, exchange_value);
```

```
8.     }
9.     // encode exchange and compare value from oop to T
10.    narrowOop val = encode_heap_oop(exchange_value);
11.    narrowOop cmp = encode_heap_oop(compare_value);
12.    narrowOop old = (narrowOop) Atomic::cmpxchg(val, (narrowOop*)dest, cmp);
13.    // decode old from T to oop
14.    return decode_heap_oop(old);
15.  } else {
16.    if (prebarrier) {
17.      update_barrier_set_pre((oop*)dest, exchange_value);
18.    }
19.    return (oop)Atomic::cmpxchg_ptr(exchange_value, (oop*)dest, compare_value);
20.  }
21. }
```

park 方法的本地实现代码如代码清单 5.48 所示。我们只关注 thread→parker()→park (isAbsolute != 0, time); 这一行，这是实现 park 的核心方法，其他代码直接忽略掉。每个 thread 都有一个 parker 与之对应，由它来实现 park 操作。

代码清单 5.48　park 方法的实现代码

```
1.  UNSAFE_ENTRY(void, Unsafe_Park(JNIEnv *env, jobject unsafe, jboolean isAbsolute, jlong time)) {
2.     EventThreadPark event;
3.     HOTSPOT_THREAD_PARK_BEGIN((uintptr_t) thread->parker(), (int) isAbsolute, time);
4.     JavaThreadParkedState jtps(thread, time != 0);
5.     thread->parker()->park(isAbsolute != 0, time);
6.     HOTSPOT_THREAD_PARK_END((uintptr_t) thread->parker());
7.     if (event.should_commit()) {
8.        oop obj = thread->current_park_blocker();
9.        event.set_parkedClass((obj != NULL) ? obj->klass() : NULL);
10.       event.set_timeout(time);
11.       event.set_address((obj != NULL) ? (TYPE_ADDRESS) cast_from_oop<uintptr_t>(obj) : 0);
12.       event.commit();
13.    }
14. }UNSAFE_END
```

此外，由于不同操作系统的实现不一样，所以需要各个系统各自实现。下面分别看 Linux 和 Windows 的实现。

先看 Linux 的实现，如代码清单 5.49 所示。它的核心实际上使用了 pthread 库，通过它提供的互斥锁和条件等待等函数来实现 park 功能。_counter 变量用于表示信号量，有许可时为 1，无许可时为 0。如果为 1 则直接返回，因为已经有许可了，也就是先调用过 unpark 了。接着对时间 time 进行转换，然后尝试获取互斥锁，只有成功获取互斥锁后才能往下执行。当 time 为 0 时，调用不带超时的 pthread_cond_wait 进入阻塞而等待唤醒信号，否则调用带超时的 pthread_cond_timedwait 进入阻塞而等待唤醒信号。最终释放互斥锁。

代码清单 5.49　park 方法的的 Linux 版本

```
1.    void Parker::park(bool isAbsolute, jlong time) {
2.      if (Atomic::xchg(0, &_counter) > 0) return;
3.      Thread* thread = Thread::current();
4.      assert(thread->is_Java_thread(), "Must be JavaThread");
5.      JavaThread *jt = (JavaThread *)thread;
6.      if (Thread::is_interrupted(thread, false)) {
7.        return;
8.      }
9.      timespec absTime;
10.     if (time < 0 || (isAbsolute && time == 0)) { // don't wait at all
11.       return;
12.     }
13.     if (time > 0) {
14.       unpackTime(&absTime, isAbsolute, time);
15.     }
16.     if (Thread::is_interrupted(thread, false) || pthread_mutex_trylock(_mutex) != 0) {
17.       return;
18.     }
19. 
20.     int status;
21.     if (_counter > 0)  { // no wait needed
22.       _counter = 0;
23.       status = pthread_mutex_unlock(_mutex);
24.       return;
25.     }
26.     if (time == 0) {
27.       _cur_index = REL_INDEX; // arbitrary choice when not timed
28.       status = pthread_cond_wait(&_cond[_cur_index], _mutex);
29.     } else {
30.       _cur_index = isAbsolute ? ABS_INDEX : REL_INDEX;
31.       status = pthread_cond_timedwait(&_cond[_cur_index], _mutex, &absTime);
32.     }
33.     _cur_index = -1;
34.     _counter = 0;
35.     status = pthread_mutex_unlock(_mutex);
36.   }
```

对于 Windows 的实现，则是直接通过 Windows 提供的 WaitForSingleObject 函数进行 park 操作，如代码清单 5.50 所示。

代码清单 5.50　park 方法的 Windows 版本

```
1.    void Parker::park(bool isAbsolute, jlong time) {
2.   
3.        ...
4.   
5.        WaitForSingleObject(_ParkEvent, time);
```

```
 6.         ResetEvent(_ParkEvent);
 7.
 8.         ...
 9.
10. }
```

unpark 实现对应的函数如代码清单 5.51 所示。我们只保留了核心的几行代码，其中只关注 p→unpark();这一行。它的核心是调用 parker 的 unpark 函数。我们来看 Linux 和 Windows 对应的实现。

代码清单 5.51　unpark 实现对应的函数

```
 1. UNSAFE_ENTRY(void, Unsafe_Unpark(JNIEnv *env, jobject unsafe, jobject jthread)) {
 2.
 3.     ...
 4.
 5.     if (p != NULL) {
 6.         HOTSPOT_THREAD_UNPARK((uintptr_t) p);
 7.         p->unpark();
 8.     }
 9.
10.     ...
11.
12. }UNSAFE_END
```

在 Linux 系统中，unpark 的实现逻辑是：先通过 pthread_mutex_lock 获取互斥锁，然后将 _counter 的值修改为 1，即表示许可为 1。最后通过 pthread_cond_signal 去唤醒前面 park 中进入阻塞的线程。相应的代码如代码清单 5.52 所示。

代码清单 5.52　unpark 方法的 Linux 版本

```
 1. void Parker::unpark() {
 2.     int status = pthread_mutex_lock(_mutex);
 3.     const int s = _counter;
 4.     _counter = 1;
 5.     int index = _cur_index;
 6.     status = pthread_mutex_unlock(_mutex);
 7.     if (s < 1 && index != -1) {
 8.        status = pthread_cond_signal(&_cond[index]);
 9.     }
10. }
```

在 Windows 系统中，unpark 则简单调用 SetEvent 函数来设置许可信号，park 操作中的 WaitForSingleObject 函数得到信号后才能往下执行。相应的代码如代码清单 5.53 所示。

代码清单 5.53　unpark 方法的 Windows 版本

```
 1. void Parker::unpark() {
 2.     guarantee(_ParkEvent != NULL, "invariant");
```

```
3.      SetEvent(_ParkEvent);
4.  }
```

5.10　AQS 的自旋锁

AQS 同步器以 CLH 锁为基础，同时为了让 CLH 锁更容易实现取消与超时功能，它从两方面对 CLH 锁进行了改造：节点的结构与节点等待机制。在结构上引入了头节点和尾节点，它们分别指向队列的头和尾，尝试获取锁、入队列、释放锁等实现都与头尾节点相关，并且每个节点都引入前驱节点和后继节点。在等待机制上由原来的自旋改为阻塞唤醒。在图 5.31 中，通过前驱节点和后继节点的引用将节点连接起来形成一个链表队列。

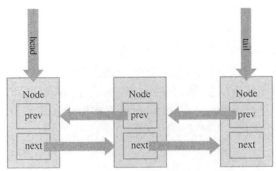

图 5.31　通过引用前驱节点和后继节点形成链表队列

下面来看一下入队、检测节点、出队、判断超时、取消节点等操作。

入队操作的逻辑其实是用一个无限循环进行 CAS 操作，即用自旋方式竞争直到成功。将尾节点 tail 的旧值赋予新节点 Node 的前驱节点，并尝试通过 CAS 操作将新节点 Node 赋予尾节点 tail，原先的尾节点的后继节点指向新建节点 Node。上面步骤完成后就建立起一条链表队列。简化的入队操作如代码清单 5.54 所示。

代码清单 5.54　入队操作的简化代码

```
1.  for(;;){
2.      Node t = tail;
3.      node.prev = t;
4.      if (compareAndSetTail(t,node)){
5.          t.next = node;
6.          return node;
7.      }
8.  }
```

上面说到，节点等待机制已经被 AQS 作者由自旋机制改造成阻塞机制。一个新建的节点在完成入队操作后，如果是自旋则直接进入循环检测前驱节点是否为头节点即可。如果改为阻塞机制，则当前线程将先检测前驱节点是否为头节点并尝试获取锁。如果当前节点的前驱节点为头节点并成功获取锁，则直接返回，当前线程不进入阻塞，否则将当前线程阻塞。简化的检测操作如代码清单 5.55 所示。

代码清单 5.55　检测新建节点是否为前驱节点的简化代码

```
1.   for(;;){
2.       if(node.prev == head)
3.           if(尝试获取锁成功){
4.               head = node;
5.               node.next = null;
6.               return;
7.           }
8.       阻塞线程
9.   }
```

出队的主要工作是负责唤醒等待队列中的后继节点，让所有等待节点环环相扣，每条线程都有序地往下执行。在共享模式下出队工作将变得异常复杂，它针对释放时的竞争优化引入了另外一种状态 PROPAGATE。多条线程并发执行出队操作时可将头节点状态改为 PROPAGATE，当下一节点被唤醒时，可根据此状态将继续往下唤醒而不用去执行尝试获取，以达到优化效果。下面只给出独占模式下出队的简化代码。

```
1.   Node s = node.next;
2.   唤醒节点 s 包含的线程
```

超时模式下需要用到 LockSupport 类的 parkNanos 方法，线程在阻塞一段时间后会自动唤醒。每次循环将累加消耗时间，当总消耗时间大于等于自定义的超时时间时就直接返回。简化的判断超时如代码清单 5.56 所示。

代码清单 5.56　判断总消耗时间是否大于自定义的超时时间的简化代码

```
1.   for(;;){
2.       尝试获取锁
3.       if(nanosTimeout <= 总消耗时间)
4.           return;
5.       LockSupport.parkNanos(this,nanosTimeout);
6.   }
```

队列中等待锁的线程可能因为中断或超时而涉及取消操作，这种情况下被取消的节点将不再进行锁竞争，此过程主要完成的工作是将取消的节点移除，简化代码如代码清单 5.57 所示。先将节点 node 状态设置成取消状态，再将前驱节点 prev 的后继节点指向 node 的后继节点。这里由于涉及竞争，必须通过 CAS 进行操作。CAS 操作即使失败也不必理会，因为已经改了节点的状态，在尝试获取锁操作中会循环对节点的状态判断。

代码清单 5.57　移除取消的节点的简化代码

```
1.    node.waitStatus = Node.CANCELLED;
2.    Node pred = node.prev;
3.    Node predNext = pred.next;
4.    Node next = node.next;
5.    compareAndSetNext(pred, predNext, next);
```

5.11　AQS 的公平性

所谓公平性，是指所有线程向临界资源申请访问权限时成功率都一样，不会让某些线程拥有优先权。我们知道，AQS 的锁是基于 CLH 锁进行优化的，而后者使用了 FIFO 队列，也就是说，等待队列是一个先进先出的队列。那么是否可以说每个线程在获取锁时就是公平的呢？关于公平性，严格来说应该分成 3 个点来看：入队阶段、唤醒阶段以及闯入策略。

在入队阶段，主要关注的是线程准备加入到等待队列时产生的竞争是否公平。对于这些准备入队列的线程节点（见图 5.32），在尝试获取锁失败后将加入到等待队列中，而此时每个线程都通过自旋操作将节点加入队列。所有线程在自旋过程中是无法保证其公平性的，后来的线程可能会比早到的线程先进入队列，所以节点入队阶段不具有公平性。

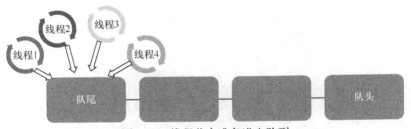

图 5.32　线程节点准备进入队列

唤醒阶段的公平性是指所有加入到等待队列中的节点能够按照加入等待队列时的先后顺序被唤醒。我们知道，当线程节点成功加入等待队列后便成为等待队列（先进先出队列）中的节点，等待队列中的所有节点都按照顺序等待自己被前驱节点唤醒并获取锁，所以队列中的所有节点在唤醒阶段是公平的，如图 5.33 所示。

闯入策略是 AQS 同步器为了提升性能而设计的一个策略，具体是指一个新线程到达共享资源边界时，不管等待队列中是否存在其他等待节点，新线程都将优先尝试去获取锁。闯入策略破坏了公平性，AQS 对外体现的公平性也主要体现在闯入策略上。

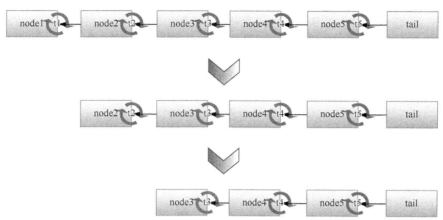

图 5.33　等待队列中的节点按序唤醒并获取锁

AQS 提供的锁获取操作运用了闯入算法，即如果有新线程到来则先进行一次获取尝试，不成功的情况下才将当前线程加入等待队列。在图 5.34 中，等待队列中节点线程按照顺序一个接一个尝试去获取共享资源的使用权。而在某一时刻，头节点线程在准备尝试获取的同时，另外一个线程闯入，新线程并非直接加入等待队列的尾部，而是先跟头节点线程竞争获取资源。闯入线程如果成功获取共享资源则直接执行，头节点线程则继续等待下一次尝试。如此一来，闯入线程成功插队，后来的线程比早到的线程先执行。这说明 AQS 锁获取算法是不严格公平的。

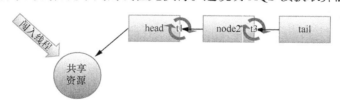

图 5.34　进入了闯入线程

为什么要使用闯入策略呢？闯入策略通常可以提升总吞吐量。一般同步器的颗粒度比较小，也可以说共享资源的范围较小，而线程从阻塞状态到被唤醒所消耗的时间可能是占用共享资源时间的几倍甚至几十倍。如此一来，线程唤醒过程中将存在一个很大的时间周期空窗期，导致资源没有得到充分利用。同时，如果每个线程都先入队再唤醒，也会导致效率低下。为了避免没必要的线程挂起和唤醒，以及为了提高吞吐量，于是引入了这种闯入策略。

闯入机制的实现对外提供一种竞争调节机制，开发人员可以在自定义同步器中定义闯入尝试获取的次数。假设次数为 n，则不断重复获取，直到 n 次都获取不成功时才把线程加入等待队列中，而且随着次数 n 的增加，成功闯入的概率也得以增加。

同时，这种闯入策略可能导致等待队列中的线程饥饿，因为锁可能一直被闯入的线程获取。在实际情况中要根据需求制定策略。在一个公平性要求很高的场景中，可以把闯入策略去除以达到公平性。在自定义同步器中可以通过 AQS 预留方法 tryAcquire 方法实现闯入策略，只需判断当前线程是否为等待队列中头节点对应的线程即可。若不是则直接返回 false，尝试获取失败。

5.12 AQS 的条件队列

条件（Condition）队列提供了阻塞与唤醒的机制。下面分析 AQS 条件队列的实现逻辑，由于完整代码较多，这里只关注核心的实现代码，如代码清单 5.58 所示。首先看 ConditionObject 类，它就是 AQS 同步器的条件队列的实现类，实现了 Condition 接口。该类中的属性主要包含等待队列中的头节点 firstWaiter 和尾节点 lastWaiter。

代码清单 5.58　AQS 条件队列的实现代码

```
1.  public class ConditionObject implements Condition, java.io.Serializable {
2.
3.      private transient Node firstWaiter;
4.      private transient Node lastWaiter;
5.
6.      public ConditionObject() {}
7.
8.      ...
9.
10. }
```

Condition 的队列由 Node 节点串联组成，Node 节点属于 AQS 同步器内部结构。每个 Node 节点的主要属性如图 5.35 所示。相较于 AQS 同步器的队列，Condition 的队列并没有使用前一节点 prev 和下一节点 next 组成的双向链表结构，而是简单地使用由下一节点 nextWaiter 组成的单向链表结构。

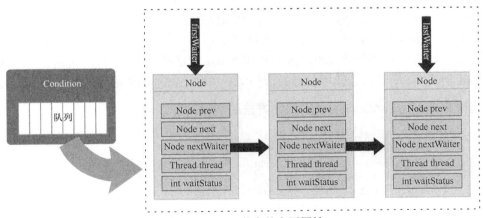

图 5.35　Node 节点的主要属性

5.12.1 await 方法

可以通过 await 和 signal 这两个核心方法来理解 Condition 的实现原理。下面先看 await 方法，如代码清单 5.59 所示。该方法用于将当前线程加入到等待集，然后释放锁且使当前线程进入阻塞等待状态，接着在接收到通知后重新进行锁竞争。

代码清单 5.59　await 方法的实现代码

```
1.  public final void await() throws InterruptedException {
2.      if (Thread.interrupted())
3.          throw new InterruptedException();
4.      Node node = addConditionWaiter();
5.      int savedState = fullyRelease(node);
6.      while (!isOnSyncQueue(node)) {
7.          LockSupport.park(this);
8.      }
9.      acquireQueued(node, savedState);
10.     if (node.nextWaiter != null)
11.         unlinkCancelledWaiters();
12. }
```

await 方法的逻辑如下：首先判断是否已经中断，如果中断则抛 InterruptedException 异常，否则往下执行，调用 addConditionWaiter 方法将当前线程添加到 Condition 队列（等待集）中。然后调用 fullyRelease 释放当前线程的锁。释放锁后开始等待下次的锁竞争操作，通过 while(!isOnSyncQueue(node)) 来实现，其中 isOnSyncQueue 方法表示该节点是否已经进入到 AQS 同步器的队列中。如果进入则表示已经开始下一次的锁竞争，否则通过 LockSupport.park(this) 将线程挂起。等该线程被唤醒后则继续往下执行，调用 acquireQueued 方法执行获取锁操作，即 CLH 锁不断检测前驱节点看是否轮到自己获得锁。最后还会调用 unlinkCancelledWaiters 来清理 Condition 队列中被取消的节点。

Condition 队列入队操作的逻辑对应的方法为 addConditionWaiter 方法，如代码清单 5.60 所示。该方法的代码逻辑为：首先检查当前线程是否持有锁，如果没有则抛出 IllegalMonitorStateException 异常。接着判断队列最后的节点状态是否为 CONDITION，如果不是则表示已经取消，需要调用 unlinkCancelledWaiters 方法将队列中所有被取消的节点清理掉。最后才创建新的 Node 节点，按单向链表结构组织队列（详细结构可参考图 5.35）。

代码清单 5.60　addConditionWaiter 方法的实现代码

```
1.  private Node addConditionWaiter() {
2.      if (!isHeldExclusively())
3.          throw new IllegalMonitorStateException();
4.      Node t = lastWaiter;
5.      if (t != null && t.waitStatus != Node.CONDITION) {
6.          unlinkCancelledWaiters();
7.          t = lastWaiter;
8.      }
9.      Node node = new Node(Node.CONDITION);
10.     if (t == null)
11.         firstWaiter = node;
12.     else
```

```
13.         t.nextWaiter = node;
14.     lastWaiter = node;
15.     return node;
16. }
```

清理队列中被取消的节点的操作对应的是 unlinkCancelledWaiters 方法，如代码清单 5.61 所示。总体而言，它的逻辑就是遍历访问单向链表。从第一个节点 firstWaiter 开始，通过 nextWaiter 引用不断获取下一个节点，判断每个节点的状态是否为 CONDITION，如果不是则将对应节点从单向链表中删除。

代码清单 5.61　unlinkCancelledWaiters 方法的实现代码

```
1.  private void unlinkCancelledWaiters() {
2.      Node t = firstWaiter;
3.      Node trail = null;
4.      while (t != null) {
5.          Node next = t.nextWaiter;
6.          if (t.waitStatus != Node.CONDITION) {
7.              t.nextWaiter = null;
8.              if (trail == null)
9.                  firstWaiter = next;
10.             else
11.                 trail.nextWaiter = next;
12.             if (next == null)
13.                 lastWaiter = trail;
14.         } else
15.             trail = t;
16.         t = next;
17.     }
18. }
```

按照条件队列的要求，当某个线程被加入到等待集后，它需要将原来持有的锁释放掉。这个操作由 fullyRelease 方法完成，如代码清单 5.62 所示。它的逻辑为：首先获取 AQS 的状态值，然后调用 release 方法进行锁释放操作。如果释放失败，则抛出 IllegalMonitorStateException 异常，意味着当前线程没有持有锁，同时将该节点的状态改为取消。

代码清单 5.62　fullyRelease 方法的实现代码

```
1.  final int fullyRelease(Node node) {
2.      try {
3.          int savedState = getState();
4.          if (release(savedState))
5.              return savedState;
6.          throw new IllegalMonitorStateException();
7.      } catch (Throwable t) {
8.          node.waitStatus = Node.CANCELLED;
9.          throw t;
10.     }
11. }
```

某个线程释放完锁后会准备重新进入到锁竞争的行列。该线程需要不断地检查其对应的节点是否已经在 AQS 同步器队列中。如果在，则说明它已经在竞争锁，否则需要继续等待，直到它进入到 AQS 同步器队列中。与之相对应的是 isOnSyncQueue 方法（见代码清单 5.63），如果状态为 CONDITION 或节点的前驱节点为 null，则说明不在 Condition 队列中（Condition 队列是不使用 prev 和 next 引用的）。如果节点的 next 引用不为 null，则说明已经在 AQS 同步器队列中。如果上面的条件都判断不出来，还需要从尾节点开始往前遍历，看是否在 AQS 队列中。

代码清单 5.63　isOnSyncQueue 方法的实现代码

```
1.   final boolean isOnSyncQueue(Node node) {
2.       if (node.waitStatus == Node.CONDITION || node.prev == null)
3.           return false;
4.       if (node.next != null)
5.           return true;
6.       return findNodeFromTail(node);
7.   }
```

以上便是 await 方法的实现涉及的核心代码。可以看到，所有方法都不必考虑多线程并发访问的问题，这是因为 Condition 的操作都是在成功获取锁后才执行的，所以完全不必担心线程安全问题。

5.12.2　signal 方法

signal 方法用于将某个线程从等待集中移除，并使之重新参与到锁竞争中，如代码清单 5.64 所示。该方法的逻辑为：首先检查当前线程是否持有锁，如果没有则抛出 IllegalMonitorStateException 异常。然后取出 Condition 队列中的第一个节点，将第一个节点从等待集中移除并唤醒该节点（具体由 doSignal 方法实现）。

代码清单 5.64　signal 方法的实现代码

```
1.   public final void signal() {
2.       if (!isHeldExclusively())
3.           throw new IllegalMonitorStateException();
4.       Node first = firstWaiter;
5.       if (first != null)
6.           doSignal(first);
7.   }
```

doSignal 方法主要将第一个节点的后继节点作为头节点 firstWaiter，然后将原来的头节点移除，如代码清单 5.65 所示。可以看到其中使用了 while 循环。之所以使用 while 循环，是因为头节点可能被取消，此时会导致状态修改失败，于是需要选择后继节点作为头节点并进行处理。

代码清单 5.65　doSignal 方法的实现代码

```
1.   private void doSignal(Node first) {
2.       do {
```

```
3.        if ((firstWaiter = first.nextWaiter) == null)
4.            lastWaiter = null;
5.        first.nextWaiter = null;
6.    } while (!transferForSignal(first) && (first = firstWaiter) != null);
7. }
```

transferForSignal 方法（见代码清单 5.66）先通过 CAS 方式修改头节点的状态，如果失败则返回 false。如果成功则调用 enq 方法将头节点加入到 AQS 同步器的等待队列中。注意这里返回的是原来的尾节点。如果原来的尾节点状态大于 0（已取消）或者修改节点的状态失败，则直接通过 unpark 方法将头节点的线程唤醒。为什么成功将状态修改为 SIGNAL 后就不用 unpark 操作了呢？原因是当前驱节点的状态为 SIGNAL 时，会在释放锁或取消时 unpark 后继节点。

代码清单 5.66　transferForSignal 方法的实现代码

```
1. final boolean transferForSignal(Node node) {
2.     if (!node.compareAndSetWaitStatus(Node.CONDITION, 0))
3.         return false;
4.     Node p = enq(node);
5.     int ws = p.waitStatus;
6.     if (ws > 0 || !p.compareAndSetWaitStatus(ws, Node.SIGNAL))
7.         LockSupport.unpark(node.thread);
8.     return true;
9. }
```

5.13　AQS 自定义同步器

5.13.1　AQS 设计思想

AQS 同步器的设计思想是通过继承的方式提供一个模板，方便开发人员根据不同场景实现一个个性化的同步器。通过前面的学习可知，同步器的核心是管理一个共享状态，通过对状态的控制来实现不同的锁机制。

在设计 AQS 时，必须考虑把繁杂且容易出错的队列管理工作统一抽象出来，并且统一控制好流程，而暴露给子类调用的方法主要就是操作共享状态的方法。我们通过继承 AbstractQueuedSynchronizer 类来实现同步器，主要使用 AQS 提供的 getState、setState、compareAndSetState 这 3 个方法。前两个为普通的 get 和 set 方法，在使用这两个方法之前必须保证不存在数据竞争，而 compareAndSetState 方法则提供了硬件级别的 CAS 原子更新。

独占模式和共享模式的锁获取与释放分别交给 acquire/release 组合与 acquireShared/releaseShared 组合，它们定义了锁获取与释放的逻辑，同时也为子类提供了获取和释放锁的接口。它们制定了同步器必要的逻辑流程模板，但也给子类预留了自定义的接口。通过对共享状态的管理，可以自定义各种各样的同步器，而无须考虑队列的管理及流程的控制，一切交给

AQS 就好了。自定义同步器的整体设计思想如图 5.36 所示。

同步器可实现任意不同的锁语义，同步器是用 AQS 类封装实现的更高层次的概念，可给开发人员提供更加形象的 API，而不是让开发人员直接接触 AQS 同步器，如图 5.37 所示。比如 ReentrantLock、Semphore、CountDownLatch 等，它们都是基于 AQS 类来实现的，这些不同的同步器更方便开发人员的理解与使用。AQS 同步器面向的是线程和状态的控制，它定义了线程获取状态的机制及线程排队等操作。这样能很好地隔离两者的关注点（更高层关注的是场景的使用，而 AQS 同步器关心的则是并发的控制）。

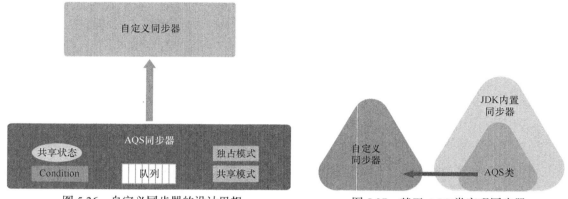

图 5.36　自定义同步器的设计思想　　　图 5.37　基于 AQS 类实现同步器

当然，假如想要实现一个自定义同步器，也是很方便的。官方推荐的做法是将继承了 AQS 类的子类作为自定义同步器的内部类，而自定义同步器中相关的操作只需间接调用子类中对应的方法即可。下面用一个简单的例子看看如何实现自己的锁。由于同步器被分为两种模式：独占模式和共享模式，所以也会给出这两种模式的例子。

5.13.2　独占模式

独占模式的一个示例是银行服务窗口应用。假如某个银行网点只有一个服务窗口，那么此银行服务窗口在一个时间内只能服务于一个人，其他人必须排队等待。这种银行窗口对应的同步器是独占模型。我们先定义一个银行窗口同步器类，如代码清单 5.67 所示。按照官方推荐的做法，使用一个继承 AQS 同步器的子类实现，该子类同样为内部类。然后重写 tryAcquire 和 tryRelease 方法，用于获得锁与释放锁的操作（主要就是维护同步状态变量，该变量只能为 0 或 1）。此外，还提供了 handle 和 unhandle 这两个方法，主要用于间接调用自定义同步器。

代码清单 5.67　独占模式下的银行服务窗口应用示例

```
1.  public class BankServiceWindow {
2.      private final Sync sync;
3.
4.      public BankServiceWindow() {
```

```
5.      sync = new Sync();
6.    }
7.
8.    private static class Sync extends AbstractQueuedSynchronizer {
9.      public boolean tryAcquire(int acquires) {
10.         if (compareAndSetState(0, 1)) {
11.             setExclusiveOwnerThread(Thread.currentThread());
12.             return true;
13.         }
14.         return false;
15.     }
16.
17.     protected boolean tryRelease(int releases) {
18.         if (getState() == 0 || Thread.currentThread() != this.getExclusiveOwnerThread())
19.             throw new IllegalMonitorStateException();
20.         setExclusiveOwnerThread(null);
21.         setState(0);
22.         return true;
23.     }
24.    }
25.
26.    public void handle() {
27.        sync.acquire(1);
28.    }
29.
30.    public void unhandle() {
31.        sync.release(1);
32.    }
33.
34. }
```

接着定义一个测试类，如代码清单 5.68 所示。形象地来说就是 3 个人到银行去办理业务，他们分别是 Tom、Jim 和 Jay。现在使用 BankServiceWindow 同步器就可以约束他们进行排队，一个一个轮着办理业务从而避免陷入混乱的局面。

代码清单 5.68　测试类（独占模式）

```
1.  public class BankServiceWindowTest {
2.      public static void main(String[] args) {
3.          final BankServiceWindow bankServiceWindow = new BankServiceWindow();
3.          Thread tom = new Thread(() -> {
4.              bankServiceWindow.handle();
5.              System.out.println("Tom 开始办理业务");
6.              try {
7.                  Thread.currentThread().sleep(5000);
8.              } catch (InterruptedException e) {
9.              }
10.             System.out.println("Tom 结束办理业务");
11.             bankServiceWindow.unhandle();
```

```
12.          });
13.          Thread jim = new Thread(() -> {
14.              bankServiceWindow.handle();
15.              System.out.println("Jim 开始办理业务");
16.              try {
17.                  Thread.currentThread().sleep(5000);
18.              } catch (InterruptedException e) {
19.              }
20.              System.out.println("Jim 结束办理业务");
21.              bankServiceWindow.unhandle();
22.          });
23.          Thread jay = new Thread(() -> {
24.              bankServiceWindow.handle();
25.              System.out.println("Jay 开始办理业务");
26.              try {
27.                  Thread.currentThread().sleep(5000);
28.              } catch (InterruptedException e) {
29.              }
30.              System.out.println("Jay 结束办理业务");
31.              bankServiceWindow.unhandle();
32.          });
33.          tom.start();
34.          jim.start();
35.          jay.start();
36.      }
37. }
```

代码清单 5.68 执行后的输出结果如下。很明显，Tom、Jim、Jay 这 3 人是排队完成业务的。但我们没有办法保证三者的顺序，它可能是 Tom、Jim、Jay，也可能是 Tom、Jay、Jim，因为在入列以前的执行顺序是无法确定的。这个同步器的语义是保证一个接一个办理。如果没有同步器的限制，输出结果将不可预测。

```
1. Jim 开始办理业务
2. Jim 结束办理业务
3. Tom 开始办理业务
4. Tom 结束办理业务
5. Jay 开始办理业务
6. Jay 结束办理业务
```

5.13.3 共享模式

共享模式的示例同样是银行服务窗口应用。随着银行网点的发展，办理业务的人越来越多，一个服务窗口已经无法满足需求。于是安排了另外一位员工负责另外一个服务窗口，这时就可以同时服务两个人了。但当两个窗口都有人占用时，同样也必须排队等待，这种多个服务窗口的同步器就是共享模式。我们先定义共享模式的同步器类，与独占模式不同的是，它的状态的初始值可以自由定义（有多少个窗口就设为多少），获取与释放就是对状态的递减和累加操作。重写 AQS 类的 tryAcquireShared 和 tryReleaseShared 方法，最终的示例代码如代码清单 5.69 所示。

代码清单 5.69 共享模式下的银行服务窗口示例

```java
1.  public class BankServiceWindowShared {
2.      private final Sync sync;
3.
4.      public BankServiceWindowShared(int count) {
5.          sync = new Sync(count);
6.      }
7.
8.      private static class Sync extends AbstractQueuedSynchronizer {
9.          Sync(int count) {
10.             setState(count);
11.         }
12.         public int tryAcquireShared(int interval) {
13.             for (;;) {
14.                 int current = getState();
15.                 int newCount = current - 1;
16.                 if (newCount < 0 || compareAndSetState(current, newCount))
17.                     return newCount;
18.             }
19.         }
20.         public boolean tryReleaseShared(int interval) {
21.             for (;;) {
22.                 int current = getState();
23.                 int newCount = current + 1;
24.                 if (compareAndSetState(current, newCount))
25.                     return true;
26.             }
27.         }
28.     }
29.
30.     public void handle() {
31.         sync.acquireShared(1);
32.     }
33.     public void unhandle() {
34.         sync.releaseShared(1);
35.     }
36.
37. }
```

接着定义一个测试类，如代码清单 5.70 所示。Tom、Jim 和 Jay 再次来到银行，现在因为有两个窗口，所以他们可以两个人同时办理，时间缩短了不少。现在使用 BankServiceWindowShared 同步器就可以同时给两个人办理业务。

代码清单 5.70　测试类（共享模式）

```java
1.  public class BankServiceWindowSharedTest {
2.      public static void main(String[] args) {
3.          final BankServiceWindowShared bankServiceWindows = new BankServiceWindowShared(2);
4.          Thread tom = new Thread(() -> {
5.              bankServiceWindows.handle();
6.              System.out.println("Tom 开始办理业务");
7.              try {
8.                  Thread.currentThread().sleep(5000);
9.              } catch (InterruptedException e) {
10.             }
11.             System.out.println("Tom 结束办理业务");
12.             bankServiceWindows.unhandle();
13.         });
14.         Thread jim = new Thread(() -> {
15.             bankServiceWindows.handle();
16.             System.out.println("Jim 开始办理业务");
17.             try {
18.                 Thread.currentThread().sleep(5000);
19.             } catch (InterruptedException e) {
20.             }
21.             System.out.println("Jim 结束办理业务");
22.             bankServiceWindows.unhandle();
23.         });
24.         Thread jay = new Thread(() -> {
25.             bankServiceWindows.handle();
26.             System.out.println("Jay 开始办理业务");
27.             try {
28.                 Thread.currentThread().sleep(5000);
29.             } catch (InterruptedException e) {
30.             }
31.             System.out.println("Jay 结束办理业务");
32.             bankServiceWindows.unhandle();
33.         });
34.         tom.start();
35.         jim.start();
36.         jay.start();
37.     }
38. }
```

代码清单 5.70 执行可能的输出结果如下。Jim 和 Jay 几乎同时开始办理业务，而当 Jay 结束后，一有空闲的服务窗口，Tom 就马上过去办理业务。

```
1.  Jim 开始办理业务
2.  Jay 开始办理业务
3.  Jay 结束办理业务
4.  Jim 结束办理业务
5.  Tom 开始办理业务
6.  Tom 结束办理业务
```

第 6 章

常见的同步器

6.1 常见的同步器

JDK 提供了一些常用的同步器,它们都基于 AQS 同步器来实现。AQS 同步器的核心功能围绕着其 32 位同步状态变量来实现。一般认为同步状态变量表示锁的数量,通过对同步状态的控制就可以实现不同的同步工具,比如闭锁、信号量、循环屏障、相位器、交换器等。为了保证线程的可见性,同步状态变量需被声明为 volatile,以保证每次的原子更新都能及时同步到每个线程上。同步状态的模式大体可以分为两类:独占模式和共享模式。通过对这两类模式的灵活使用,可以实现各种不同的同步器。

图 6.1 所示为独占模式的同步器。同步器是通过对同步状态 state 的控制来实现的,此时的同步器可以看成一个管道,管道的大小决定了同时通过管道的线程数量。独占模式就好比只有一个线程大小的管道,在这种模式下线程只能逐一通过管道,任意时刻管道内都只能存在一个线程。在图 6.1 中,线程 1 首先进入同步器,其他两个线程只能一直等待,直到线程 1 离开同步器后,线程 2 才进入同步器。类似地,等到线程 2 离开同步器后线程 3 才进入。

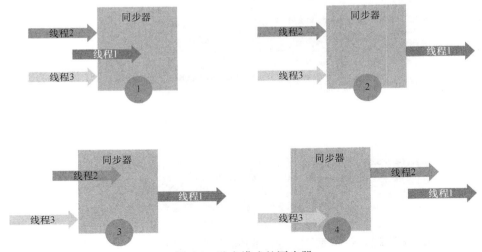

图 6.1 独占模式的同步器

在独占模式中如何控制同步状态 state 呢？我们只需要两个状态，即 0 和 1。0 表示通道没有被占用，而 1 表示通道正在被占用。我们来看一下 tryAcquire 和 tryRelease 这两个方法，如代码清单 6.1 所示。多个线程通过 tryAcquire 方法尝试把 state 变量改为 1，由于有自旋和 CAS 的保证，在竞争条件下有且仅有一个线程能成功修改 state。修改成功的线程代表获取锁成功，因此拥有进入管道往下执行的权利，而其他线程则不断地循环检测 state 值是否改回 0。当线程执行完毕且退出管道时，通过 tryRelease 方法把 state 变量改为 0，从而让出管道，让其他线程继续对该锁竞争。

代码清单 6.1　独占模式中的 tryAcquire 和 tryRelease 方法

```
1.   public class ExclusiveModel {
2.       private static Unsafe unsafe = null;
3.       private static final long stateOffset;
4.       private volatile int state = 0;
5.       static {
6.           try {
7.               unsafe = getUnsafeInstance();
8.               stateOffset = unsafe.objectFieldOffset(ExclusiveModel.class.getDeclaredField("state"));
9.           } catch (Exception ex) {
10.              throw new Error(ex);
11.          }
12.      }
13.
14.      private static Unsafe getUnsafeInstance() throws Exception {
15.          Field theUnsafeInstance = Unsafe.class.getDeclaredField("theUnsafe");
16.          theUnsafeInstance.setAccessible(true);
17.          return (Unsafe) theUnsafeInstance.get(Unsafe.class);
18.      }
19.
20.      public void tryAcquire() {
21.          for (;;) {
22.              int newV = state + 1;
23.              if (newV == 1)
24.                  if (unsafe.compareAndSwapInt(this, stateOffset, 0, newV)) {
25.                      return;
26.                  }
27.          }
28.      }
29.
30.      public void tryRelease() {
31.          unsafe.compareAndSwapInt(this, stateOffset, 1, 0);
32.      }
33.
34.  }
```

共享模式的同步器就是管道宽度大于 1 的同步器，它可以同时让 n 个线程通过管道。它与独占模式不同的地方主要是对同步状态 state 的管理及判断。独占模式中 state 的值只能为 0 或 1，而共享模式的 state 则可以被初始化为任意整数。初始值 n 用来表示同步器管道的宽度为 n，即可供 n 个线程同时通过管道。如图 6.2 所示，假设同步器的管道宽度为 3，则意味着最多可以有 3 个线程在管道内。线程 1 首先进入同步器，线程 2 跟着也进入同步器，此时线程 1 还未离开同步器。然后线程 3 也进入同步器，这时 3 个线程同时在同步器中，最终 3 个线程都顺利通过。

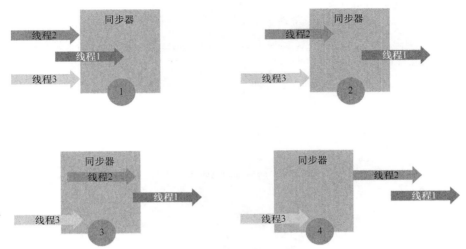

图 6.2　共享模式的同步器

在共享模式中又该如何控制同步状态 state 呢？多个线程通过 tryAcquireShared 方法尝试将 state 的值减去 1，自旋及 CAS 能保证线程安全地修改 state 的值。只有当新值大于等于 0 才表示获取锁成功，并拥有进入管道往下执行的权利。在线程执行完毕时将调用 tryReleaseShared 方法尝试修改 state 值，使之增加 1，表示已经执行完毕并让出同步器的管道空间，如代码清单 6.2 所示。需要说明的是，共享模式的同步器可能存在多个线程并发地释放锁，所以此处必须使用自旋和 CAS 来保证线程的安全。

代码清单 6.2　共享模式中的 tryAcquireShared 和 tryReleaseShared 方法

```
1.   public class SharedModel {
2.       private static Unsafe unsafe = null;
3.       private static final long stateOffset;
4.       private volatile int state = 10;
5.       static {
6.           try {
7.               unsafe = getUnsafeInstance();
8.               stateOffset = unsafe.objectFieldOffset(SharedModel.class.getDeclaredField("state"));
9.           } catch (Exception ex) {
10.              throw new Error(ex);
11.          }
12.      }
13.
```

```java
14.    private static Unsafe getUnsafeInstance() throws Exception {
15.        Field theUnsafeInstance = Unsafe.class.getDeclaredField("theUnsafe");
16.        theUnsafeInstance.setAccessible(true);
17.        return (Unsafe) theUnsafeInstance.get(Unsafe.class);
18.    }
19.
20.    public int tryAcquireShared() {
21.        for (;;) {
22.            int newCount = state - 1;
23.            if (newCount >= 0
24.                    && unsafe.compareAndSwapInt(this, stateOffset, newCount + 1, newCount)) {
25.                return newCount;
26.            }
27.        }
28.    }
29.
30.    public int tryReleaseShared() {
31.        for (;;) {
32.            int newCount = state + 1;
33.            if (unsafe.compareAndSwapInt(this, stateOffset, newCount - 1, newCount)) {
34.                return newCount;
35.            }
36.        }
37.    }
38. }
```

AQS 同步器是 JDK 中常见同步器的基础，它提供了对同步状态 state 的基础管理，同时也提供了独占和共享两种模式，基于这两种模式可以很方便地实现各种同步器。实际上，AQS 同步器在提供 state 状态管理的同时也会维护等待队列。这两项工作被封装成一个模板，从而规定了同步器的基本流程。该流程包括在什么条件下将线程加入等待队列、在什么条件下移除等待线程、如何操作等待队列、是否需要阻塞、是否支持中断等。AQS 同步器对外仅仅提供 state 状态操作接口，而队列的维护工作已经绑定在模板中，无须动手便能实现高效的同步器。

6.2 闭锁

闭锁（CountDownLatch）是 Java 多线程并发中的一种很常见的同步器，它是基于 AQS 同步器来实现的，如图 6.3 所示。通过它可以定义一个倒计数器，当倒计数器的值大于 0 时，所有调用 await 方法的线程都会等待。而调用 countDown 方法则可以让倒计数器的值减 1，当倒计数器值为 0 时，所有等待的线程都将继续往下执行。

闭锁的主要应用场景是让某个或某些线程在某个运行节点上等待 n 个条件都满足后，才让所有线程继续往下执行，其中倒计数器的值为 n，每满足一个条件，倒计数器就减 1。比如在图 6.4 中，倒计数器初始值为 3，然后 3 个线程调用 await 方法后都在等待。随后倒计数器减 1

为 2，再减 1 为 1，最后减 1 为 0，然后所有等待的线程都往下继续执行。

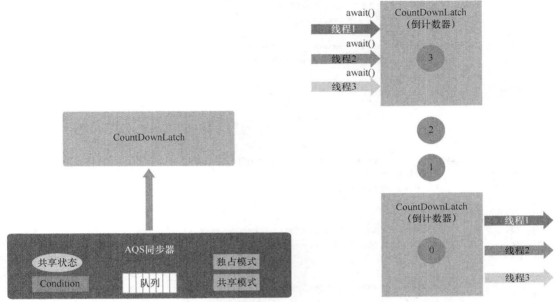

图 6.3　CountDownLatch 基于 AQS 同步器来实现　　图 6.4　闭锁的应用

闭锁的三要素为倒计数器的初始值、await 方法以及 countDown 方法。倒计数器的初始值在构建 CountDownLatch 对象时指定，它表示需要等待的条件个数。await 方法能让线程进入等待状态，等待的条件是倒计数器的值大于 0。countDown 方法用于将倒计数器的值减 1。

闭锁的实现代码如代码清单 6.3 所示。CountDownLatch 类的构造函数需要传入一个整型参数，表示倒计数器的初始值，对应着 AQS 的 state 状态变量。如果想将继承了 AQS 类的子类 Sync 作为 CountDownLatch 类的内部类，在 CountDownLatch 同步器中只需间接调用子类中对应的方法即可。比如 await 方法和 countDown 方法分别调用 Sync 子类的 acquireSharedInterruptibly 方法和 releaseShared 方法。

Sync 子类中需要实现的两个方法是 tryAcquireShared 和 tryReleaseShared，分别用于获取共享锁和释放共享锁。获取共享锁的逻辑是，如果状态变量（倒计数器的值）等于 0 则返回 1，表示线程可以得到共享锁；而当倒计数器的值为非 0 时则返回 -1，表示获取锁失败并对线程进行入队管理。释放共享锁的逻辑是，通过自旋来进行减 1 操作，使用 getState 方法获取状态变量，将其值减 1 后使用 compareAndSetState 方法通过 CAS 修改状态值。

代码清单 6.3　闭锁的实现代码

```
1.  public class CountDownLatch {
2.      private final Sync sync;
3.
4.      public CountDownLatch(int count) {
```

```
5.         if (count < 0)
6.             throw new IllegalArgumentException("count < 0");
7.         this.sync = new Sync(count);
8.     }
9.
10.    public void await() throws InterruptedException {
11.        sync.acquireSharedInterruptibly(1);
12.    }
13.
14.    public void countDown() {
15.        sync.releaseShared(1);
16.    }
17.
18.    private static final class Sync extends AbstractQueuedSynchronizer {
19.        Sync(int count) {
20.            setState(count);
21.        }
22.
23.        protected int tryAcquireShared(int acquires) {
24.            return (getState() == 0) ? 1 : -1;
25.        }
26.
27.        protected boolean tryReleaseShared(int releases) {
28.            for (;;) {
29.                int c = getState();
30.                if (c == 0)
31.                    return false;
32.                int nextc = c - 1;
33.                if (compareAndSetState(c, nextc))
34.                    return nextc == 0;
35.            }
36.        }
37.    }
38. }
```

在闭锁的第一个应用示例（见代码清单 6.4）中，首先创建一个 CountDownLatch 对象作为倒计数器，其值为 2。然后，线程 1 调用 await 方法进行等待，线程 2 调用 countDown 方法将倒计数器的值减 1 并往下执行。接着，线程 3 再调用 countDown 方法，将倒计数器的值再减 1 并往下执行，此时倒计数器的值为 0，线程 1 停止等待并往下执行。整个流程如图 6.5 所示。

代码清单 6.4 将 CountDownLatch 对象作为倒计数器的应用示例 1

```
1. public class CountDownLatchDemo {
2.     static CountDownLatch latch = new CountDownLatch(2);
3.
4.     public static void main(String[] args) {
5.         Thread thread1 = new Thread(() -> {
6.             System.out.println("thread1 is waiting");
```

```
7.          try {
8.              latch.await();
9.              System.out.println("thread1 go");
10.         } catch (InterruptedException e) {
11.         }
12.     });
13.     Thread thread2 = new Thread(() -> {
14.         try {
15.             Thread.sleep(2000);
16.         } catch (InterruptedException e) {
17.         }
18.         System.out.println("thread2 count down");
19.         latch.countDown();
20.         System.out.println("thread2 goes");
21.     });
22.     Thread thread3 = new Thread(() -> {
23.         try {
24.             Thread.sleep(4000);
25.         } catch (InterruptedException e) {
26.         }
27.         System.out.println("thread3 count down");
28.         latch.countDown();
29.         System.out.println("thread3 goes");
30.     });
31.     thread1.start();
32.     thread2.start();
33.     thread3.start();
34. }
35. }
```

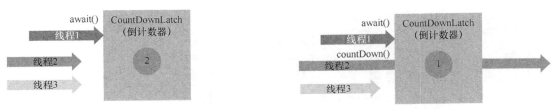

图 6.5 将 CountDownLatch 对象作为倒计数器的应用示例 1

代码清单 6.4 执行后的输出结果如下所示。首先，thread1 调用 await 方法后进入等待状态，

thread2 睡眠 2s 后调用 countDown 方法并往下执行。接着，thread3 睡眠 4s 后调用 countDown 方法并往下执行。最后，thread1 才停止等待并继续往下执行。

```
1.    thread1 is waiting
2.    thread2 count down
3.    thread2 goes
4.    thread3 count down
5.    thread3 goes
6.    thread1 go
```

闭锁的第二个应用示例（见代码清单 6.5）中，首先创建一个 CountdownLatch 对象作为倒计数器，其值为 2。然后，线程 1 和线程 2 都分别调用 await 方法进行等待。接着，线程 3 调用两次 countDown 方法，将倒计数器的值减 2 并往下执行，此时倒计数器的值为 0，线程 1 和线程 2 停止等待并往下执行。整个流程如图 6.6 所示。

代码清单 6.5　将 CountDownLatch 对象作为倒计数器的应用示例 2

```java
1.    public class CountDownLatchDemo2 {
2.        static CountDownLatch latch = new CountDownLatch(2);
3.
4.        public static void main(String[] args) {
5.            Thread thread1 = new Thread(() -> {
6.                System.out.println("thread1 is waiting");
7.                try {
8.                    latch.await();
9.                    System.out.println("thread1 goes");
10.               } catch (InterruptedException e) {
11.               }
12.           });
13.
14.           Thread thread2 = new Thread(() -> {
15.               System.out.println("thread2 is waiting");
16.               try {
17.                   latch.await();
18.                   System.out.println("thread2 goes");
19.               } catch (InterruptedException e) {
20.               }
21.           });
22.
23.           thread1.start();
24.           thread2.start();
25.           latch.countDown();
26.           try {
27.               Thread.currentThread().sleep(2000);
28.           } catch (InterruptedException e) {
29.               e.printStackTrace();
30.           }
31.           latch.countDown();
32.       }
```

```
33.  }
```

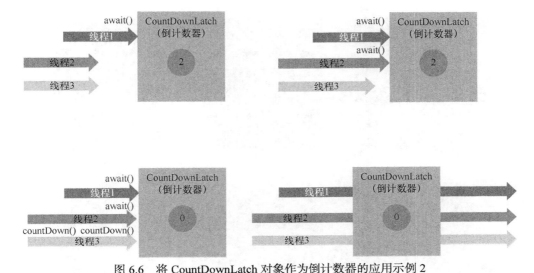

图 6.6　将 CountDownLatch 对象作为倒计数器的应用示例 2

代码清单 6.5 执行后的输出结果如下所示。首先，thread1 和 thread2 调用 await 方法后进入等待状态。然后，主线程调用 countDown 方法后休眠 2s。此时的倒计数器值为 1，所以 thread1 和 thread2 继续等待，直到主线程休眠结束后再次调用 countDown 方法后，thread1 和 thread2 才能继续往下执行。

```
1.  thread1 is waiting
2.  thread2 is waiting
3.  thread2 goes
4.  thread1 goes
```

6.3　信号量

信号量（Semaphore）可以实现对公共资源的并发访问控制。一个线程在进入公共资源时，需要先获取许可，否则要等待其他线程释放许可。每个线程在离开公共资源时都会释放许可。可以把 Semaphore 看成一个计数器，当计数器的值小于许可最大值时，线程可以得到一个许可并往下执行。

信号量的主要应用场景是控制最多 n 个线程同时访问资源，其中计数器的最大值是许可的最大值 n。比如在一个停车场中（见图 6.7），假设停车场一共有 8 个车位，其中 6 个车位已停放车辆，然后又来了两辆汽车，此时因为刚好剩下两个车位，所以这两辆车都能停放。接着又来了一辆车，由于现在已经没有空位，所以只能等待其他车辆离开。此时刚好一辆红色汽车离开停车场，黄车就可以停进去。假如又有一辆汽车进来，则该车又得等待，如此往复。在这个过程中，停车场就是公共资源，车位数就是信号量的最大许可数，车辆就好比线程。

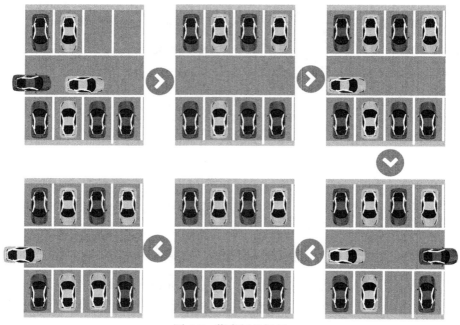

图 6.7　停车场的场景

信号量的四要素为最大许可数、公平/非公平模式、acquire 方法以及 release 方法。最大许可数和公平模式在构建 Semaphore 对象时指定，分别表示公共资源中最多可以有多少个线程同时访问以及获取许可时是否使用公平模式。acquire 方法用于获取许可，如果许可不够则进入等待状态。release 方法用于释放许可。

6.3.1　非公平模式的实现

Semaphore 类是基于 AQS 同步器来实现的（见图 6.8）。不管是公平模式还是非公平模式，都是基于 AQS 的共享模式，只是在获取许可的操作逻辑上有差异。Semaphore 的默认模式为非公平模式。我们先看非公平模式的实现。

Semaphore 类的几个主要方法如代码清单 6.6 所示。其中提供了两个构造函数，相关的两个参数分别为许可最大数和是否使用公平模式，FairSync 是公平模式的同步器，而 NonfairSync 则是非公平模式的同步器。其中有两个 acquire 方法，无参时默认是一次获取一个许可，而如果传入整型参数，则表示一次获取若干个许可。既然有两个 acquire 方法，则对应的有两个 release 方法，无参时表示释放一个许可，而如果传入整型参数，则表示一次释放若干个许可。

6.3 信号量

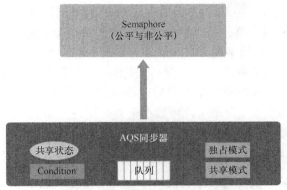

图 6.8 Semaphore 类基于 AQS 同步器实现

代码清单 6.6　Semaphore 类的几个主要方法

```java
public class Semaphore implements java.io.Serializable {
    private final Sync sync;

    public Semaphore(int permits) {
        sync = new NonfairSync(permits);
    }

    public Semaphore(int permits, boolean fair) {
        sync = fair ? new FairSync(permits) : new NonfairSync(permits);
    }

    public void acquire() throws InterruptedException {
        sync.acquireSharedInterruptibly(1);
    }

    public void release() {
        sync.releaseShared(1);
    }

    public void acquire(int permits) throws InterruptedException {
        if (permits < 0)
            throw new IllegalArgumentException();
        sync.acquireSharedInterruptibly(permits);
    }

    public void release(int permits) {
        if (permits < 0)
            throw new IllegalArgumentException();
        sync.releaseShared(permits);
    }

    .......
}
```

Semaphore 内部的 Syn 子类是公平模式 FairSync 类和非公平模式 NonfairSync 类的抽象父类，许可最大数与 AQS 同步器的状态变量对应。代码清单 6.7 提供了非公平的许可获取方法 nonfairTryAcquireShared。非公平模式其实就是在许可数量允许的情况下，让所有线程都进行自旋操作，而不管它们到来的顺序，将全部线程放到一起去竞争许可。其中 compareAndSetState 方法提供了 CAS 算法，从而能够保证并发修改许可值，而剩余许可数等于当前可用许可值减去消耗许可数。需要注意的是，当剩余许可数小于 0 时则返回负数，表示线程会进入等待队列。tryReleaseShared 方法提供了释放许可的操作。释放许可的逻辑相同，即不管是不是公平模式都使用该方法，通过自旋操作将释放的许可数增加到当前剩余许可数中。

代码清单 6.7　非公平的许可获取方法 nonfairTryAcquireShared

```
1.   abstract static class Sync extends AbstractQueuedSynchronizer {
2.
3.       Sync(int permits) {
4.           setState(permits);
5.       }
6.
7.       final int getPermits() {
8.           return getState();
9.       }
10.
11.      final int nonfairTryAcquireShared(int acquires) {
12.         for (;;) {
13.             int available = getState();
14.             int remaining = available - acquires;
15.             if (remaining < 0 || compareAndSetState(available, remaining))
16.                 return remaining;
17.         }
18.      }
19.
20.      protected final boolean tryReleaseShared(int releases) {
21.         for (;;) {
22.             int current = getState();
23.             int next = current + releases;
24.             if (next < current)
25.                 throw new Error("Maximum permit count exceeded");
26.             if (compareAndSetState(current, next))
27.                 return true;
28.         }
29.      }
30.  }
```

非公平模式 NonfairSync 类的实现主要通过 tryAcquireShared 方法，直接调用父类 Sync 的 nonfairTryAcquireShared 方法，如代码清单 6.8 所示。

代码清单 6.8　tryAcquireShared 方法

```
1.    static final class NonfairSync extends Sync {
2.        NonfairSync(int permits) {
3.            super(permits);
4.        }
5.
6.        protected int tryAcquireShared(int acquires) {
7.            return nonfairTryAcquireShared(acquires);
8.        }
9.    }
```

6.3.2　公平模式的实现

公平模式与非公平模式的主要差异是获取许可时的机制。非公平模式直接通过自旋操作让所有线程竞争许可，从而导致非公平。而公平模式则通过队列来实现公平机制。它们的区别就在 tryAcquireShared 方法中（见代码清单 6.9），即增加了 hasQueuedPredecessors() 判断，它会检查是否已经存在等待队列。如果已经有等待队列，则返回 −1，表示让 AQS 同步器将当前线程放进等待队列中（有了队列则意味着公平）。要注意的是，在未达到最大许可数的情况下，所有线程都不会进入等待队列。

代码清单 6.9　tryAcquireShared 方法中的区别

```
1.    static final class FairSync extends Sync {
2.
3.        FairSync(int permits) {
4.            super(permits);
5.        }
6.
7.        protected int tryAcquireShared(int acquires) {
8.            for (;;) {
9.                if (hasQueuedPredecessors())
10.                   return -1;
11.               int available = getState();
12.               int remaining = available - acquires;
13.               if (remaining < 0 || compareAndSetState(available, remaining))
14.                   return remaining;
15.           }
16.       }
17.   }
```

6.3.3 信号量的使用示例

信号量的第一个使用示例如图 6.9 所示，代码如代码清单 6.10 所示。首先实例化一个拥有 5 个许可的信号量对象；然后一共有 10 个线程一同尝试获取 5 个许可，得到许可的线程将 value 进行累加 1，接着睡眠 5s；最后释放许可。

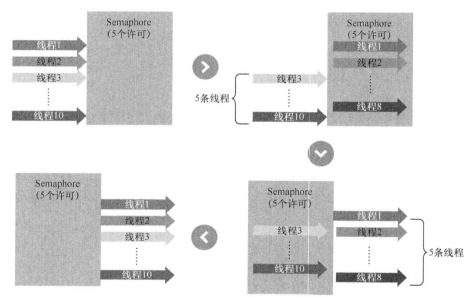

图 6.9　Semaphore 的使用示例

代码清单 6.10　Semaphore 的使用示例 1

```
1.   public class SemaphoreDemo {
2.       static Semaphore semaphore = new Semaphore(5);
3.       static AtomicInteger value = new AtomicInteger(0);
4.
5.       public static void main(String[] args) {
6.
7.           for (int i = 0; i < 10; i++) {
8.               Thread thread = new Thread(() -> {
9.                   try {
10.                      semaphore.acquire();
11.                      System.out.println("counting number : " + value.incrementAndGet());
12.                      Thread.sleep(5000);
13.                      semaphore.release();
14.                  } catch (InterruptedException e) {
15.                  }
16.              });
```

```
17.            thread.start();
18.        }
19.    }
20. }
```

代码清单 6.10 执行后的输出结果如下。其中 2 个线程输出 "counting number：xx" 后其他线程开始等待。大概等待 5s 后获得许可的 5 个线程执行释放许可操作，然后其他线程才能获得许可并往下执行。

```
1.  counting number : 2
2.  counting number : 1
3.  counting number : 5
4.  counting number : 3
5.  counting number : 4
6.  counting number : 6
7.  counting number : 7
8.  counting number : 8
9.  counting number : 9
10. counting number : 10
```

信号量的第二个使用示例如代码清单 6.11 所示。代码清单 6.11 与代码清单 6.10 很相似，不同的地方在于每次获取许可时会消耗 2 个许可，释放时也释放 2 个许可。这里实例化 1 个拥有 6 个许可的信号量对象，然后 10 个线程一同尝试获取许可。但这次最多只能有 3 个线程同时得到许可，也就是 3 个线程得到许可后对 value 值进行累加 2，然后睡眠 5s 后释放许可。接着另外 3 个线程又获得许可往下执行，直到 10 个线程都执行完毕。

代码清单 6.11　Semaphore 的使用示例 2

```
1.  public class SemaphoreDemo2 {
2.      static Semaphore semaphore = new Semaphore(6);
3.      static AtomicInteger value = new AtomicInteger(0);
4.
5.      public static void main(String[] args) {
6.
7.          for (int i = 0; i < 10; i++) {
8.              Thread thread = new Thread(() -> {
9.                  try {
10.                     semaphore.acquire(2);
11.                     System.out.println("counting number : " + value.incrementAndGet());
12.                     Thread.sleep(5000);
13.                     semaphore.release(2);
14.                 } catch (InterruptedException e) {
15.                 }
16.             });
17.             thread.start();
18.         }
19.     }
20. }
```

6.4 循环屏障

循环屏障（CyclicBarrier）是 JDK 提供的另一种同步器。该同步器的规则是，只有在到达屏障的线程数量达到指定值时，屏障才会放行。可以将 CyclicBarrier 看成一个倒计数器。倒计数器的最大值是屏障的大小，每个线程调用 await 方法时都会让倒计数器的值减 1。当倒计数器的值为 0 时，会让所有等待的线程往下执行。

循环屏障主要的应用场景是在某些节点约束 n 个线程，比如让指定数量的线程共同到达某个节点后，这些线程才能一起往下执行。在图 6.10 中，对于一个倒计数器最大值为 3 的循环屏障，初始时 3 个线程都未调用 await 方法。首先，线程 1 调用 await 方法后，倒计数器的值变为 2。接着，线程 2 继续调用 await 方法，倒计数器的值变为 1。然后，线程 3 也调用 await 方法，此时倒计数器的值为 0，3 个线程都通过屏障继续往下执行。最后，倒计数器的值又重新恢复到最大值 3，这就是称为循环屏障的原因。

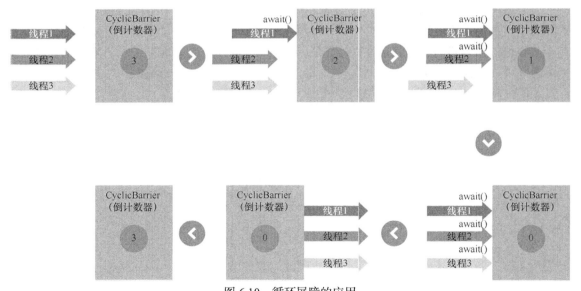

图 6.10　循环屏障的应用

循环屏障的三要素为倒计数器的最大值、await 方法以及触发点 Runnable 任务。倒计数器的最大值在构建 CyclicBarrier 对象时指定，它表示需要等待的线程数。await 方法能让倒计数器的值减 1，并且让线程进入等待状态。触发点 Runnable 任务指的是当指定数量的线程到达屏障后会触发执行的任务。

与闭锁和信号量的实现不同的是，循环屏障是通过 ReentrantLock 来实现的。如果往更底层追究，则会发现循环屏障也使用了 AQS 同步器，因为 ReentrantLock 的实现就基于 AQS 同步器，如图 6.11 所示。

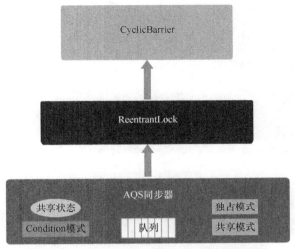

图 6.11　ReentrantLock 基于 AQS 同步器实现

为了方便阅读，这里将循环屏障的源码整理成两部分，并分别讲解。第一部分是循环屏障的核心源码，如代码清单 6.12 所示。其中，CyclicBarrier 类中的 ReentrantLock 对象和 Condition 对象用于控制线程，parties 变量表示倒计数器的最大值，count 变量表示倒计数器当前值，而 Runnable 对象为触发点任务。这里主要有两个构造函数，一个传入倒计数器最大值，另一个传入倒计数器的最大值和触发点任务。nextGeneration 方法表示已经达到屏障倒计数器的最大值，准备进行下一轮，它会将屏障中所有的线程放行，而且将倒计数器的当前值重置为最大值。getNumberWaiting 方法用于获取当前屏障中等待的线程数，其值为倒计数器的最大值减去倒计数器的当前值。

代码清单 6.12　循环屏障的核心源码

```
1.  public class CyclicBarrier {
2.      private final ReentrantLock lock = new ReentrantLock();
3.      private final Condition trip = lock.newCondition();
4.      private final int parties;
5.      private final Runnable barrierCommand;
6.      private int count;
7.      private int flag;
8.
9.      public CyclicBarrier(int parties) {
10.         this(parties, null);
11.     }
12.
13.     public CyclicBarrier(int parties, Runnable barrierAction) {
14.         if (parties <= 0)
15.             throw new IllegalArgumentException();
```

```
16.            this.parties = parties;
17.            this.count = parties;
18.            this.barrierCommand = barrierAction;
19.        }
20.
21.        private void nextGeneration() {
22.            trip.signalAll();
23.            count = parties;
24.            flag++;
25.        }
26.
27.        public int getNumberWaiting() {
28.            final ReentrantLock lock = this.lock;
29.            lock.lock();
30.            try {
31.                return parties - count;
32.            } finally {
33.                lock.unlock();
34.            }
35.        }
36.
37.        ......
38.
39.    }
```

第二部分是循环屏障最核心的部分——await 方法,如代码清单 6.13 所示。它会调用 dowait 方法,所以这里主要看 dowait 方法。为了线程安全,它首先会通过 lock.lock()进行加锁,然后判断当前线程是否被中断,如果被中断,则往上抛出 InterruptedException 异常。线程调用 await 方法时会让倒计数器减 1,所以接下来会将当前倒计数器的当前值减 1。如果倒计数器的当前值为 0,则需要执行一个 Runnable 对象,它就是前面构造函数传入的触发点任务,然后调用 nextGeneration 方法进入下一轮。如果倒计数器当前值不为 0,则调用 Condition 对象的 await 方法进入等待状态。如果设置了超时,则使用 awaitNanos 方法,期间如果发生中断异常则通过 Thread.currentThread().interrupt()设置当前线程的中断标识。此外,如果等待超过指定时间,则抛出 TimeoutException 异常。最后调用 lock.unlock()释放锁。

代码清单 6.13　CyclicBarrier 中的 await 方法

```
1.    public int await() throws InterruptedException, BrokenBarrierException {
2.        try {
3.            return dowait(false, 0L);
4.        } catch (TimeoutException toe) {
5.            throw new Error(toe);
6.        }
7.    }
8.
9.    private int dowait(boolean timed, long nanos)
10.           throws InterruptedException, BrokenBarrierException, TimeoutException {
```

```
11.        final ReentrantLock lock = this.lock;
12.        lock.lock();
13.        try {
14.            int f = flag;
15.            if (Thread.interrupted())
16.                throw new InterruptedException();
17.            int index = --count;
18.            if (index == 0) {
19.                final Runnable command = barrierCommand;
20.                if (command != null)
21.                    command.run();
22.                nextGeneration();
23.                return 0;
24.            }
25.            for (;;) {
26.                try {
27.                    if (!timed)
28.                        trip.await();
29.                    else if (nanos > 0L)
30.                        nanos = trip.awaitNanos(nanos);
31.                } catch (InterruptedException ie) {
32.                    Thread.currentThread().interrupt();
33.                }
34.                if (f != flag)
35.                    return index;
36.                if (timed && nanos <= 0L)
37.                    throw new TimeoutException();
38.            }
39.        } finally {
40.            lock.unlock();
41.        }
42.    }
```

循环屏障的两个使用示例分别如代码清单 6.14 和代码清单 6.15 所示。在代码清单 6.14 中，创建了一个 CyclicBarrier 对象，倒计数器最大值为 3。然后创建 3 个线程，线程会在不同时间调用 await 方法。

代码清单 6.14 循环屏障的使用示例 1

```
1.  public class CyclicBarrierDemo {
2.      static CyclicBarrier barrier = new CyclicBarrier(3);
3.
4.      public static void main(String[] args) {
5.          Thread thread1 = new Thread(() -> {
6.              try {
7.                  System.out.println("thread1 is waiting");
8.                  barrier.await();
9.                  System.out.println("thread1 goes");
10.             } catch (InterruptedException | BrokenBarrierException e) {
11.             }
```

```
12.        });
13.        Thread thread2 = new Thread(() -> {
14.            try {
15.                Thread.sleep(2000);
16.                System.out.println("thread2 is waiting");
17.                barrier.await();
18.                System.out.println("thread2 goes");
19.            } catch (InterruptedException | BrokenBarrierException e) {
20.            }
21.        });
22.        Thread thread3 = new Thread(() -> {
23.            try {
24.                Thread.sleep(4000);
25.                System.out.println("thread3 is waiting");
26.                barrier.await();
27.                System.out.println("thread3 goes");
28.            } catch (InterruptedException | BrokenBarrierException e) {
29.            }
30.        });
31.        thread1.start();
32.        thread2.start();
33.        thread3.start();
34.    }
35. }
```

代码清单 6.14 运行后的输出结果如下。线程 1 启动后输出 "thread1 is waiting"，然后调用 await 方法进入等待状态，倒计数器的值减 1 后为 2。线程 2 启动后先睡眠 2s，然后输出 "thread2 is waiting"，再调用 await 方法进入等待状态，倒计数器的值再减 1 后为 1。线程 3 启动后先睡眠 4s，并输出 "thread3 is waiting"，再调用 await 方法进入等待状态，倒计数器的值减 1 后为 0。此时所有等待的线程都将被放行往下执行，随机输出 "*** goes"。

```
1. thread1 is waiting
2. thread2 is waiting
3. thread3 is waiting
4. thread3 goes
5. thread1 goes
6. thread2 goes
```

代码清单 6.15 用于演示触发点 Runnable 任务（这里以等女朋友吃饭为场景）。假设我与女朋友约好去饭馆吃饭，我准时到达饭馆后开始等女朋友。而女朋友半个小时后到达饭馆。两个人都到齐后开始点餐。

代码清单 6.15　循环屏障的使用示例 2

```
1. public class CyclicBarrierDemo2 {
2.     static CyclicBarrier barrier = new CyclicBarrier(2, new Runnable() {
3.         public void run() {
4.             System.out.println("我和女朋友都到饭馆了，开始点餐");
```

```
5.         }
6.     });
7.
8.     public static void main(String[] args) {
9.         Thread me = new Thread(() -> {
10.            try {
11.                System.out.println("我到达饭馆等女朋友");
12.                barrier.await();
13.            } catch (InterruptedException | BrokenBarrierException e) {
14.            }
15.        });
16.        Thread girlfriend = new Thread(() -> {
17.            try {
18.                System.out.println("女朋友晚到半小时");
19.                Thread.sleep(30 * 60 * 1000);
20.                System.out.println("女朋友到达饭馆");
21.                barrier.await();
22.            } catch (InterruptedException | BrokenBarrierException e) {
23.            }
24.        });
25.        me.start();
26.        girlfriend.start();
27.    }
28. }
```

代码清单 6.15 的最终输出如下。

```
1. 我到达饭馆等女朋友
2. 女朋友晚到半小时
3. 女朋友到达饭馆
4. 我和女朋友都到饭馆了，开始点餐
```

本节所讲的循环屏障与前面讲到的闭锁有点类似，它们都用于对多线程并发进行控制，都类似于一种倒计数器。它们之间不同的地方是，闭锁针对的是倒计数器的值，而循环屏障针对的是线程数。这句话如何理解呢？假如倒计数器的值为 5，那么对于闭锁来说只要调用 5 次 countDown 方法便能让等待的线程往下执行，而不管是一个线程调 5 次 countDown 方法还是 5 个线程分别调用一次 countDown 方法。对于循环屏障来说，必须要有 5 个线程分别调用 await 方法才能使得等待的线程往下执行。闭锁和循环屏障都是等倒计数器的值为 0 后让所有等待的线程通过并往下执行，而循环屏障规定只能有不同的线程来将倒计数器的值减 1。

代码清单 6.16 是循环屏障和闭锁对比的例子。主线程启动线程 1 后调用闭锁的 await 方法进入等待状态，此时线程 1 睡眠 2s 后连续调用两次 countDown 方法将倒计数器的值减为 0，使得主线程得以往下执行。接着主线程启动线程 2，线程 2 睡眠 2s 后调用两次循环屏障的 await 方法让倒计数器的值变为 0，但失败了，因为第一次执行 barrier.await() 时该线程就已经进入等待，所以无法往下执行。这个程序只能输出 "CountDownLatch" 和 "CyclicBarrier"，并不能输出 "can you get here"。此外，循环屏障能在一轮倒计数完后重置倒计数器，闭锁则不能。

代码清单6.16　循环屏障与闭锁对比的例子

```java
1.  public class CountDownLatchVSCyclicBarrier {
2.      static CyclicBarrier barrier = new CyclicBarrier(2);
3.      static CountDownLatch latch = new CountDownLatch(2);
4.
5.      public static void main(String[] args) throws InterruptedException, BrokenBarrierException {
6.          Thread thread1 = new Thread(() -> {
7.              try {
8.                  System.out.println("CountDownLatch");
9.                  Thread.sleep(2000);
10.                 latch.countDown();
11.                 latch.countDown();
12.             } catch (InterruptedException e) {
13.             }
14.         });
15.         thread1.start();
16.         latch.await();
17.         Thread thread2 = new Thread(() -> {
18.             try {
19.                 System.out.println("CyclicBarrier");
20.                 Thread.sleep(2000);
21.                 barrier.await();
22.                 barrier.await();
23.                 System.out.println("can you get here");
24.             } catch (InterruptedException | BrokenBarrierException e) {
25.             }
26.         });
27.         thread2.start();
28.     }
29. }
```

6.5　相位器

相位器（Phaser）能够让多个线程分别在不同阶段进行同步，而且支持动态调整注册的线程数量。相位器与前面讲解的循环屏障类似，但拥有更加灵活的功能，而且更侧重于多阶段屏障。我们通过图 6.12 所示的流程来理解相位器的基本功能。假如我们创建了一个相位器，4 个线程准备通过该相位器进行同步。起初 4 个线程各自运行，然后必须都到达时阶段 1 位置时才能同步，并进阶到阶段 2。接下来 4 个线程继续各自运行，也必须都到达阶段 2 位置时才能进行同步。往下的阶段 3 也是类似的过程。

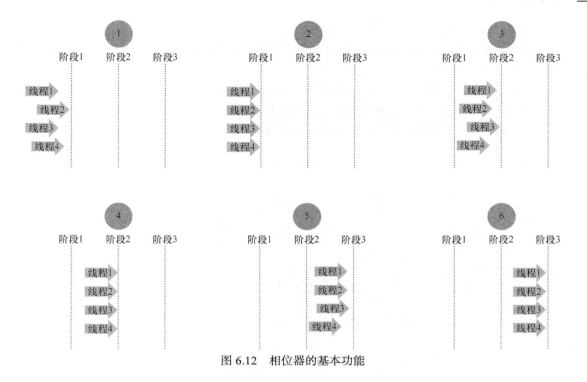

图 6.12　相位器的基本功能

6.5.1　相位器的主要概念及方法

下面介绍相位器的主要概念及方法，如图 6.13 所示。

- **阶段（phase）**：每个相位器都包含若干个同步点，这些同步点称之为阶段。任意时刻，相位器只能处于某个阶段。图 6.13 中的相位器包含了 3 个阶段，所有线程在阶段 1、阶段 2 和阶段 3 进行同步。
- **参与者（party）**：表示参与到相位器中进行同步的线程。可以在构造函数中指定参与者数量，也可以在运行过程中动态更改参与者的数量。参与者会在相位器中进行同步。图 6.13 中的 4 个线程就是参与者。
- **到达（arrived）**：表示参与者到达某个阶段时的状态，已到达某个阶段的参与者会等待其他参与者的到来。图 6.13 中的线程 2 和线程 4 就是处于到达状态。
- **进阶（advance）**：表示所有参与者从某个阶段进入另外一个阶段。需要注意的是，只有所有参与者都到达某个阶段后才能进入下一个阶段。图 6.13 中的 4 个线程都到达阶段 1 后就开始进阶到阶段 2。
- **终止（terminated）**：表示相位器进入终止状态，所有阶段都已运行完成。
- **register 方法**：该方法用于注册一个参与者到相位器中。
- **arriveAndAwaitAdvance 方法**：该方法表示到达某个阶段后等待其他参与者的到来，

然后一起进阶。
- **arriveAndDeregister 方法**：该方法表示到达某个阶段后等待其他参与者的到来，然后将自己进行反注册。
- **onAdvance 方法**：该方法表示当所有参与者都到达某个阶段时会触发这个方法，然后根据它的返回值决定是否让相位器进入终止状态。
- **isTerminated 方法**：判断相位器是否已经终止。

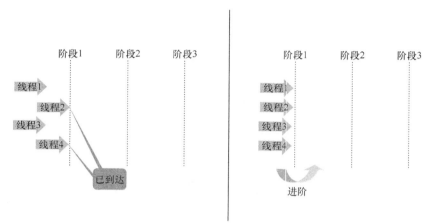

图 6.13　相位器中的主要概念及方法

6.5.2　相位器的 3 个例子

我们先通过一个简单的例子来理解相位器的基本功能，如代码清单 6.17 所示。该代码清单创建了一个相位器，并且通过 3 个 arriveAndAwaitAdvance 方法来实现 3 个阶段的同步操作。执行默认构造函数后调用 4 次 register 方法进行注册操作，也就是说相位器的参与者的数量为 4，这个操作等同于在构造函数中直接指定数量，即 new Phaser(4)。

代码清单 6.17　相位器基本功能的演示示例 1

```
1.   public class PhaserDemo {
2.       public static void main(String[] args) throws IOException {
3.
4.           Phaser phaser = new Phaser();
5.           for (int i = 0; i < 4; i++) {
6.               phaser.register();
7.               new Thread(() -> {
8.                   System.out.println("第一阶段" + Thread.currentThread().getName()
     + ": 执行完任务");
9.                   phaser.arriveAndAwaitAdvance();
10.                  System.out.println("第二阶段" + Thread.currentThread().getName()
     + ": 执行完任务");
```

```
11.                    phaser.arriveAndAwaitAdvance();
12.                    System.out.println("第三阶段" + Thread.currentThread().getName()
    + ": 执行完任务");
13.                    phaser.arriveAndAwaitAdvance();
14.                }).start();
15.            }
16.
17.        }
18.    }
```

代码清单 6.17 执行的输出结果大致如下。可以看到 4 个线程在 3 个阶段都进行了同步。

```
1.  第一阶段Thread-2: 执行完任务
2.  第一阶段Thread-3: 执行完任务
3.  第一阶段Thread-0: 执行完任务
4.  第一阶段Thread-1: 执行完任务
5.  第二阶段Thread-1: 执行完任务
6.  第二阶段Thread-0: 执行完任务
7.  第二阶段Thread-3: 执行完任务
8.  第二阶段Thread-2: 执行完任务
9.  第三阶段Thread-1: 执行完任务
10. 第三阶段Thread-2: 执行完任务
11. 第三阶段Thread-0: 执行完任务
12. 第三阶段Thread-3: 执行完任务
```

再来看第二个例子，如代码清单 6.18 所示。这里重点看相位器如何通过重写 onAdvance 方法来实现多阶段的同步。它的逻辑是若当前阶段值大于等于 3，则返回 true，并终止相位器。该方法如果返回 true，则表示相位器处于终止状态，反之则表示处于非终止状态。一共有 4 个线程分别调用 Phaser 对象的 register 方法进行注册，每个线程都通过 while (!phaser.isTerminated()) 来不断执行每个阶段，直到相位器终止后才结束。其中调用 arriveAndAwaitAdvance 方法会使得每个线程在每个阶段都会等待其他线程，同步后一起进阶到下一阶段。

代码清单 6.18 相位器基本功能的演示示例 2

```
1.  public class PhaserDemo2 {
2.      public static void main(String[] args) throws IOException {
3.          int phaseNum = 3;
4.          Phaser phaser = new Phaser() {
5.              protected boolean onAdvance(int phase, int registeredParties) {
6.                  System.out.println("当前处于第" + phase + "阶段，当前参与线程数为"
    + registeredParties + "。");
7.                  return phase + 1 >= phaseNum || registeredParties == 0;
8.              }
9.          };
10.         for (int i = 0; i < 4; i++) {
11.             phaser.register();
12.             new Thread(() -> {
```

```
13.            while (!phaser.isTerminated()) {
14.                phaser.arriveAndAwaitAdvance();
15.                System.out.println(Thread.currentThread().getName() + ": 执行完任务");
16.            }
17.        }).start();
18.    }
19.
20. }
21. }
```

代码清单 6.18 某次运行的输出结果如下。可以看到 4 个线程在每个阶段都进行了同步。

```
1.  当前处于第 0 阶段，当前参与线程数为 4。
2.  Thread-1: 执行完任务
3.  Thread-2: 执行完任务
4.  Thread-3: 执行完任务
5.  Thread-0: 执行完任务
6.  当前处于第 1 阶段，当前参与线程数为 4。
7.  Thread-2: 执行完任务
8.  Thread-3: 执行完任务
9.  Thread-1: 执行完任务
10. Thread-0: 执行完任务
11. 当前处于第 2 阶段，当前参与线程数为 4。
12. Thread-1: 执行完任务
13. Thread-2: 执行完任务
14. Thread-3: 执行完任务
15. Thread-0: 执行完任务
```

相位器基本功能的第三个例子与选手比赛有关，如代码清单 6.19 所示。这里重点看 arriveAndDeregister 方法的使用。首先，5 位选手各自花费若干时间做预备，arriveAndAwaitAdvance 方法用于保证 5 位选手都同步就位，枪响后正式比赛。然后，5 位选手陆续到达终点，调用 arriveAndDeregister 方法可使所有参与者到达终点后进行反注册操作，反注册操作会将相位器的参与者数量减 1。最后，相位器的参与者数量为 0，主线程的 phaser.isTerminated()返回 true，即相位器处于终止状态，比赛结束。

代码清单 6.19　相位器基本功能的演示示例 3

```
1.  public class PhaserDemo3 {
2.      public static void main(String[] args) throws InterruptedException {
3.          Phaser phaser = new Phaser();
4.          System.out.println("比赛即将开始");
5.          for (int index = 0; index < 5; index++) {
6.              phaser.register();
7.              new Thread(() -> {
8.                  try {
9.                      Thread.sleep((long) (Math.random() * 5000));
10.                     System.out.println(Thread.currentThread().getName() + "选手已就位");
11.                     phaser.arriveAndAwaitAdvance();
12.                     //比赛枪响，正式比赛
```

```
13.                    Thread.sleep((long) (Math.random() * 1000));
14.                    System.out.println(Thread.currentThread().getName() + "选手到达终点");
15.                    phaser.arriveAndDeregister();
16.                } catch (InterruptedException e) {
17.                    e.printStackTrace();
18.                }
19.            }).start();
20.        }
21.        while (!phaser.isTerminated()) {
22.        }
23.        System.out.println("比赛结束");
24.    }
25. }
```

代码清单 6.19 执行后的输出结果如下。

```
1.  比赛即将开始
2.  Thread-2 选手已就位
3.  Thread-3 选手已就位
4.  Thread-5 选手已就位
5.  Thread-4 选手已就位
6.  Thread-6 选手已就位
7.  Thread-2 选手到达终点
8.  Thread-4 选手到达终点
9.  Thread-6 选手到达终点
10. Thread-3 选手到达终点
11. Thread-5 选手到达终点
12. 比赛结束
```

6.5.3 相位器的状态示意图

我们通过图 6.14 来理解相位器的内部状态。首先，通过 new Phaser(3) 创建一个相位器对象，该对象有 3 个参与者。假如又调用了 register() 方法，则该相位器一共有 4 个参与者，这些参与者都处于未到达状态。然后，某个线程执行 arriveAndAwaitAdvance 方法后变为到达状态，接着第二个和第三个参与者都变为到达状态。假如第四个线程执行了 arriveAndDeregister 方法，则第四个参与者到达后将进行反注册，此时只剩 3 个参与者。接着进入下一阶段，相位器会自动调用 onAdvance 方法判断是否终止相位器，如果还未终止则进入下一阶段，此时 3 个参与者又处于未到达状态。

相位器的底层实现需要对共享变量进行维护，包括终止状态、当前阶段数、参与者数量以及未到达数量这 4 个属性。这就要求我们要保证相位器内部属性的一致性。可以定义 4 个变量来分别表示相位器的 4 个属性，但为了方便维护共享变量，可以使用一个 long 类型作为共享变量，这样就能很方便地进行硬件基别的原子更新。所以现在的重点工作就是对 64 位的 long 类型进行划分，0~15 位表示未到达数量，16~31 位表示参与者数量，32~62 位表示当前阶段数，63 位表示终止状态，如图 6.15 所示。

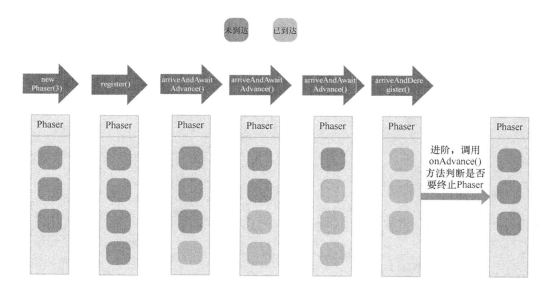

图 6.14　相位器的内部状态示意图

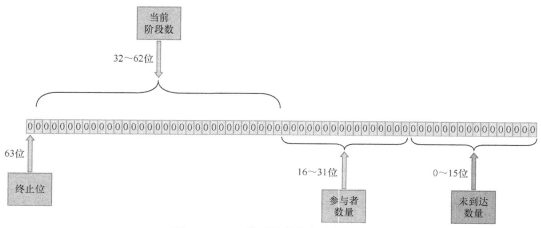

图 6.15　long 类型的共享变量的划分

6.5.4　相位器的实现原理

相位器的实现原理如代码清单 6.20 所示。该代码清单出自 JDK 源码，但并非完全相同，其中去掉了大量的非核心代码，只保留了最核心的代码，以便能使读者更好地理解实现原理。

我们先看 Phaser 类包含的一些属性以及构造方法。state 变量表示共享状态，它是一个 long 类型，我们会使用不同的位范围来表示相位器的内部属性。VarHandle 对象用于进行 CAS 操作，保证能原子地维护共享状态 state。MAX_PHASE=Integer.MAX_VALUE 表示允许的最大阶段数,PARTIES_SHIFT=16 表示参与者数量偏移,PHASE_SHIFT=32 表示当前阶段偏移,UNARRIVED_MASK=0xffff 表示未到达掩码,PARTIES_MASK=0xffff0000L 表示参与者掩码,TERMINATION_BIT=1L<<63 表示终止位,EMPTY=1 表示相位器是空的。该类提供了不带参数和带参数的两种构造函数,当不带参数时表示参与者数量为 0,当带参数时则可以通过参数值来指定参与者数量。两种构造函数都需要将当前阶段设为 0,并且通过偏移将当前阶段和参与者数量存放到共享变量 state 中。

代码清单 6.20　相位器的实现原理

```
1.   public class Phaser {
2.       private static final VarHandle STATE;
3.       static {
4.           try {
5.               MethodHandles.Lookup l = MethodHandles.lookup();
6.               STATE = l.findVarHandle(Phaser.class, "state", long.class);
7.           } catch (ReflectiveOperationException e) {
8.               throw new ExceptionInInitializerError(e);
9.           }
10.      }
11.      private volatile long state;
12.
13.      private static final int MAX_PHASE = Integer.MAX_VALUE;
14.      private static final int PARTIES_SHIFT = 16;
15.      private static final int PHASE_SHIFT = 32;
16.      private static final int UNARRIVED_MASK = 0xffff;
17.      private static final long PARTIES_MASK = 0xffff0000L;
18.      private static final long TERMINATION_BIT = 1L << 63;
19.      private static final int EMPTY = 1;
20.
21.      public Phaser() {
22.          this(0);
23.      }
24.
25.      public Phaser(int parties) {
26.          int phase = 0;
27.          this.state = (parties == 0) ? (long) EMPTY
28.              : ((long) phase << PHASE_SHIFT) | ((long) parties << PARTIES_SHIFT)
29.                          | ((long) parties);
30.      }
31.
32.      ......
33.
34.  }
```

相位器的 register 方法的核心是通过自旋来更新 state,需要按照不同的位范围进行状态更新,如代码清单 6.21 所示。其中 counts 是 state 的低 32 位,首先通过偏移和掩码得到参与者数量和未到达数量,而 state 右移 32 位则得到当前阶段数。然后通过 counts != EMPTY 来判断相位器是否为空，为空时则设置参与者数量为 1。最后通过 compareAndSet 设置 state 变量，成功则退出自旋。不为空时则分两种情况：如果未到达数量等于 0，则调用 internalAwaitAdvance 方法等待下一个阶段，其中 Thread.onSpinWait() 是 JVM 提供的等待方法；将参与者数量和未到达数量都加 1，然后通过 compareAndSet 方法设置 state 变量。

代码清单 6.21　相位器的 register 方法

```java
1.   public int register() {
2.       long adjust = ((long) 1 << PARTIES_SHIFT) | 1;
3.       int phase;
4.       for (;;) {
5.           int counts = (int) state;
6.           int parties = counts >>> PARTIES_SHIFT;
7.           int unarrived = counts & UNARRIVED_MASK;
8.           phase = (int) (state >>> PHASE_SHIFT);
9.           if (counts != EMPTY) {
10.              if (unarrived == 0)
11.                  this.internalAwaitAdvance(phase);
12.              else if (STATE.compareAndSet(this, state, state + adjust))
13.                  break;
14.          } else {
15.              long next = ((long) phase << PHASE_SHIFT) | adjust;
16.              if (STATE.compareAndSet(this, state, next))
17.                  break;
18.          }
19.      }
20.      return phase;
21.  }
22.
23.  private int internalAwaitAdvance(int phase) {
24.      int p;
25.      while ((p = (int) (state >>> PHASE_SHIFT)) == phase)
26.          Thread.onSpinWait();
27.      return p;
28.  }
```

相位器的 arriveAndAwaitAdvance 方法如代码清单 6.22 所示。其核心逻辑是通过自旋将未到达数量减 1 并等待其他参与者到来后一起进阶。首先通过偏移得到当前阶段数量和未到达数量，然后将未到达数减 1 并通过 compareAndSet 设置 state。如果未到达数量大于 1，则调用 internalAwaitAdvance 等待其他参与者一起进阶。接着调用 onAdvance 方法判断是否要终止相位器，前面的 onAdvance 例子就是在这里触发的。根据情况分别设置终止位和未到达状态（主要通过或运算来实现），最后将当前阶段数进行加 1，并尝试通过 compareAndSet 设置 state，最终的返回值为新的当前阶段数。

代码清单 6.22　相位器的 arriveAndAwaitAdvance 方法

```java
1.   public int arriveAndAwaitAdvance() {
```

```
2.      for (;;) {
3.          long s = state;
4.          int phase = (int) (s >>> PHASE_SHIFT);
5.          int counts = (int) s;
6.          int unarrived = (counts == EMPTY) ? 0 : (counts & UNARRIVED_MASK);
7.          if (STATE.compareAndSet(this, s, s -= 1)) {
8.              if (unarrived > 1)
9.                  return this.internalAwaitAdvance(phase);
10.             long n = s & PARTIES_MASK;
11.             int nextUnarrived = (int) n >>> PARTIES_SHIFT;
12.             if (onAdvance(phase, nextUnarrived))
13.                 n |= TERMINATION_BIT;
14.             else if (nextUnarrived == 0)
15.                 n |= EMPTY;
16.             else
17.                 n |= nextUnarrived;
18.             int nextPhase = (phase + 1) & MAX_PHASE;
19.             n |= (long) nextPhase << PHASE_SHIFT;
20.             if (!STATE.compareAndSet(this, s, n))
21.                 return (int) (state >>> PHASE_SHIFT);
22.             return nextPhase;
23.         }
24.     }
25. }
```

相位器的 arriveAndDeregister 方法（见代码清单 6.23）的核心逻辑是通过自旋将未到达数量和参与者数量都减 1。注意，此方法不进行阻塞等待。首先通过偏移及掩码得到当前阶段数量、未到达数量，然后将 state 的参与者数量和未到达数量分别减 1，并通过 compareAndSet 设置 state。如果未到达数量等于 1，则说明该线程是最后一个到达的参与者，此时会调用 onAdvance 方法判断是否要终止相位器。最后通过或运算设置终止位、未到达数量、当前阶段数量，并通过 compareAndSet 设置 state。

代码清单 6.23　相位器的 arriveAndDeregister 方法

```
1.  public int arriveAndDeregister() {
2.      for (;;) {
3.          long s = state;
4.          int phase = (int) (s >>> PHASE_SHIFT);
5.          if (phase < 0)
6.              return phase;
7.          int counts = (int) s;
8.          int unarrived = (counts == EMPTY) ? 0 : (counts & UNARRIVED_MASK);
9.          if (STATE.compareAndSet(this, s, s -= (1 | 1 << PARTIES_SHIFT))) {
10.             if (unarrived == 1) {
11.                 long n = s & PARTIES_MASK;
12.                 int nextUnarrived = (int) n >>> PARTIES_SHIFT;
13.                 if (onAdvance(phase, nextUnarrived))
14.                     n |= TERMINATION_BIT;
15.                 else if (nextUnarrived == 0)
```

```
16.                        n |= EMPTY;
17.                    else
18.                        n |= nextUnarrived;
19.                    int nextPhase = (phase + 1) & MAX_PHASE;
20.                    n |= (long) nextPhase << PHASE_SHIFT;
21.                    boolean result = STATE.compareAndSet(this, s, n);
22.                }
23.                return phase;
24.            }
25.        }
26.    }
```

相位器的 onAdvance 方法和 isTerminated 方法如代码清单 6.24 所示。onAdvance 方法的逻辑是判断注册的参与者数量是否为 0，返回 true 表示需要终止相位器。isTerminated 方法的逻辑是直接判断 state 的值是否小于 0，小于 0 则最高位为 1，也就是 long 的符号位为 1，即小于 0。

代码清单 6.24　相位器的 onAdvance 和 isTerminated 方法

```
1.    protected boolean onAdvance(int phase, int registeredParties) {
2.        return registeredParties == 0;
3.    }
4.
5.    public boolean isTerminated() {
6.        return this.state < 0L;
7.    }
```

6.6　交换器

在并发的场景中，有时候想要在两个线程之间互相交换信息，此时会发现实现起来并不容易，原因是两个线程并发执行的先后顺序不好控制。假如两个线程在运行过程中的某个节点要互相交换信息（见图 6.16），那么应该如何来实现呢？

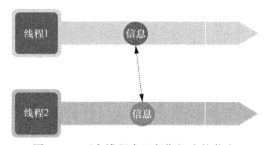

图 6.16　两个线程在运行期间交换信息

为了更好地理解，我们来模拟在两个线程之间进行通信。在代码清单 6.25 中，两个变量分别用于保存来自线程 1 和线程 2 的消息，其中 message2 对应来自线程 1 的消息，而 message1

对应来自线程 2 的消息。由于无法预知两个线程的先后顺序，所以需要线程 1 和线程 2 通过一个 while 循环来确保给变量赋值，以完成消息的交换。

代码清单 6.25　在两个线程之间进行通信

```
1.  public class ThreadExchangeInfoDemo {
2.      private static String message1;
3.      private static String message2;
4.
5.      public static void main(String[] args) {
6.          Thread thread1 = new Thread(() -> {
7.              while (message1 == null || message2 == null) {
8.                  if (message2 == null)
9.                      message2 = "message from thread1";
10.             }
11.             System.out.println(Thread.currentThread().getName() + ":" + message1);
12.         });
13.         thread1.setName("thread1");
14.         Thread thread2 = new Thread(() -> {
15.             while (message2 == null || message1 == null) {
16.                 if (message1 == null)
17.                     message1 = "message from thread2";
18.             }
19.             System.out.println(Thread.currentThread().getName() + ":" + message2);
20.         });
21.         thread2.setName("thread2");
22.         thread1.start();
23.         thread2.start();
24.     }
25. }
```

代码清单 6.25 某次运行的输出结果如下。可以看到线程 1 和线程 2 之间的信息已经完成交换。

```
1.  thread2:message from thread1
2.  thread1:message from thread2
```

交换器（Exchanger）是 JDK 提供的在两个线程之间进行信息交换的工具，它的完整类名为 java.util.concurrent.Exchanger。可以将 Exchanger 看成一个同步点，在这个点上一对线程能够进行信息交换。图 6.17 所示为两个线程的信息交换过程。首先，两个线程都未到达 Exchanger，接着线程 1 进入 Exchanger，它发现线程 2 还没有到，于是会停下来等线程 2 到来。然后，线程 2 进入 Exchanger，它发现线程 1 已经准备就绪，于是两个线程之间互相交换信息。最后，两个线程携带着交换后的信息走出 Exchanger。需要注意的是，Exchanger 类用于线程两两之间的信息交换，但我们无法指定哪个线程与哪个线程进行交换。比如一共有 10 个线程，则先到达的线程两两之间进行交换，最终 5 对线程完成交换。

第 6 章 常见的同步器

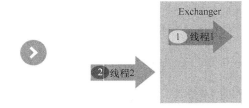

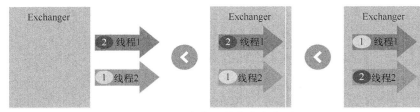

图 6.17 两个线程的信息交换过程

在深入分析 Exchanger 的实现原理之前，我们来看一个例子。假设 Tom 和 Jack 约好在某个地点交易，他们将进行一手交钱、一手交货的操作。这个例子的实现如代码清单 6.26 所示。

代码清单 6.26　模拟一手交钱、一手交货的示例

```
1.   public class ExchangerDemo {
2.
3.       public static void main(String[] args) {
4.           Exchanger<Message> exchanger = new Exchanger<>();
5.           Thread thread1 = new Thread(() -> {
6.               try {
7.                   Message goods = new Message("good");
8.                   Message receivedMessage = exchanger.exchange(goods);
9.                   System.out.println(Thread.currentThread().getName() + ":" +
     receivedMessage.data);
10.              } catch (InterruptedException e) {
11.              }
12.          });
13.          thread1.setName("Tom");
14.          Thread thread2 = new Thread(() -> {
15.              try {
16.                  Message money = new Message("money");
17.                  Message receivedMessage = exchanger.exchange(money);
18.                  System.out.println(Thread.currentThread().getName() + ":" +
     receivedMessage.data);
19.              } catch (InterruptedException e) {
20.              }
21.          });
```

```
22.            thread2.setName("Jack");
23.            thread1.start();
24.            thread2.start();
25.        }
26.
27.        public static class Message {
28.            public String data;
29.
30.            public Message(String s) {
31.                this.data = s;
32.            }
33.        }
34.    }
```

首先创建一个 Exchanger 对象，然后分别创建两个线程，线程 1 和线程 2 分别表示 Tom 和 Jack。Tom 的手里拿着货物，而 Jack 的手里拿着钱，他们通过 Exchanger 进行交换。最后 Tom 得到钱，Jack 得到货物。代码清单 6.26 执行后的输出如下。

```
1.    Tom:money
2.    Jack:good
```

6.6.1 交换器的实现原理

通过上面的例子我们已经对交换器有了大致的了解，接下来看一下交换器是如何实现的，如代码清单 6.27 所示。交换器实现原理的核心有两个：对两个线程在同步点的控制；信息的交换及交换后保证两个线程继续往下执行。为了更好地理解核心思想，我们去掉了一些非核心代码，比如自旋的逻辑、多槽数组动态伸缩的逻辑。其中自旋主要是为了优化等待操作而引入的策略，实际上是优先自旋而不进入系统等待，只有超过自旋次数后才进入系统等待，这样可以有效地减少系统切换成本。而多槽数组动态伸缩则是为了优化竞争而引入的策略，数组的大小会根据竞争的程度动态调整。

代码清单 6.27　交换器的实现原理

```
1.    public class Exchanger<V> {
2.
3.        private static final int ASHIFT = 5;
4.        private static final int MMASK = 0xff;
5.        private static final int NCPU = Runtime.getRuntime().availableProcessors();
6.        static final int FULL = (NCPU >= (MMASK << 1)) ? MMASK : NCPU >>> 1;
7.        private static final Object NULL_ITEM = new Object();
8.        private static final Object TIMED_OUT = new Object();
9.        private final Participant participant;
10.       private volatile Node[] arena;
11.       private volatile Node slot;
12.       private volatile int bound;
13.
```

```
14.    static final class Node {
15.        Object item;
16.        volatile Object match;
17.        volatile Thread parked;
18.    }
19.
20.    static final class Participant extends ThreadLocal<Node> {
21.        public Node initialValue() {
22.            return new Node();
23.        }
24.    }
25.
26.    public Exchanger() {
27.        participant = new Participant();
28.    }
29.
30.    ......
31.
32. }
```

我们先看 Exchanger 类包含的一些属性和内部类。ASHIFT 表示多槽数组中两个有效槽之间的移位数（值为 5），MMASK 表示多槽模式下可支持的最大数组索引（值为 255），NCPU 为 CPU 数（这里假设为 4），FULL 为实际的多槽数组的最大索引值（值与 CPU 数有关，CPU 数为 4 时值为 2），NULL_ITEM 表示交换信息为空，TIMED_OUT 表示信息交换超时。

Participant 是一个内部类，它继承了 ThreadLocal 类，用来保存线程本地变量。注意，这里在 initialValue 方法中创建了 Node 对象，该方法会在首次调用 get 方法时被调用。arena 为 Node 对象数组（多槽模式），slot 为 Node 对象（单槽模式），bound 表示多槽数组的最大索引。Node 是一个内部类，用于封装保存交换信息。Node 类中的 item 为当前线程所携带的信息，match 为另一个线程所携带的信息，parked 为阻塞线程。Exchanger 类只提供一个构造函数，仅仅是创建了 Participant 对象。

此外，还提供了 4 个 VarHandle 类型对象，它们用于 bound、slot、match、arena 等变量的操作，如代码清单 6.28 所示。有了 VarHandle 对象，就能非常方便地进行 CAS 操作。该类型对象是在 JDK 9 后才出现的，主要是为了替代 Unsafe 类。

代码清单 6.28　4 个 VarHandle 类型对象

```
1.  private static final VarHandle BOUND;
2.  private static final VarHandle SLOT;
3.  private static final VarHandle MATCH;
4.  private static final VarHandle AA;
5.  static {
6.      try {
7.          MethodHandles.Lookup l = MethodHandles.lookup();
8.          BOUND = l.findVarHandle(Exchanger.class, "bound", int.class);
9.          SLOT = l.findVarHandle(Exchanger.class, "slot", Node.class);
10.         MATCH = l.findVarHandle(Node.class, "match", Object.class);
```

```
11.              AA = MethodHandles.arrayElementVarHandle(Node[].class);
12.         } catch (ReflectiveOperationException e) {
13.             throw new Error(e);
14.         }
15.     }
```

Exchanger 类的核心方法为 exchange，如代码清单 6.29 所示。两个线程之间也是通过该方法进行信息交换的，该方法默认永不超时，但也支持传入超时参数从而实现超时机制。

代码清单 6.29　exchange 方法的实现

```
1.  public V exchange(V x) throws InterruptedException {
2.      Object v;
3.      Node[] a;
4.      Object item = (x == null) ? NULL_ITEM : x;
5.      if (((a = arena) != null || (v = slotExchange(item, false, 0L)) == null)
6.              && ((Thread.interrupted() || (v = arenaExchange(item, false, 0L)) == null)))
7.          throw new InterruptedException();
8.      return (v == NULL_ITEM) ? null : (V) v;
9.  }
10.
11. public V exchange(V x, long timeout, TimeUnit unit)
12.         throws InterruptedException, TimeoutException {
13.     Object v;
14.     Object item = (x == null) ? NULL_ITEM : x;
15.     long ns = unit.toNanos(timeout);
16.     if ((arena != null || (v = slotExchange(item, true, ns)) == null)
17.             && ((Thread.interrupted() || (v = arenaExchange(item, true, ns)) == null)))
18.         throw new InterruptedException();
19.     if (v == TIMED_OUT)
20.         throw new TimeoutException();
21.     return (v == NULL_ITEM) ? null : (V) v;
22. }
```

我们先看永不超时的情况。如果交换信息为 null，则 item 为 NULL_ITEM，表示空消息对象。如果多槽数组为空，则先尝试调用 slotExchange 方法进行单槽模式信息交换，如果返回为 null，则看是否已经被中断，如果中断则向上抛出 InterruptedException 异常，否则调用 arenaExchange 方法进行多槽模式信息交换。

在支持超时的情况下，先通过 unit.toNanos(timeout) 将超时时长转换为纳秒数值，然后分别传入 slotExchange 方法和 arenaExchange 方法（注意永不超时的情况是直接传入 0）。此外，如果超时，则要向上抛出 TimeoutException 异常。两种情况都是调用 slotExchange 方法或 arenaExchange 方法进行信息交换，分别对应单槽模式和多槽模式。下面分别看看这两种模式的实现。

6.6.2 交换器的单槽模式

slotExchange 方法包含了单槽模式的逻辑,我们先通过图 6.18 来了解什么是单槽模式。现在有 4 个线程,任意两个之间可以进行信息交换,交换过程中只有一个槽可供使用。当线程 1 先进入槽后则占有了该槽,此时线程 2 可以进入槽与线程 1 互相交换信息。但线程 3 却无法再进入槽,它只有等到线程 1 和线程 2 完成信息交换并让出槽后才能进入该槽。线程 3 在进入槽后,线程 4 才能进入与之进行信息交换。

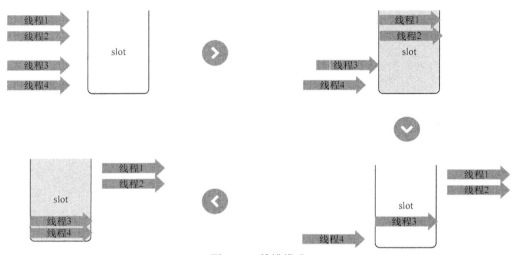

图 6.18　单槽模式

slotExchange 方法的实现如代码清单 6.30 所示。该方法的逻辑是,首先通过 participant.get() 获取当前线程的本地变量 Node 对象,然后判断线程是否已经被中断。接着通过 for(Node q;;) 循环实现自旋操作(因为涉及多线程的并发修改,所以需要自旋)。先看 if((q = slot) != null) 条件,槽不为空说明另一个线程已经占有该槽,尝试通过 CAS 将 slot 设为 null 后让出槽。并且对两个线程的信息进行交换,如果占有槽的线程处于等待状态,就要通过 unpark 唤醒它,然后返回交换得到的对象。如果 CAS 失败,说明已经产生竞争,此时就创建多槽数组,即 arena = new Node[(FULL + 2) << ASHIFT]。需要注意的是,一旦创建了多槽数组,后面的信息交换就将通过多槽模式来实现。而 if(arena != null)条件则表示已启用了多槽模式,直接返回 null,从而会在 exchange 中调用 arenaExchange 多槽模式方法。最后,如果槽为空,则该线程尝试通过 CAS 将 slot 设为当前线程的本地变量 p(即占有槽),如果成功则跳出自旋。

代码清单 6.30　slotExchange 方法的实现

```
1.    private final Object slotExchange(Object item, boolean timed, long ns) {
2.        Node p = participant.get();
3.        Thread t = Thread.currentThread();
```

```
4.        if (t.isInterrupted())
5.            return null;
6.        for (Node q;;) {
7.            if ((q = slot) != null) {
8.                if (SLOT.compareAndSet(this, q, null)) {
9.                    Object v = q.item;
10.                   q.match = item;
11.                   Thread w = q.parked;
12.                   if (w != null)
13.                       LockSupport.unpark(w);
14.                   return v;
15.               }
16.               if (NCPU > 1 && bound == 0 && BOUND.compareAndSet(this, 0, FULL))
17.                   arena = new Node[(FULL + 2) << ASHIFT];
18.           } else if (arena != null)
19.               return null;
20.           else {
21.               p.item = item;
22.               if (SLOT.compareAndSet(this, null, p))
23.                   break;
24.               p.item = null;
25.           }
26.       }
27.       long end = timed ? System.nanoTime() + ns : 0L;
28.       Object v;
29.       while ((v = p.match) == null) {
30.           if (!t.isInterrupted() && arena == null
31.                   && (!timed || (ns = end - System.nanoTime()) > 0L)) {
32.               p.parked = t;
33.               if (slot == p) {
34.                   if (ns == 0L)
35.                       LockSupport.park(this);
36.                   else
37.                       LockSupport.parkNanos(this, ns);
38.               }
39.               p.parked = null;
40.           } else if (SLOT.compareAndSet(this, p, null)) {
41.               v = timed && ns <= 0L && !t.isInterrupted() ? TIMED_OUT : null;
42.               break;
43.           }
44.       }
45.       MATCH.setRelease(p, null);
46.       p.item = null;
47.       return v;
48.   }
```

跳出自旋后还有一部分逻辑，用于等待另外一个线程将交换信息传递过来，同时提供超时机制。这里重要的循环条件是 while((v = p.match) == null)，表示还未获得另外一个线程的交换信息，它将一直等待下去。如果线程被中断，也会跳出循环。(ns = end - System.nanoTime()) > 0L 条件用于判断是否超时，如果没有超时，则会通过 LockSupport.parkNanos 进入系统级别的等待状态,到时间后会被自动唤醒且进行下一个循环。if(ns == 0L)表示永不超时,会通过 LockSupport.park 方法进入系统级别的等待状态,且一直不会超时。该方法最终会返回 Node 节点的 match 变量值,即交换得到的信息。当然,如果返回 TIMED_OUT,则表示交换信息操作超时,此时要通过 SLOT.compareAndSet(this, p, null)将占用的槽让出来。

6.6.3 交换器的多槽模式

为了解决单槽模式下存在的槽竞争问题,Exchanger 引入了槽数组,即多槽模式,从而能够同时供多对线程进行信息交换。多槽模式如图 6.19 所示,这是 4 个线程的情况,线程两两之间进行信息交换。假如线程 1 占用了第一个槽并与线程 2 进行信息交换,那么线程 3 发现该槽无法使用后则会去尝试另外一个槽。结果发现还未被占用,所以线程 3 成功占有另外一个槽,与线程 4 在该槽中进行信息交换。

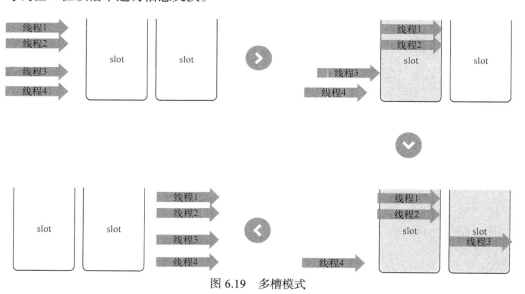

图 6.19 多槽模式

为了避免伪共享问题,Exchanger 类将多槽数组设计成填充槽和有效槽。虽然槽数组是连续的,但是真正使用的槽却是相隔一定距离的,这个距离是 32。图 6.20 所示为一个连续的槽数组,其中大部分都是填充槽,有效槽只有 3 个,对应的索引分别为 31、63、95。从有效槽的索引来看,它们分别为 0、1、2。伪共享说的是高速缓存与内存之间的问题,即高度缓存在加载内存数据时一般会将相邻数据也一起加载进去,而数组的结构在内存中是连续的,所以如果不使用填充槽,则全部有效槽都被加载到缓存行中,当缓存行上的某个槽被更新后就会导致所有数据失效(即所有槽都无法再使用),需要重新从内存中加载,从而产生性能问题。

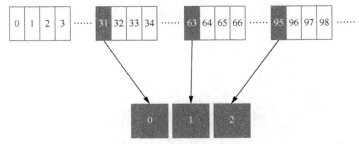

图 6.20 一个连续的槽数组

图 6.21 所示为多槽数组的逻辑结构示意图，其中 bound 变量表示多槽数组的最大索引，MMASK 表示多槽模式下可支持的最大数组索引。

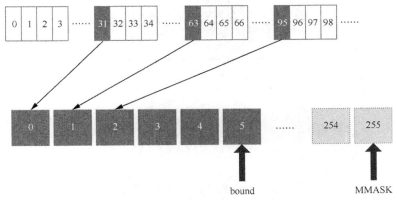

图 6.21 多槽数据的逻辑结构示意图

arenaExchange 方法实现了多槽模式的逻辑，它的实现思想如图 6.22 所示。其中一共有 6 个槽，在图 6.22 最上方的第一个槽正在被线程 1 和线程 2 用于信息交换。线程 3 发现第一个槽被占用后，尝试在第二个槽中自旋等待与之交换的线程，线程 4 的到来将会与之相配对。类似地，线程 5 使用了第三个槽。需要注意的是，只有第一个槽中的线程才会进入系统等待状态，而第二、第三及后面的槽所对应的线程都不会进入系统等待，它们使用自旋来替代等待，从而减少线程上下文开销。继续看图 6.22 的中间部分，假如第一、第二个槽已经使用完毕并释放，此时线程 5 已经自旋了一段时间，但是仍然未等来与之进行信息交换的线程，那么线程 5 将会倾向于向第一个槽去靠拢。最终线程 5 进入到第一个槽中，并且在该槽中进入系统等待状态，等待线程 6 的到来。

代码清单 6.31 所示为 arenaExchange 方法的代码逻辑，其中 for (int i = p.index;;) 为主体逻辑，通过自旋方式尝试在槽数组中进行信息交换操作。前面已经介绍过，为了避免伪共享问题而引入了填充槽和有效槽，所以我们需要通过(i << ASHIFT) + ((1 << ASHIFT) - 1)将有效槽的索引转为实际数组索引。其中 if(q != null && AA.compareAndSet(a, j, q, null))条件表示槽不为空，即已经有线程在等待交换，所以代码块 1 负责信息交换及唤醒线程。而 if(i <= FULL

&& q == null)条件表示槽为空,线程可以尝试去占用该槽,所以代码块 2 负责尝试在槽数组中占有槽。

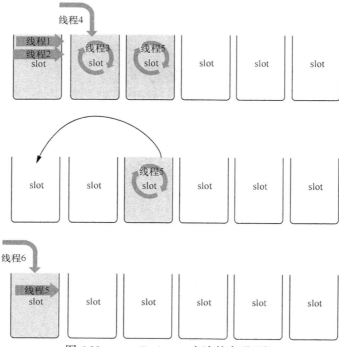

图 6.22 arenaExchange 方法的实现思想

代码清单 6.31 arenaExchange 方法的代码逻辑

```
1.    private final Object arenaExchange(Object item, boolean timed, long ns) {
2.        Node[] a = arena;
3.        int alen = a.length;
4.        Node p = participant.get();
5.        for (int i = 0;;) {
6.            int j = (i << ASHIFT) + ((1 << ASHIFT) - 1);
7.            if (j < 0 || j >= alen)
8.                j = alen - 1;
9.            Node q = (Node) AA.getAcquire(a, j);
10.           if (q != null && AA.compareAndSet(a, j, q, null)) {
11.
12.               //代码块 1
13.
14.           } else if (i <= FULL && q == null) {
15.
16.               //代码块 2
17.
18.           } else {
```

```
19.              if (i == FULL)
20.                  i = 0;
21.              else
22.                  i = i + 1;
23.          }
24.      }
25. }
```

代码块 1 负责信息交换及唤醒线程,其逻辑如代码清单 6.32 所示。首先分别交换信息,然后通过 LockSupport.unpark(w)唤醒线程。

代码清单 6.32　代码块 1 的逻辑

```
1.  Object v = q.item;
2.  q.match = item;
3.  Thread w = q.parked;
4.  if (w != null)
5.      LockSupport.unpark(w);
6.  return v;
```

下面看代码块 2 的逻辑,如代码清单 6.33 所示。AA.compareAndSet(a, j, null, p)表示尝试去占有第 i 个有效槽,成功占用后则会通过 for (int spins = 1024;;)进入到一个自旋操作中。在自旋过程中,if(v != null)表示其他某个线程跟该线程进行交换,于是可以清空并返回交换到的数据。然后 if (spins > 0)则说明还在自旋中,通过 Thread.yield()让出 CPU 时间。接着 if (!t.isInterrupted() && i == 0 && (!timed || (ns = end - System.nanoTime()) > 0L))条件用于判断是否占有第一个槽,如果是第一个槽则允许线程进入系统等待状态。最后的 if (AA.getAcquire(a, j) == p && AA.compareAndSet(a, j, p, null))是中断、超时、非第一个槽的情况,中断及超时则直接返回对应的指示,而非第一个槽的情况则会通过 i = i >>>= 1 来向第一个槽靠拢。

代码清单 6.33　代码块 2 的逻辑

```
1.  p.item = item;
2.  if (AA.compareAndSet(a, j, null, p)) {
3.      long end = (timed && i == 0) ? System.nanoTime() + ns : 0L;
4.      Thread t = Thread.currentThread();
5.      for (int spins = 1024;;) {
6.          Object v = p.match;
7.          if (v != null) {
8.              MATCH.setRelease(p, null);
9.              p.item = null;
10.             return v;
11.         } else if (spins > 0) {
12.             spins--;
13.             Thread.yield();
14.         } else if (!t.isInterrupted() && i == 0
15.                 && (!timed || (ns = end - System.nanoTime()) > 0L)) {
16.             p.parked = t;
```

```
17.              if (AA.getAcquire(a, j) == p) {
18.                  if (ns == 0L)
19.                      LockSupport.park(this);
20.                  else
21.                      LockSupport.parkNanos(this, ns);
22.              }
23.              p.parked = null;
24.          } else if (AA.getAcquire(a, j) == p && AA.compareAndSet(a, j, p, null)) {
25.              p.item = null;
26.              i = i >>>= 1;
27.              if (Thread.interrupted())
28.                  return null;
29.              if (timed && i == 0 && ns <= 0L)
30.                  return TIMED_OUT;
31.              break;
32.          }
33.      }
34. } else
35.      p.item = null;
36.
```

至此，整个多槽模式的逻辑演示介绍完毕。

第 7 章

原子类

7.1 原子整型

原子整型，顾名思义就是具备原子性的整型。在介绍原子整型之前，我们先了解一下什么是原子性。原子性主要是指一个操作是原子的、不可分割的，它要么全部执行，要么全部不执行。之所以能表达这个意思，是因为在人类认知的某个阶段，公认原子为最小的粒子，它不能够再被分割。从计算机并发的角度来看，原子性就是某个操作是不可分割的整体，整个操作可以看成一个不可拆分的执行指令。

7.1.1 一行代码等于原子性吗

对采用 Java 语言编写的程序来说，如果不做额外处理，我们都应当把它看成是非原子性的。比如很多代码看起来只有一行，直觉上看它具备原子性，但实际上却并非如此。这是因为 Java 代码会编译成字节码指令，一行代码可能会被编译成若干条指令。另外，即使只有一条字节码指令，JVM 在解析运行这条字节码时也可能变成多步操作。

我们来看看代码清单 7.1。其中 main 方法只有 count++;这一行代码，如果从 Java 语言层面来看，很容易认为这是一个不可分割的操作，但实际情况是怎样的呢？从编译后的字节码指令可以看到，由 getstatic、iconst_1、iadd 和 putstatic 这 4 个指令来完成 count++ 运算，如代码清单 7.2 所示。

代码清单 7.1　一行代码并不等于原子性代码的示例

```
1.  public class AtomicIntegerDemo {
2.      static int count = 0;
3.      public static void main(String[] args) {
4.          count++;
5.      }
6.  }
```

代码清单 7.2　编译后的字节码指令

```
public static void main(java.lang.String[]);
```

```
descriptor: ([Ljava/lang/String;]V
flags: (0x0009) ACC_PUBLIC, ACC_STATIC
Code:
  Stack=2, locals=1, args_size=1
    0: getstatic      #10            // Field count:I
    3: iconst_1
    4: iadd
    5: putstatic      #10            // Field count:I
    8: return
```

整个流程为先获取 count 静态变量的值并推入栈中，然后将 1 推入栈中，接着对栈中的两个数执行相加操作，最终将结果赋值回 count 静态变量。所以从字节码指令的角度来看，Java 语言层的一行代码被分解成了 4 个操作，如图 7.1 所示。此外，JVM 在执行每个指令时同样也可能需要多个操作来完成。综合来看，对于 Java 常规代码，我们都不应该认为其具有原子性。

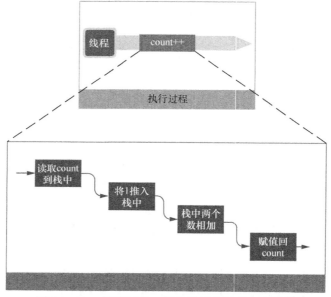

图 7.1　Java 语言层的一行代码被分解成 4 个操作

7.1.2　volatile 能保证原子性吗

我们看一下将变量声明为 volatile 能否保证原子性。直接将 count 变量声明为 volatile，某次运行得到的结果为 count = 50665。可以看到运算的结果仍然是错误的，为什么呢？下面通过 volatile 的语义来分析整个过程，如代码清单 7.3 所示。

代码清单 7.3　将变量声明为 volatile

```
1.   public class AtomicIntegerDemo3 {
2.       static volatile int count = 0;
```

```
3.      public static void main(String[] args) throws InterruptedException {
4.          for (int i = 0; i < 10; i++)
5.              new Thread(() -> {
6.                  for (int j = 0; j < 10000; j++)
7.                      count++;
8.              }).start();
9.          Thread.sleep(3000);
10.         System.out.println("count = " + count);
11.     }
12. }
```

为了提升 CPU 效率，JVM 的每个线程都会保存一个变量副本，这样就能提升数据的读取速度，从而使 CPU 的整体执行效率更高。在 Java 中，volatile 主要用于保证变量的可见性，被声明为 volatile 的变量表示每次读取时必须读取主存中的变量，而且每次修改时都必须同步修改主存中的变量。在图 7.2 中，线程 1、线程 2 和线程 3 都有各自的 count 副本。假如线程 1 要读取 count 的值，它必须将主存的 count 变量的值同步到高速缓冲中的 count 变量。假如线程 1 修改了 count 的值，它也必须将修改同步到主存中的 count 变量。如果所有线程都按这样的规定来读写变量，那么就可以保证任何线程对变量的修改都能够及时反映到其他线程，从而保证了变量的可见性。

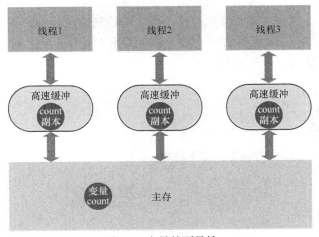

图 7.2　变量的可见性

但需要注意的是，可见性并不等于原子性。对于图 7.2 所示的例子，假如线程 1 和线程 2 同时读取主存的 count 变量，此时都为 0，线程 1 执行完 count++ 后将结果写回主存，而线程 2 也做同样的操作，最终主存的 count 变量值为 1，而实际上已经执行了两次自加操作。

7.1.3　synchronized 能解决问题吗

既然 volatile 不能有效解决问题，那么 synchronized 关键词能否解决问题呢？我们使用 synchronized 来将 count++ 操作括起来（见代码清单 7.4），表示这个区域会使用互斥锁来控制访问，也就是通过锁机制使得 count++ 变成一个原子操作。此时运行的结果是正确的，结果为 count = 100000。

代码清单 7.4　使用 synchronized 进行加锁确保操作是原子性的

```
1.   public class AtomicIntegerDemo4 {
2.       static int count = 0;
3.       public static void main(String[] args) throws InterruptedException {
4.           for (int i = 0; i < 10; i++)
5.               new Thread(() -> {
6.                   for (int j = 0; j < 10000; j++)
7.                       synchronized (AtomicIntegerDemo4.class) {
8.                           count++;
9.                       }
10.              }).start();
11.          Thread.sleep(3000);
12.          System.out.println("count = " + count);
13.      }
14.  }
```

使用 synchronized 加锁后的逻辑可以通过图 7.3 来理解。此时 count++ 的边界就好比有了两个屏障，每次只能有一个线程进入该屏障中执行 count++。所以不管 count++ 最终会被分解为多少个操作，在整个过程中都只能同时有一个线程执行该段逻辑。synchronized 保证了操作的原子性，同时 synchronized 也具备可见性，从而保证了执行结果的正确性。可见原子性和可见性共同保证了并发执行结果的正确性。

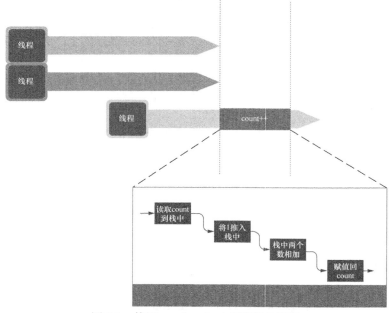

图 7.3　使用 synchronized 加锁后的逻辑

7.1.4 AtomicInteger

除了上述加锁的实现方式外，还可通过 AtomicInteger 提供更优的实现方式，实现代码如代码清单 7.5 所示。只需要创建一个 AtomicInteger 对象，然后调用 addAndGet 方法进行加法操作，就能保证执行结果是正确的。AtomicInteger 使用了一种无锁的方式来实现，通过 volatile 和 CAS 的组合来保证可见性和原子性。相较于锁机制，自旋的无锁方式能够提供更高的吞吐量。

代码清单 7.5　通过 AtomicInteger 提供更优的实现方式

```
1.   public class AtomicIntegerDemo5 {
2.       static AtomicInteger count = new AtomicInteger(0);
3.       public static void main(String[] args) throws InterruptedException {
4.           for (int i = 0; i < 10; i++) {
5.               new Thread(() -> {
6.                   for (int j = 0; j < 10000; j++)
7.                       count.getAndAdd(1);
8.               }).start();
9.           Thread.sleep(3000);
10.          System.out.println("count = " + count.get());
11.      }
12.  }
```

AtomicInteger 是 JDK 提供的原子整型，可以拆分为 Atomic 和 Integer 两个关键词，即能保证原子性的整型操作。通过它能在多线程的场景下安全并发地操作某个整型变量，保证对整型变量操作的正确性。实际上，AtomicInteger 类只包含了一个 volatile 整型变量，它通过 CAS 算法和自旋来实现原子性。比如在图 7.4 中有 3 个线程，它们都通过一个无限循环来不断执行 CAS 操作，而跳出循环的条件就是 CAS 操作执行成功。CAS 会判断内存值 value 有没有被其他线程更改，如果没有则本线程能够安全地更新。

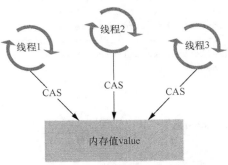

图 7.4　通过 CAS 算法和自旋实现原子性

7.1.5 实现原理

AtomicInteger 类处于 java.util.concurrent.atomic 包下，为了方便理解，我们只对其核心源码进行分析，而且也不保证与 JDK 源码一致。从整体来看，AtomicInteger 就是对普通的整型进行封装，从而达到原子性效果。所以肯定需要有一个 volatile 整型变量，硬件级别的原子性则由 Unsafe 提供。其中 static 块用于创建 Unsafe 对象，因为 Java 为了保证安全而限制了 Unsafe 的使用，必须通过反射技巧来绕开限制。valueoffset 是 value 变量的偏移量，这个变量供 Unsafe 来定位。

接着看 AtomicInteger 的主要方法。构造函数可传入整型参数或者不传参数，get 方法用于获取 value 值，如代码清单 7.6 所示。我们主要关注 getAndSet 和 getAndAdd 这两个方法，掌握其中一个方法的实现机制便能完全掌握该类的实现。我们看 getAndAdd 方法，它通过 do-while 循环来实现自旋，U.getIntVolatile 用于获取 value 变量值，U.compareAndSwapInt 则是 CAS 操作，其中 v 是期望值，而 v+delta 则是新值。总体来说，就是使用了自旋+CAS，CAS 返回 true 时表示修改成功并跳出循环，而如果修改失败则继续下一个循环，直到 CAS 成功为止。这就是 AtomicInteger 相关操作的核心，该类的其他方法如果涉及并发修改，也是通过自旋+CAS 来实现的。

代码清单 7.6　AtomicInteger 的主要方法

```
1.   public class AtomicInteger {
2.   
3.       private static final Unsafe U;
4.       private static final long valueoffset;
5.       private volatile int value;
6.   
7.       static {
8.           try {
9.               Field theUnsafeInstance = Unsafe.class.getDeclaredField("theUnsafe");
10.              theUnsafeInstance.setAccessible(true);
11.              U = (Unsafe) theUnsafeInstance.get(Unsafe.class);
12.              valueoffset = U.objectFieldOffset(AtomicInteger.class.getDeclaredField("value"));
13.          } catch (ReflectiveOperationException e) {
14.              throw new Error(e);
15.          }
16.      }
17.  
18.      public AtomicInteger() {
19.      }
20.  
21.      public AtomicInteger(int initialValue) {
22.          value = initialValue;
23.      }
24.  
```

```
25.     public final int get() {
26.         return value;
27.     }
28.
29.     public final int getAndSet(int newValue) {
30.         int v;
31.         do {
32.             v = U.getIntVolatile(this, valueoffset);
33.         } while (!U.compareAndSwapInt(this, valueoffset, v, newValue));
34.         return v;
35.     }
36.
37.     public final int getAndAdd(int delta) {
38.         int v;
39.         do {
40.             v = U.getIntVolatile(this, valueoffset);
41.         } while (!U.compareAndSwapInt(this, valueoffset, v, v + delta));
42.         return v;
43.     }
44.
45. }
```

本节分析了 AtomicInteger 原子类的实现原理，为了让读者更好理解，我们从原子性开始讲起，并且应该意识到并非一行代码就具有原子性，很多时候它会被底层分解成多个步骤去执行。比如在 JVM 中被分解成多个字节码，还比如更底层的编译器会分解成多个机器指令。我们还分析了 volatile 和 synchronized 在并发过程中的效果，从而知道了并发结果的正确性需要同时由可见性和原子性来保证。最后还分析了 AtomicInteger 类的实现原理，JDK 提供的原子类还有 AtomicBoolean 和 AtomicLong，它们的实现原理类似，这里不再赘述。

7.2 原子引用

上一节介绍了原子整型 AtomicInteger，通过它可以对整型变量进行原子操作。然而，如果想要对浮点型变量进行原子操作，却没有办法在 JDK 中找到对应的 AtomicDouble 类或 AtomicFloat 类。一般对普通的 Java 对象进行原子操作，又该如何进行呢？这时就可以使用 JDK 提供的原子引用（AtomicReference）类，该类是 JDK 提供的可以用来原子更新 Java 对象的辅助类，它能解决多个线程在并发地更新对象时所带来的线程安全问题。该类的核心方法是 compareAndSet，它的原理是通过对象地址的比较来确定是否可以更新（这其实就是 CAS 算法）。

我们先看普通对象在多线程情况下进行更新时会不会产生线程安全问题。代码清单 7.7 所示为一个报数示例，Tom、Jack 和 Lucy 这 3 个人分别进行报数。

代码清单 7.7　报数示例

```java
public class AtomicReferenceDemo {

    static String[] names = { "Tom", "Jack", "Lucy" };
    static Baoshu bs = new Baoshu();

    public static void main(String[] args) throws InterruptedException {
        for (String name : names) {
            new Thread(() -> {
                Baoshu newbs = new Baoshu();
                newbs.name = name;
                newbs.num = bs.num + 1;
                bs = newbs;
                System.out.println(bs);
            }).start();
        }
        Thread.sleep(3000);
    }

    static class Baoshu {
        public String name;
        public int num = 0;
        public String toString() {
            return name + " : " + num;
        }
    }
}
```

我们创建了 3 个线程进行报数，每个报数的人要将当前数加 1。正确情况下是 3 人按 1、2、3 的顺序进行报数，某次的输出如下。

```
Jack : 2
Jack : 2
Lucy : 3
```

可以看出，明显产生了线程安全问题，这是 3 个线程并发地修改 Baoshu 对象导致的。

对象安全问题最简单的解决办法就是通过加锁来解决，在代码清单 7.7 的基础上，我们创建了一个 lock 对象作为锁，然后通过 synchronized(lock) 来指定同步块以提供锁机制来保证线程安全，如代码清单 7.8 所示。

代码清单 7.8　通过锁机制保证线程安全

```java
public class AtomicReferenceDemo2 {

    static String[] names = { "Tom", "Jack", "Lucy" };
    static Baoshu bs = new Baoshu();
    static Object lock = new Object();
```

```
6.
7.    public static void main(String[] args) throws InterruptedException {
8.        for (String name : names) {
9.            new Thread(() -> {
10.               synchronized (lock) {
11.                   Baoshu newbs = new Baoshu();
12.                   newbs.name = name;
13.                   newbs.num = bs.num + 1;
14.                   bs = newbs;
15.                   System.out.println(bs);
16.               }
17.           }).start();
18.        }
19.        Thread.sleep(3000);
20.    }
21.
22.    static class Baoshu {
23.        public String name;
24.        public int num = 0;
25.        public String toString() {
26.            return name + " : " + num;
27.        }
28.    }
29. }
```

锁只允许一个线程进入锁区域，它能保证以串行的方式执行，最终输出如下。

```
1. Tom : 1
2. Lucy : 2
3. Jack : 3
```

synchronized 是一种悲观锁，虽然它简单有效，但我们有时想通过无锁的方式来得到更好的执行性能，此时可以使用 AtomicReference，如代码清单 7.9 所示。通过 for 自旋且使用 AtomicReference 来封装 Baoshu 对象以提供 CAS 操作，然后通过 get 方法获取 Baoshu 对象，并且通过 compareAndSet 方法尝试修改 Baoshu 对象，如果当前对象地址与期望地址一致则成功更新，否则更新失败并进入下一次更新尝试。

代码清单 7.9　使用 AtomicReference 来获得更好的执行性能

```
1.  public class AtomicReferenceDemo3 {
2.      Baoshu bs = new Baoshu();
3.      AtomicReference<Baoshu> atomicBS = new AtomicReference<Baoshu>(bs);
4.
5.      public void next(String name) {
6.          for (;;) {
7.              Baoshu newbs = new Baoshu();
8.              newbs.name = name;
9.              newbs.num = atomicBS.get().num + 1;
10.             if (atomicBS.compareAndSet(atomicBS.get(), newbs)) {
```

```
11.                    System.out.println(newbs.toString());
12.                    break;
13.                }
14.            }
15.        }
16.
17.     static class Baoshu {
18.         private String name;
19.         private int num = 0;
20.         public String toString() {
21.             return name + " : " + num;
22.         }
23.     }
24.
25.     public static void main(String[] args) throws InterruptedException {
26.         AtomicReferenceDemo3 demo = new AtomicReferenceDemo3();
27.         String[] names = { "Tom", "Jack", "Lucy" };
28.         for (String name : names) {
29.             new Thread(() -> {
30.                 demo.next(name);
31.             }).start();
32.         }
33.         Thread.sleep(3000);
34.     }
35.
36. }
```

AtomicReference 的核心思想还是 CAS。对于 Java 对象来说，如果属性一个一个地更新，那么即使每个属性是原子更新，也没办法达到整体原子性的效果。所以需要另外一种更新的方式，那就是原子更新对象地址，这样一来，不管它有多少个属性，都能够通过地址更新来达到整体原子性效果（AtomicReference 通过 CAS 比较的正是对象的地址）。

比如图 7.5 中的两个线程同时拿到了指向对象 1 的地址（也就是变量 a），两个线程并发地更新该对象。假如线程 1 和线程 2 都创建了新的对象 2 和对象 3，那么这两个线程都通过 CAS 来比较对象的地址，且只有变量 a 指向对象 1 时才会被成功地进行更新。假设线程 1 成功更新，则此时变量 a 指向对象 2，而线程 2 的 CAS 发现变量 a 已经不是执行对象 1 的地址了，于是更新失败，它可以进行下一轮的尝试更新。

JDK 提供的 AtomicReference 类位于 java.util.concurrent.atomic 包下。为了将对象封装成原子引用，这里通过泛型来声明需要封装的对象，具体的对象则由 AtomicReference 的构造函数传入，如代码清单 7.10 所示。AtomicReference 内部使用了 value 变量来保存对象，该对象可以通过 get 方法来获得。我们来看 compareAndSet 核心方法，它的两个参数分别为期望值和新值，也就是期望的对象地址和新的对象地址。这里底层的 CAS 由 VarHandle 来实现，这是 JDK 9 后提供的方法，可替代 Unsafe 类，这里只要知道这两种方法都可行即可。

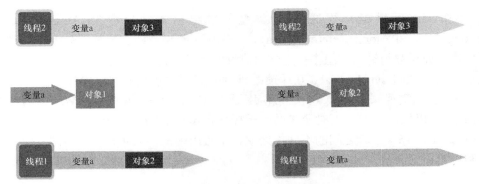

图 7.5 通过 CAS 来比较对象的地址

代码清单 7.10 AtomicReference 类的实现

```
1.  public class AtomicReference<V> {
2.      private static final VarHandle VALUE;
3.      private volatile V value;
4.      static {
5.          try {
6.              MethodHandles.Lookup l = MethodHandles.lookup();
7.              VALUE = l.findVarHandle(AtomicReference.class, "value", Object.class);
8.          } catch (ReflectiveOperationException e) {
9.              throw new Error(e);
10.         }
11.     }
12.
13.     public AtomicReference() {
14.     }
15.
16.     public AtomicReference(V initialValue) {
17.         value = initialValue;
18.     }
19.
20.     public final V get() {
21.         return value;
22.     }
23.
24.     public final boolean compareAndSet(V expectedValue, V newValue) {
25.         return VALUE.compareAndSet(this, expectedValue, newValue);
26.     }
27.
28. }
```

7.3 原子数组

AtomicInteger、AtomicLong 和 AtomicReference 原子类分别用于整型、长整型和对象引用

的原子操作。下面继续介绍这 3 种类型的数组形式，分别为 AtomicIntegerArray、AtomicLongArray 和 AtomicReferenceArray，分别用于整型数组、长整型数组和对象引用数组的原子操作。原子数组针对数组中指定索引位置的元素提供了原子操作。

当然也可以通过 synchronized 互斥锁的机制来实现对数组的原子操作，但如果想通过更高效的 CAS 机制实现原子更新，就必须使用原子数组。在图 7.6 中，对数组中的元素提供了 CAS 机制，可以保证多线程下对元素的原子操作。AtomicIntegerArray、AtomicLongArray 和 AtomicReferenceArray 三者之间的不同之处就在于数组中保存的元素类型不同，分别为整型、长整型和对象引用。

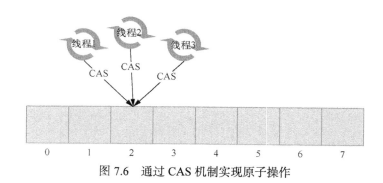

图 7.6 通过 CAS 机制实现原子操作

7.3.1 AtomicIntegerArray

在讲解 AtomicIntegerArray 的实现原理前，先看一个简单的例子，如代码清单 7.11 所示。其中有一个长度为 2 的整型数组，每个元素都用于计数。代码清单 7.11 启动 10 个线程进行计算，每个线程负责对两个计数器进行 10 000 次累加操作，在作者本地某次运算的结果为 counts[0] = 56791、counts[1] = 57533。显然这是有问题的，正确的结果应该是 100 000。

代码清单 7.11 进行计数的简单示例

```
1.   public class AtomicIntegerArrayDemo {
2.       static int[] counts = {0,0};
3.       public static void main(String[] args) throws InterruptedException {
4.           for (int i = 0; i < 10; i++)
5.               new Thread(() -> {
6.                   for (int j = 0; j < 10000; j++) {
7.                       counts[0]++;
8.                       counts[1]++;
9.                   }
10.              }).start();
11.          Thread.sleep(3000);
12.          System.out.println("counts[0] = " + counts[0]);
13.          System.out.println("counts[1] = " + counts[1]);
```

```
14.    }
15. }
```

然后通过 AtomicIntegerArray 来实现计数器，如代码清单 7.12 所示。同样启动 10 个线程进行计算，此时通过 incrementAndGet 方法来指定对数组中的元素进行累加操作。最终的执行结果为 counts[0] = 100000、counts[1] = 100000，此时的结果是正确的。

代码清单 7.12　通过 AtomicIntegerArray 来实现计数器

```
1.  public class AtomicIntegerArrayDemo2 {
2.      static AtomicIntegerArray counts = new AtomicIntegerArray(2);
3.      public static void main(String[] args) throws InterruptedException {
4.          for (int i = 0; i < 10; i++) {
5.              new Thread(() -> {
6.                  for (int j = 0; j < 10000; j++) {
7.                      counts.incrementAndGet(0);
8.                      counts.incrementAndGet(1);
9.                  }
10.             }).start();
11.         Thread.sleep(3000);
12.         System.out.println("counts[0] = " + counts.get(0));
13.         System.out.println("counts[1] = " + counts.get(1));
14.     }
15. }
```

实际上也可以通过 AtomicInteger[]的方式来实现上述 AtomicIntegerArray 的计数器，如代码清单 7.13 所示。在该代码中，将两个 AtomicInteger 对象作为计数器，然后分别调用 getAndAdd 方法，最终结果输出为 counts[0] = 100000、counts[1] = 100000。同样能保证两个计数器的准确性。

代码清单 7.13　使用 AtomicInteger[]方式实现计数器

```
1.  public class AtomicIntegerArrayDemo3 {
2.      static AtomicInteger[] counts = new AtomicInteger[2];
3.      public static void main(String[] args) throws InterruptedException {
4.          counts[0] = new AtomicInteger();
5.          counts[1] = new AtomicInteger();
6.          for (int i = 0; i < 10; i++)
7.              new Thread(() -> {
8.                  for (int j = 0; j < 10000; j++) {
9.                      counts[0].getAndAdd(1);
10.                     counts[1].getAndAdd(1);
11.                 }
12.             }).start();
13.         Thread.sleep(3000);
14.         System.out.println("counts[0] = " + counts[0].get());
15.         System.out.println("counts[1] = " + counts[1].get());
```

```
16.     }
17. }
```

AtomicIntegerArray 的实现思想如图 7.7 所示。它对一个整型数组的每个元素提供 CAS 操作，同时结合自旋操作来保证原子性。

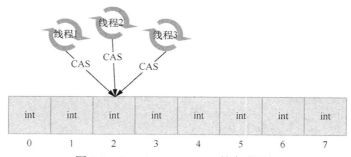

图 7.7　AtomicIntegerArray 的实现思想

接着看 AtomicIntegerArray 的源码实现核心（见代码清单 7.14），JDK 提供的 AtomicIntegerArray 类位于 java.util.concurrent.atomic 包下。构造函数的参数可为数组长度或者数组对象，对于数组长度则按照长度创建数组，而对于数组对象则直接复制一个副本。除此之外，其他的核心方法都通过 VarHandle 对象来完成。该类中数组对应 array 变量，而 AA 则是 VarHandle 对象，通过 VarHandle 对象能够实现各种原子操作。

代码清单 7.14　AtomicIntegerArray 的源码实现核心

```java
1.  public class AtomicIntegerArray {
2.      private static final VarHandle AA = MethodHandles.arrayElementVarHandle(int[].class);
3.      private final int[] array;
4.
5.      public AtomicIntegerArray(int length) {
6.          array = new int[length];
7.      }
8.
9.      public AtomicIntegerArray(int[] array) {
10.         this.array = array.clone();
11.     }
12.
13.     public final int get(int i) {
14.         return (int) AA.getVolatile(array, i);
15.     }
16.
17.     public final boolean compareAndSet(int i, int expectedValue, int newValue) {
18.         return AA.compareAndSet(array, i, expectedValue, newValue);
19.     }
20.
21.     public final int incrementAndGet(int i) {
22.         return (int) AA.getAndAdd(array, i, 1) + 1;
```

```
23.     }
24.
25.     public final int decrementAndGet(int i) {
26.         return (int) AA.getAndAdd(array, i, -1) - 1;
27.     }
28.
29. }
```

7.3.2 AtomicLongArray

掌握了 AtomicIntegerArray 的原理后，AtomicLongArray 和 AtomicReferenceArray 就很简单了，它们的基本原理都类似。AtomicLongArray 的核心方法实现如代码清单 7.15 所示。可以看到它与 AtomicIntegerArray 几乎一样，只是类型变成了长整型（见图 7.8）。

代码清单 7.15　AtomicLongArray 的核心方法实现

```
1.  public class AtomicLongArray {
2.      private static final VarHandle AA = MethodHandles.arrayElementVarHandle(long[].class);
3.      private final long[] array;
4.
5.      public AtomicLongArray(int length) {
6.          array = new long[length];
7.      }
8.
9.      public AtomicLongArray(long[] array) {
10.         this.array = array.clone();
11.     }
12.
13.     public final long get(int i) {
14.         return (long) AA.getVolatile(array, i);
15.     }
16.
17.     public final boolean compareAndSet(int i, long expectedValue, long newValue) {
18.         return AA.compareAndSet(array, i, expectedValue, newValue);
19.     }
20.
21.     public final long incrementAndGet(int i) {
22.         return (long) AA.getAndAdd(array, i, 1L) + 1L;
23.     }
24.
25.     public final long decrementAndGet(int i) {
26.         return (long) AA.getAndAdd(array, i, -1L) - 1L;
27.     }
28.
29. }
```

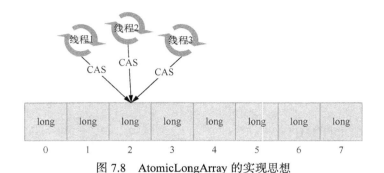

图 7.8 AtomicLongArray 的实现思想

7.3.3 AtomicReferenceArray

接着看 AtomicReferenceArray 的核心方法实现，如代码清单 7.16 所示。此时数组对应的是 Object 类型，通过泛型来声明数组的类型。构造函数可以传入整型或泛型指定的数组，使用 Arrays.copyOf 对数组进行复制。compareAndSet 是核心方法，它是通过对象地址来进行 CAS（见图 7.9），这在上一节的 AtomicReference 中已经分析过，这里不再赘述。

代码清单 7.16　AtomicReferenceArray 的核心方法实现

```
1.  public class AtomicReferenceArray<E> {
2.      private static final VarHandle AA = MethodHandles.arrayElementVarHandle
   (Object[].class);
3.      private final Object[] array;
4.
5.      public AtomicReferenceArray(int length) {
6.          array = new Object[length];
7.      }
8.
9.      public AtomicReferenceArray(E[] array) {
10.         this.array = Arrays.copyOf(array, array.length, Object[].class);
11.     }
12.
13.     public final E get(int i) {
14.         return (E) AA.getVolatile(array, i);
15.     }
16.
17.     public final boolean compareAndSet(int i, E expectedValue, E newValue) {
18.         return AA.compareAndSet(array, i, expectedValue, newValue);
19.     }
20.
21. }
```

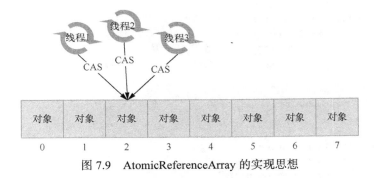

图 7.9 AtomicReferenceArray 的实现思想

本节主要对原子数组进行了分析，包括 AtomicIntegerArray、AtomicLongArray 和 AtomicReferenceArray 这 3 个类，分别对应的是整型原子数组、长整型原子数组和对象引用原子数组。

7.4 原子变量更新器

本节介绍 AtomicIntegerFieldUpdater、AtomicLongFieldUpdater 以及 AtomicReferenceFieldUpdater 这 3 个原子类，我们称为原子变量更新器，如图 7.10 所示。它们的功能与 AtomicInteger、AtomicLong 和 AtomicReference 这 3 个原子变量类的功能是一样的，都是提供 CAS+自旋模式的原子更新操作。但是，在有了原子变量类后为什么还要原子变量更新器呢？原子变量更新器的主要优势是能够避免或者减少对变量所在类代码的修改。比如，要对 A 类中的某个变量进行原子操作，那么通过原子变量更新器则可以不用对 A 类进行修改，因此侵入性小。这得益于原子变量更新器使用了反射的机制。

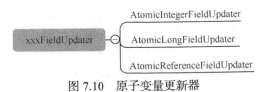

图 7.10 原子变量更新器

我们从代码清单 7.16 所示的示例开始，分别使用原子变量类和原子变量更新器对整型变量进行改造，从而实现变量的原子操作。先看不具备原子性的情况，如代码清单 7.17 所示。此时创建的 Counter 类是非线程安全的，当 10 个线程并发地执行 count++操作时，某次执行的结果为 count = 48250。这个结果明显是错误的。

代码清单 7.17　不具备原子性的代码

```
1.  public class AtomicIntegerFieldUpdaterDemo {
2.      public static void main(String[] args) throws InterruptedException {
3.          Counter counter = new Counter();
```

```
4.         for (int i = 0; i < 10; i++)
5.             new Thread(() -> {
6.                 for (int j = 0; j < 10000; j++)
7.                     counter.count++;
8.             }).start();
9.         Thread.sleep(3000);
10.        System.out.println("count = " + counter.getCount());
11.    }
12.
13.    static class Counter {
14.        volatile int count = 0;
15.        public int getCount() {
16.            return count;
17.        }
18.    }
19. }
```

现在先使用原子变量类对其进行改造,这时要将 Counter 类的 count 变量改为 AtomicInteger 对象,同时还要额外提供 atomicAdd 原子操作方法,如代码清单 7.18 所示。再次运行该程序,每次的执行结果都为 count = 100000,这说明该操作已经满足原子性,这段代码是线程安全的。

代码清单 7.18 使用原子变量类进行改造后的代码

```
1.  public class AtomicIntegerFieldUpdaterDemo2 {
2.      public static void main(String[] args) throws InterruptedException {
3.          Counter counter = new Counter();
4.          for (int i = 0; i < 10; i++)
5.              new Thread(() -> {
6.                  for (int j = 0; j < 10000; j++)
7.                      counter.atomicAdd(1);
8.              }).start();
9.          Thread.sleep(3000);
10.         System.out.println("count = " + counter.getCount());
11.     }
12.
13.     static class Counter {
14.         AtomicInteger count = new AtomicInteger(0);
15.
16.         public int getCount() {
17.             return count.get();
18.         }
19.
20.         public void atomicAdd(int delta) {
21.             count.getAndAdd(delta);
22.         }
23.     }
24. }
```

接着看使用原子变量更新器的改造方案,如代码清单 7.19 所示。可以看到,这里没有对 Counter 类做任何改造,仅仅是在使用的时候通过 AtomicIntegerFieldUpdater 对 Counter 类进行包装，从而实现原子操作。具体实现是通过 AtomicIntegerFieldUpdater.newUpdater 方法创建 AtomicIntegerFieldUpdater 对象，创建时需要传入待原子化的类以及变量名，然后通过调用 AtomicIntegerFieldUpdater 的 incrementAndGet 方法便能够实现原子累加操作。

代码清单 7.19　使用原子变量更新器改造后的代码

```java
1.   public class AtomicIntegerFieldUpdaterDemo3 {
2.       public static void main(String[] args) throws InterruptedException {
3.           AtomicIntegerFieldUpdater<Counter> countFieldUpdater = AtomicIntegerFieldUpdater
4.                   .newUpdater(Counter.class, "count");
5.           Counter test = new Counter();
6.           for (int i = 0; i < 10; i++)
7.               new Thread(() -> {
8.                   for (int j = 0; j < 10000; j++)
9.                       countFieldUpdater.incrementAndGet(test);
10.              }).start();
11.          Thread.sleep(3000);
12.          System.out.println("count = " + test.getCount());
13.      }
14.
15.      static class Counter {
16.          static volatile int count = 0;
17.          public int getCount() {
18.              return count;
19.          }
20.      }
21.  }
```

原子变量更新器的实现思想的本质与其他的原子类是一样的，它们都是通过 CAS+自旋的方式来实现原子操作。但是原子变量更新器与其他原子类不同的地方在于，它使用了反射机制，而反射能够减少代码的入侵。如图 7.11 所示，AtomicXXXFieldUpdater 是原子变量更新器，假如有 3 个线程通过它对某个 Java 对象进行原子操作，其实就是进行 CAS 和自旋操作，而对 Java 对象的访问则通过反射进行。

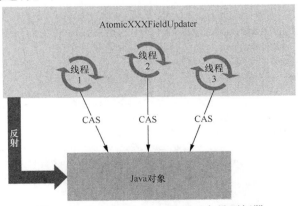

图 7.11　使用了反射机制的原子变量更新器

JDK 提供的原子变量更新器位于 java.util.concurrent.atomic 包下，一共有 3 个类，分别是 AtomicIntegerFieldUpdater、AtomicLongFieldUpdater 以及 AtomicReferenceFieldUpdater。鉴于这三者的实现比较类似，我们以 AtomicIntegerFieldUpdater 为例进行分析，如代码清单 7.20 所示。

代码清单 7.20　AtomicIntegerFieldUpdater 类

```java
 1.  public abstract class AtomicIntegerFieldUpdater<T> {
 2.      private static final Unsafe U;
 3.
 4.      static {
 5.          try {
 6.              Field theUnsafeInstance = Unsafe.class.getDeclaredField("theUnsafe");
 7.              theUnsafeInstance.setAccessible(true);
 8.              U = (Unsafe) theUnsafeInstance.get(Unsafe.class);
 9.          } catch (ReflectiveOperationException e) {
10.              throw new Error(e);
11.          }
12.      }
13.
14.      public static <U> AtomicIntegerFieldUpdater<U> newUpdater(Class<U> tclass, String fieldName) {
15.          return new AtomicIntegerFieldUpdaterImpl<U>(tclass, fieldName);
16.      }
17.
18.      public abstract boolean compareAndSet(T obj, int expect, int update);
19.
20.      public abstract int get(T obj);
21.
22.      public int incrementAndGet(T obj) {
23.          int prev, next;
24.          do {
25.              prev = get(obj);
26.              next = prev + 1;
27.          } while (!compareAndSet(obj, prev, next));
28.          return next;
29.      }
30.
31.      private static final class AtomicIntegerFieldUpdaterImpl<T>
32.              extends AtomicIntegerFieldUpdater<T> {
33.          private final long offset;
34.          final int modifiers;
35.
36.          AtomicIntegerFieldUpdaterImpl(final Class<T> tclass, final String fieldName) {
37.              final Field field;
38.              try {
39.                  field = AccessController.doPrivileged(new PrivilegedExceptionAction<Field>() {
40.                      public Field run() throws NoSuchFieldException {
41.                          return tclass.getDeclaredField(fieldName);
42.                      }
43.                  });
44.                  modifiers = field.getModifiers();
```

```
45.            } catch (Exception ex) {
46.                throw new RuntimeException(ex);
47.            }
48.            if (!Modifier.isVolatile(modifiers))
49.                throw new IllegalArgumentException("Must be volatile type");
50.            this.offset = U.objectFieldOffset(field);
51.        }
52.
53.        public final boolean compareAndSet(T obj, int expect, int update) {
54.            return U.compareAndSwapInt(obj, offset, expect, update);
55.        }
56.
57.        public final int get(T obj) {
58.            return U.getIntVolatile(obj, offset);
59.        }
60.
61.    }
62. }
```

首先可以看到 AtomicIntegerFieldUpdater 通过泛型来声明需要封装的类型，然后还有一个提供 CAS 操作的 Unsafe 对象，其中 static 块用于创建 Unsafe 对象。需要注意的是，AtomicIntegerFieldUpdater 是一个抽象类，该类通过 newUpdater 方法构建对象（真正创建的是 AtomicIntegerFieldUpdaterImpl 对象），它是具体的实现类。最后还声明了 compareAndSet 和 get 两个抽象方法，它们由实现类去实现。此外，incrementAndGet 方法为默认的实现，在实现类中可以重写该方法，也可以直接使用该方法。

具体的实现类为 AtomicIntegerFieldUpdaterImpl，该类属于 AtomicIntegerFieldUpdater 的内部类，而且还继承了 AtomicIntegerFieldUpdater 类。AtomicIntegerFieldUpdater 构造函数中传入了类名和字段名，表示要对哪个类的哪个字段进行操作，这些都是使用反射时必须知道的信息。构造函数中还会判断变量的声明，如果不为 volatile 则会报错。Unsafe 的 objectFieldOffset 方法则用于获取变量的偏移。而 compareAndSet 方法和 get 方法的实现都是间接调用了 Unsafe 的方法，这里不再赘述。

原子变量更新器对变量的操作是有要求的，不管是 AtomicIntegerFieldUpdater、AtomicLongFieldUpdater，还是 AtomicReferenceFieldUpdater，都需要满足以下 3 个条件才能正常执行。

- 待操作的变量必须对原子变量更新器可见。因为原子变量更新器是通过反射来得到该变量的，因此不可见会导致反射失败。比如变量声明为 private，则 AtomicXXXFieldUpdater 会执行失败。
- 待操作的变量必须被声明为 volatile，这用于保证变量的可见性。如果不声明为 volatile，则 AtomicXXXFieldUpdater 会报错。
- 待操作的变量不能声明为 static，这是因为 AtomicXXXFieldUpdater 在进行 CAS 操作时是通过 Unsafe 类进行的，Unsafe 类的 objectFieldOffset 用于获取变量的偏移，而该方法不支持静态变量。如果声明为 static，则 AtomicXXXFieldUpdater 会报错。

第 8 章

阻塞队列

8.1 阻塞队列概述

在分析阻塞队列（BlockingQueue）之前，先看一下生产者/消费者模式。这是一个常见的模式，生产者负责数据的生产，而消费者则负责数据的消费。一般来说，生产者与消费者的数量比例是 $m:n$，该模式最大的好处就是将数据生产方与消费方进行解耦，使得它们互相之间不会影响。为了将生产者和消费者连接起来，我们需要一个特殊的容器，该容器能存储生产者生产的数据，而消费者则能从该容器中取出数据。

我们可以通过厨师、桌子、顾客来说明生产者/消费者模式，如图 8.1 所示。厨师就好比生产者，他们生产出美食并放到桌子这个容器中；顾客则好比消费者，他们从桌子上获取美食进行消费享用。

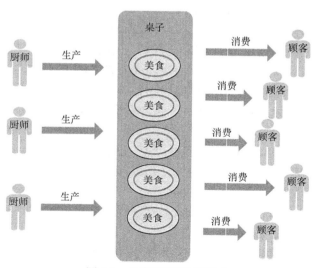

图 8.1 生产者/消费者模式

生产者/消费者模式的核心部分就是生产者和消费者之间的那个特殊容器。我们通过一个线程安全且具有一定策略的容器来连接生产者和消费者。这个容器可以具有队列性质，也可以具有栈性质，抑或是其他数据结构。最常见的就是阻塞队列——队列能保证先进的数据先出，而阻塞则是队列已满时或队列为空时的处理策略。

图 8.2 所示为阻塞队列工作示意图。线程 1、线程 2、线程 3 生产的数据通过 put 操作入队，线程 4、线程 5 通过 take 操作出队，当队列满时 put 操作会产生阻塞直到消费者将队列的元素拿走，当队列为空时 take 操作会产生阻塞直到生产者将数据入队。

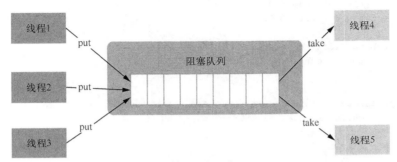

图 8.2　阻塞队列工作示意图

根据前面对阻塞队列的介绍，我们尝试模拟实现一个简单的阻塞队列。先看数据结构的设计。可以使用一个数组来存放队列的元素，并通过 head 和 tail 指针来约束先进先出的规则。入队操作使用 tail 指向的位置，而出队操作则使用 head 指向的位置，且指针一旦到达数组尾部就重新从头开始，如图 8.3 所示。

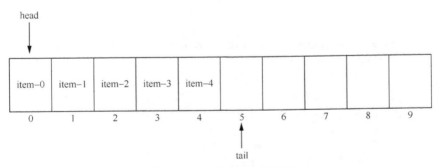

图 8.3　一个简单的阻塞队列

下面看看具体的实现，如代码清单 8.1 所示。其中，Object 数组用于保存元素，size 表示队列的大小，此外还有 head 和 tail 指针。通过构造函数可指定阻塞队列的大小。生产者生产的数据调用 put 方法进行入队，如果 size 等于队列最大长度，则调用 wait 阻塞（此时队列已经满了），否则将元素保存到队列中，同时维护 size 和 tail。最后，如果 size 等于 1，要调用 notifyAll 方法通知消费者可以进行消费。消费者通过调用 take 方法消费数据，如果 size 等于 0，则调用 wait 阻塞（队列为空，需要等待），否则通过 head 获取队列头的元素，同时维护 size 和 head。最后，如果 size 等于 queue.length-1，调用 notifyAll 方法通知生产者可以进行生产。

代码清单 8.1　简单阻塞队列的代码实现

```
1.   public class BlockingQueueSimulater {
2.
3.       Object[] queue;
4.       public int size;
5.       private int head = 0;
6.       private int tail = 0;
7.
8.       public BlockingQueueSimulater(int maxSize) {
9.           this.queue = new Object[maxSize];
10.      }
11.
12.      public synchronized void put(Object elem) throws InterruptedException {
13.          while (size == queue.length) {
14.              wait();
15.          }
16.          size++;
17.          queue[tail] = elem;
18.          if (tail == queue.length - 1)
19.              tail = 0;
20.          else
21.              tail++;
22.          if (size == 1) {
23.              notifyAll();
24.          }
25.      }
26.
27.      public synchronized Object take() throws InterruptedException {
28.          while (size == 0) {
29.              wait();
30.          }
31.          size--;
32.          Object elem = queue[head];
33.          queue[head] = null;
34.          if (head == queue.length - 1)
35.              head = 0;
36.          else
37.              head++;
38.          if (size == queue.length - 1) {
39.              notifyAll();
40.          }
41.          return elem;
42.      }
43.
44.  }
```

下面通过代码清单 8.2 来看模拟阻塞队列。

代码清单 8.2　模拟阻塞队列

```
1.  public class BlockingQueueSimulaterDemo {
2.      public static void main(String[] args) throws InterruptedException {
3.          BlockingQueueSimulater bq = new BlockingQueueSimulater(10);
4.          new Thread(() -> {
5.              try {
6.                  System.out.println("take " + bq.take() + " from queue!");
7.                  Thread.sleep(2000);
8.                  System.out.println("queue size : " + bq.size);
9.                  int size = bq.size;
10.                 for (int i = 0; i < size; i++)
11.                     System.out.print(bq.take() + "|");
12.             } catch (InterruptedException e) {
13.             }
14.         }).start();
15.
16.         Thread.sleep(2000);
17.
18.         new Thread(() -> {
19.             try {
20.                 for (int i = 0; i < 10; i++)
21.                     bq.put("item-" + i);
22.                 System.out.println("10 items is produced!");
23.             } catch (InterruptedException e) {
24.             }
25.         }).start();
26.     }
27. }
```

输出结果如下，一个线程进行数据生产，另一个线程进行数据消费。

```
1.  take item-0 from queue!
2.  10 items is produced!
3.  queue size : 9
4.  item-1|item-2|item-3|item-4|item-5|item-6|item-7|item-8|item-9|
```

接下来看一下 JDK 的阻塞队列接口 BlockingQueue（见代码清单 8.3），该接口早在 Java 5 时就已经引入。JDK 提供了多种阻塞队列的实现，比如 ArrayBlockingQueue、LinkedBlockingQueue、PriorityBlockingQueue 等，它们都实现了 BlockingQueue 接口。该接口主要的核心方法都支持中断，其中 put 方法和 take 方法我们已经很熟悉，另外 offer 和 poll 两个方法对应的是 put 和 take 两个方法，只是前者提供了阻塞超时机制。

代码清单 8.3　BlockingQueue 接口的实现

```
1.  public interface BlockingQueue<E> {
2.      void put(E e) throws InterruptedException;
3.      E take() throws InterruptedException;
```

```
4.      boolean offer(E e, long timeout, TimeUnit unit) throws InterruptedException;
5.      E poll(long timeout, TimeUnit unit) throws InterruptedException;
6.      int size();
7.  }
```

8.2 数组阻塞队列

数组阻塞队列 ArrayBlockingQueue 位于 java.util.concurrent 包下，从名字可以看出它的存储结构是数组，即基于数组实现了一个 FIFO 的阻塞队列，如图 8.4 所示。在数组阻塞队列中，新元素插入队列尾部，最先入队的元素在队列头部，而最后入队的元素则在队列尾部。由于数组是有界的，所以构造时需要指定数组的大小。此外，该阻塞队列还提供公平和非公平两种模式。

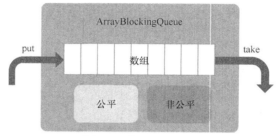

图 8.4　数组阻塞队列

在分析 ArrayBlockingQueue 的实现原理之前，我们先通过几个例子来了解它的使用。在代码清单 8.4 所示的示例 1 中，创建了一个 ArrayBlockingQueue 对象，然后通过 5 个线程去生产数据并 put 到阻塞队列，而主线程则充当消费者，不断将阻塞队列中的数据取出消费。

代码清单 8.4　数组阻塞队列使用示例 1

```
1.  public class BlockingQueueDemo {
2.      static BlockingQueue<String> blockingQueue = new ArrayBlockingQueue<String>(10);
3.      public static void main(String[] args) throws InterruptedException {
4.          for (int i = 0; i < 5; i++) {
5.              final int index = i;
6.              new Thread(() -> {
7.                  try {
8.                      System.out.println("producing string" + index);
9.                      blockingQueue.put("string" + index);
10.                 } catch (InterruptedException e) {
11.                 }
12.             }).start();
13.         }
14.         Thread.sleep(2000);
15.         int blockingQueueSize = blockingQueue.size();
```

```
16.        System.out.println("the size of blocking queue is " + blockingQueueSize);
17.        for (int i = 0; i < blockingQueueSize; i++)
18.            System.out.println(blockingQueue.take());
19.    }
20. }
```

代码清单 8.4 的输出结果如下。

```
1.  producing string3
2.  producing string2
3.  producing string0
4.  producing string4
5.  producing string1
6.  the size of blocking queue is 5
7.  string0
8.  string3
9.  string2
10. string4
11. string1
```

代码清单 8.5 所示的示例 2 主要展示了阻塞队列空间已满而产生阻塞的情况。首先创建一个最大长度为 4 的阻塞队列，然后通过 5 个线程各自生产一个数据 put 到阻塞队列中，但由于阻塞队列最大长度为 4，所以在 put 这 4 个数据后会产生阻塞。

代码清单 8.5　数组阻塞队列使用示例 2

```
1.  public class BlockingQueueDemo2 {
2.      static BlockingQueue<String> blockingQueue = new ArrayBlockingQueue<String>(4);
3.      public static void main(String[] args) throws InterruptedException {
4.          for (int i = 0; i < 5; i++) {
5.              final int index = i;
6.              new Thread(() -> {
7.                  try {
8.                      blockingQueue.put("string" + index);
9.                      System.out.println("produced string" + index);
10.                 } catch (InterruptedException e) {
11.                 }
12.             }).start();
13.         }
14.     }
15. }
```

代码清单 8.5 的输出结果如下。

```
1.  produced string3
2.  produced string0
3.  produced string4
4.  produced string2
```

代码清单 8.6 所示的示例 3 主要展示了阻塞队列为空时产生阻塞的情况。首先创建一个最大长度为 10 的阻塞队列，但没有生产者生产数据，此时阻塞队列为空，所以调用 take 方法时产生了阻塞。

代码清单 8.6　数组阻塞队列使用示例 3

```java
1.  public class BlockingQueueDemo3 {
2.      static BlockingQueue<String> blockingQueue = new ArrayBlockingQueue<String>(10);
3.      public static void main(String[] args) throws InterruptedException {
4.          System.out.println(blockingQueue.take());
5.      }
6.  }
```

我们先看 ArrayBlockingQueue 类的属性和构造函数，如代码清单 8.7 所示。其中 count 变量表示队列的长度大小，takeIndex 和 putIndex 分别表示队列入队和出队的索引，items 是用于保存队列元素的 Object 数组，lock 是并发控制锁，notEmpty 和 notFull 分别表示队列非空时和非满时的条件。ArrayBlockingQueue 类提供两个构造函数，第一个构造函数只需指定队列的大小，另一个构造函数则需要分别指定队列大小和公平模式。锁对象使用的是 ReentrantLock 对象，它负责公平机制。notEmpty 和 notFull 条件对象通过锁对象的 newCondition 来创建。

代码清单 8.7　ArrayBlockingQueue 类的属性和构造函数

```java
1.  public class ArrayBlockingQueue<E> implements BlockingQueue<E> {
2.  
3.      int count;
4.      int takeIndex;
5.      int putIndex;
6.      final Object[] items;
7.      final ReentrantLock lock;
8.      private final Condition notEmpty;
9.      private final Condition notFull;
10. 
11.     public ArrayBlockingQueue(int capacity) {
12.         this(capacity, false);
13.     }
14. 
15.     public ArrayBlockingQueue(int capacity, boolean fair) {
16.         if (capacity <= 0)
17.             throw new IllegalArgumentException();
18.         this.items = new Object[capacity];
19.         lock = new ReentrantLock(fair);
20.         notEmpty = lock.newCondition();
21.         notFull = lock.newCondition();
22.     }
23. 
24.     ......
25. 
26. }
```

enqueue 和 dequeue 这两个方法是该类内部使用的方法，负责维护数组的队列功能，如代码清单 8.8 所示。enqueue 是入队操作，通过 putIndex 索引定位插入的位置。如果索引超过了队列的最大长度，则索引从头开始，同时让长度加 1 并向 notEmpty 条件发送信号。类似地，dequeue 是出队操作，通过 takeIndex 索引定位取出的位置。如果索引超过了队列的最大长度，则索引也需要从头开始，同时让长度减 1 并向 notFull 条件发送信号。

代码清单 8.8　enqueue 和 dequeue 方法

```
1.  private void enqueue(E e) {
2.      final Object[] items = this.items;
3.      items[putIndex] = e;
4.      if (++putIndex == items.length)
5.          putIndex = 0;
6.      count++;
7.      notEmpty.signal();
8.  }
9.
10. private E dequeue() {
11.     final Object[] items = this.items;
12.     E e = (E) items[takeIndex];
13.     items[takeIndex] = null;
14.     if (++takeIndex == items.length)
15.         takeIndex = 0;
16.     count--;
17.     notFull.signal();
18.     return e;
19. }
```

put 和 take 方法是阻塞队列的入队和出队方法，它们会间接调用 enqueue 和 dequeue 方法，如代码清单 8.9 所示。其中 put 方法会检查入队的元素（该元素不能为 null），然后会先获取锁再执行 enqueue 方法维护数组。如果数组长度已经达到最大，则调用 notFull 条件的 await 方法等待。take 方法会先获取锁后才执行 dequeue 方法来维护数组。如果数组长度为 0，则调用 notEmpty 条件的 await 方法等待。

代码清单 8.9　put 和 take 方法

```
1.  public void put(E e) throws InterruptedException {
2.      Objects.requireNonNull(e);
3.      final ReentrantLock lock = this.lock;
4.      lock.lockInterruptibly();
5.      try {
6.          while (count == items.length)
7.              notFull.await();
8.          enqueue(e);
9.      } finally {
10.         lock.unlock();
11.     }
```

```
12.     }
13.
14.     public E take() throws InterruptedException {
15.         final ReentrantLock lock = this.lock;
16.         lock.lockInterruptibly();
17.         try {
18.             while (count == 0)
19.                 notEmpty.await();
20.             return dequeue();
21.         } finally {
22.             lock.unlock();
23.         }
24.     }
```

offer 和 poll 方法是支持超时的阻塞队列的入队和出队方法，如代码清单 8.10 所示。它们主要的区别就在于分别调用了 notFull 和 notEmpty 条件的 awaitNanos 方法进行等待。入队时如果超过指定的时间，则返回 false，表示入队超时导致失败。而出队时如果超过指定的时间，则返回 null，表示出队超时导致失败。

代码清单 8.10 offer 和 poll 方法

```
1.  public boolean offer(E e, long timeout, TimeUnit unit) throws InterruptedException {
2.      Objects.requireNonNull(e);
3.      long nanos = unit.toNanos(timeout);
4.      final ReentrantLock lock = this.lock;
5.      lock.lockInterruptibly();
6.      try {
7.          while (count == items.length) {
8.              if (nanos <= 0L)
9.                  return false;
10.             nanos = notFull.awaitNanos(nanos);
11.         }
12.         enqueue(e);
13.         return true;
14.     } finally {
15.         lock.unlock();
16.     }
17. }
18.
19. public E poll(long timeout, TimeUnit unit) throws InterruptedException {
20.     long nanos = unit.toNanos(timeout);
21.     final ReentrantLock lock = this.lock;
22.     lock.lockInterruptibly();
23.     try {
24.         while (count == 0) {
25.             if (nanos <= 0L)
26.                 return null;
27.             nanos = notEmpty.awaitNanos(nanos);
28.         }
29.         return dequeue();
```

```
30.        } finally {
31.            lock.unlock();
32.        }
33.    }
```

8.3 链表阻塞队列

下面继续分析 JDK 提供的另外一种阻塞队列的实现——链表阻塞队列 LinkedBlockingQueue。该类位于 java.util.concurrent 包下，如图 8.5 所示。从名字可以看出它的存储结构是一个链表，即基于链表实现了一个 FIFO 的阻塞队列。得益于链表的先天优势，构造阻塞队列时我们可以不用指定队列的长度，这也是链表结构相较于数组结构的优势。所以链表阻塞队列可以看成无界的（实际上是整型最大值 Integer.MAX_VALUE）。LinkedBlockingQueue 不提供公平模式，只提供了非公平模式。

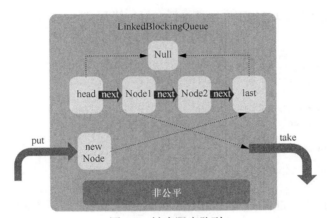

图 8.5 链表阻塞队列

LinkedBlockingQueue 的用法与 ArrayBlockingQueue 十分相似，下面通过代码清单 8.11 来了解 LinkedBlockingQueue 的使用。注意在代码清单 8.11 中，在创建 LinkedBlockingQueue 对象时可以不传入队列长度，它会使用 Integer.MAX_VALUE 作为默认值。然后通过 5 个线程去生产数据并 put 到阻塞队列，每个线程都会循环产生 1 000 个字符串。最终再调用 3 次 take 方法取出数据进行消费。

代码清单 8.11　LinkedBlockingQueue 的使用

```
1. public class LinkedBlockingQueueDemo {
2.
3.     static BlockingQueue<String> blockingQueue = new LinkedBlockingQueue<String>();
4.
5.     public static void main(String[] args) throws InterruptedException {
6.         for (int i = 0; i < 5; i++) {
```

```
7.              final int index = i;
8.              new Thread(() -> {
9.                  System.out.println("thread_" + index + " producing 1000 strings");
10.                 for (int j = 0; j < 1000; j++)
11.                     try {
12.                         blockingQueue.put("thread_" + index + "_" + j);
13.                     } catch (InterruptedException e) {
14.                     }
15.             }).start();
16.         }
17.         Thread.sleep(2000);
18.         int blockingQueueSize = blockingQueue.size();
19.         System.out.println("the size of blocking queue is " + blockingQueueSize);
20.         System.out.println(blockingQueue.take());
21.         System.out.println(blockingQueue.take());
22.         System.out.println(blockingQueue.take());
23.     }
24. }
```

代码清单 8.11 的某次输出如下。也就是说，每个线程都产生 1 000 个数据，阻塞队列的总长度为 5 000。

```
1. thread_1 producing 1000 strings
2. thread_3 producing 1000 strings
3. thread_0 producing 1000 strings
4. thread_2 producing 1000 strings
5. thread_4 producing 1000 strings
6. the size of blocking queue is 5000
7. thread_1_0
8. thread_1_1
9. thread_1_2
```

在继续分析 LinkedBlockingQueue 的实现原理之前，我们先看链表结构实现的队列是如何进行入队和出队的。入队操作如图 8.6 所示。链表结构中的节点通过一个 next 引用将所有节点链接起来，链表中包含 head 和 last 这两个特殊的节点，分别表示头节点和尾节点。链表中使用了一个特殊的 null 节点，它不保存数据，仅仅用于辅助工作，开始时，head 节点和 last 节点都指向这个 null 节点。假如有一个数据要执行入队操作，那么将新建一个 Node1 节点，然后将 last 节点（此时为 null 节点）的 next 指向 Node1 节点，最后将 last 节点也指向 Node1 节点。类似地，如果再有一个新数据执行入队操作，则创建 Node2 节点后将 last 节点（此时为 Node1 节点）的 next 指向 Node2 节点，并将 last 节点也指向 Node2 节点。

出队操作的核心步骤是先找到 head 节点所指节点的 next 节点，然后将 head 节点指向该 next 节点。如图 8.7 所示，原来有一个 Node1 和 Node2 组成的链表队列，当要执行出队操作时，只需找到 head 节点所指节点的 next 节点（即 Node1 节点），取出后将 head 节点指向 Node1 节点即可。类似地，再次执行出队操作则时，找到 head 节点所指节点的 next 节点（此时为 Node2 节点），然后将 head 节点指向 Node2 节点。

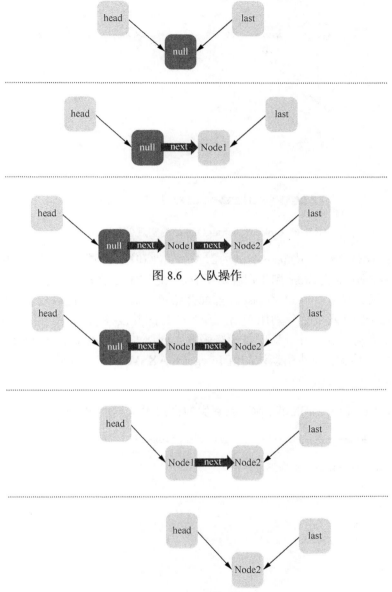

图 8.6　入队操作

图 8.7　出队操作

前面讲到，ArrayBlockingQueue 的内部实现只使用了一个锁，而 LinkedBlockingQueue 的实现则使用了两个锁。那么为什么要使用两个锁呢？实际上这是一种提升并发性能的措施。队列的入队操作和出队操作分别使用一个锁能够减少竞争，相当于将生产者和消费者分开竞争。生产者线程执行入队操作时使用 putLock 锁，而消费者线程执行出队操作时使用 takeLock 锁，如图 8.8 所示。这就是使用两个锁的主要原因。

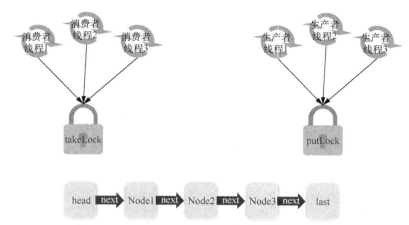

图 8.8　LinkedBlockingQueue 的实现使用了两个锁

下面先看 LinkedBlockingQueue 类的内部类、属性和构造函数，如代码清单 8.12 所示。其中定义了一个 Node 内部类，item 用于保存数据，next 的作用在前面介绍过。接着是属性，capacity 表示链表队列的最大长度，head 和 last 在前面介绍过，count 用于记录链表队列当前的大小，takeLock 和 putLock 这两个锁分别用于出队和入队时的并发控制，notEmpty 和 notFull 分别表示队列非空时和非满时的条件。最后是构造函数，创建 LinkedBlockingQueue 对象时可以指定队列大小，也可以不指定而使用默认的值 Integer.MAX_VALUE。此外还可以看到 head 和 last 在初始时都指向同一个 null 节点。

代码清单 8.12　LinkedBlockingQueue 类的内部类、属性和构造函数

```
1.   public class LinkedBlockingQueue<E> implements BlockingQueue<E> {
2.
3.       static class Node<E> {
4.           E item;
5.           Node<E> next;
6.
7.           Node(E x) {
8.               item = x;
9.           }
10.      }
11.
12.      private final int capacity;
13.      transient Node<E> head;
14.      private transient Node<E> last;
15.      private final AtomicInteger count = new AtomicInteger();
16.      private final ReentrantLock takeLock = new ReentrantLock();
17.      private final Condition notEmpty = takeLock.newCondition();
18.      private final ReentrantLock putLock = new ReentrantLock();
19.      private final Condition notFull = putLock.newCondition();
20.
```

```
21.    public LinkedBlockingQueue() {
22.        this(Integer.MAX_VALUE);
23.    }
24.
25.    public LinkedBlockingQueue(int capacity) {
26.        if (capacity <= 0)
27.            throw new IllegalArgumentException();
28.        this.capacity = capacity;
29.        last = head = new Node<E>(null);
30.    }
31.
32.    ......
33.
34. }
```

enqueue 和 dequeue 这两个方法是该类内部使用的方法，负责维护链表的队列功能，如代码清单 8.13 所示。enqueue 是入队操作，逻辑很简单，即将 last 节点的 next 指向新节点，last 节点也指向新节点。类似地，dequeue 是出队操作，它获取 head 节点的 next 所指向的节点，该节点为队列头节点 first，first 节点的 item 是保存的数据。注意 h.next=h 是为了帮助垃圾回收器进行垃圾回收，省略掉也是符合链表逻辑的。

代码清单 8.13　enqueue 和 dequeue 方法

```
1.    private void enqueue(Node<E> node) {
2.        last = last.next = node;
3.    }
4.
5.    private E dequeue() {
6.        Node<E> h = head;
7.        Node<E> first = h.next;
8.        h.next = h;
9.        head = first;
10.       E x = first.item;
11.       first.item = null;
12.       return x;
13.   }
```

put 方法是链表阻塞队列的入队方法，它会间接使用 enqueue 方法，如代码清单 8.14 所示。首先它会检查并确认入队的元素不能为 null，然后去获取 putLock 锁，成功获取后会检查队列当前长度是否已经达到最大长度，如果是，则通过 notFull 条件进入阻塞等待，否则执行入队操作并维护表示队列当前长度的 count 变量。在当前长度小于最大长度时，要向 notFull 条件发送信号，唤醒阻塞的线程。最后，如果原来队列长度为 0，则还要通知消费者的 takeLock 锁对应的 notEmpty 条件，因为队列长度为 0 时会导致消费者端的线程进入阻塞，而现在生产了一个数据，于是需要通知消费者可以消费了。

代码清单 8.14　put 方法

```
1.  public void put(E e) throws InterruptedException {
2.      if (e == null)
3.          throw new NullPointerException();
4.      final int c;
5.      final Node<E> node = new Node<E>(e);
6.      final ReentrantLock putLock = this.putLock;
7.      final AtomicInteger count = this.count;
8.      putLock.lockInterruptibly();
9.      try {
10.         while (count.get() == capacity) {
11.             notFull.await();
12.         }
13.         enqueue(node);
14.         c = count.getAndIncrement();
15.         if (c + 1 < capacity)
16.             notFull.signal();
17.     } finally {
18.         putLock.unlock();
19.     }
20.     if (c == 0) {
21.         final ReentrantLock takeLock = this.takeLock;
22.         takeLock.lock();
23.         try {
24.             notEmpty.signal();
25.         } finally {
26.             takeLock.unlock();
27.         }
28.     }
29. }
```

take 方法是链表阻塞队列的出队方法，它会间接调用 dequeue 方法，如代码清单 8.15 所示。首先它会先获取 takeLock 锁，成功获取后会检查队列当前长度是否为 0。如果是，则通过 notEmpty 条件进入阻塞等待，否则执行出队操作并维护表示队列当前长度的 count 变量。如果原来队列长度大于 1，则要向 notEmpty 条件发送信号，唤醒阻塞的线程。最后，如果原来队列长度等于最大长度，则要通知生产者的 putLock 对应的 notFull 条件，因为最大长度时会导致生产者端的线程进入阻塞，而被拿掉一个数据后就可以通知生产者继续生产数据了。

代码清单 8.15　take 方法

```
1.  public E take() throws InterruptedException {
2.      final E x;
3.      final int c;
4.      final AtomicInteger count = this.count;
5.      final ReentrantLock takeLock = this.takeLock;
6.      takeLock.lockInterruptibly();
7.      try {
```

```
8.          while (count.get() == 0) {
9.              notEmpty.await();
10.         }
11.         x = dequeue();
12.         c = count.getAndDecrement();
13.         if (c > 1)
14.             notEmpty.signal();
15.     } finally {
16.         takeLock.unlock();
17.     }
18.     if (c == capacity) {
19.         final ReentrantLock putLock = this.putLock;
20.         putLock.lock();
21.         try {
22.             notFull.signal();
23.         } finally {
24.             putLock.unlock();
25.         }
26.     }
27.     return x;
28. }
```

offer 方法是支持超时的链表阻塞队列的入队方法，如代码清单 8.16 所示。offer 方法与 put 方法的主要区别在于它调用了 notFull 条件的 awaitNanos 方法进入阻塞等待。如果等待超过指定的时间，则返回 false，表示入队操作因超时而导致失败。

代码清单 8.16　offer 方法

```
1.  public boolean offer(E e, long timeout, TimeUnit unit) throws InterruptedException {
2.      if (e == null)
3.          throw new NullPointerException();
4.      long nanos = unit.toNanos(timeout);
5.      final int c;
6.      final ReentrantLock putLock = this.putLock;
7.      final AtomicInteger count = this.count;
8.      putLock.lockInterruptibly();
9.      try {
10.         while (count.get() == capacity) {
11.             if (nanos <= 0L)
12.                 return false;
13.             nanos = notFull.awaitNanos(nanos);
14.         }
15.         enqueue(new Node<E>(e));
16.         c = count.getAndIncrement();
17.         if (c + 1 < capacity)
18.             notFull.signal();
19.     } finally {
20.         putLock.unlock();
21.     }
22.     if (c == 0) {
```

```
23.         final ReentrantLock takeLock = this.takeLock;
24.         takeLock.lock();
25.         try {
26.             notEmpty.signal();
27.         } finally {
28.             takeLock.unlock();
29.         }
30.     }
31.     return true;
32. }
```

poll 方法是支持超时的出队方法，如代码清单 8.17 所示。它与 take 方法的主要区别在于它调用了 notEmpty 条件的 awaitNanos 方法进入阻塞等待，如果超时，则返回 null，表示出队操作因超时而导致失败。

代码清单 8.17　poll 方法

```
1.  public E poll(long timeout, TimeUnit unit) throws InterruptedException {
2.      final E x;
3.      final int c;
4.      long nanos = unit.toNanos(timeout);
5.      final AtomicInteger count = this.count;
6.      final ReentrantLock takeLock = this.takeLock;
7.      takeLock.lockInterruptibly();
8.      try {
9.          while (count.get() == 0) {
10.             if (nanos <= 0L)
11.                 return null;
12.             nanos = notEmpty.awaitNanos(nanos);
13.         }
14.         x = dequeue();
15.         c = count.getAndDecrement();
16.         if (c > 1)
17.             notEmpty.signal();
18.     } finally {
19.         takeLock.unlock();
20.     }
21.     if (c == capacity) {
22.         final ReentrantLock putLock = this.putLock;
23.         putLock.lock();
24.         try {
25.             notFull.signal();
26.         } finally {
27.             putLock.unlock();
28.         }
29.     }
30.     return x;
31. }
```

8.4 优先级阻塞队列

优先级阻塞队列 PriorityBlockingQueue 是一种具备优先级功能的阻塞队列,它的内部存储核心结构是二叉堆,如图 8.9 所示。二叉堆包括最小堆和最大堆,最小堆能保证最顶层的元素为最小元素,而最大堆则相反。在理解优先级阻塞队列时,最重要的就是理解二叉堆结构,在新元素插入时会根据元素值的大小去调整二叉堆结构,从而保证堆顶一直为最大值或最小值。在出队时先拿掉堆顶最大或最小的元素,然后再调整二叉堆结构。由于二叉堆使用了数组来保存元素,而数组可以通过复制来扩容,所以 PriorityBlockingQueue 可以看成是无界的。此外,优先级阻塞队列的内部还包含一个比较器,二叉堆就是通过这个比较器构建起来的。关于二叉堆的详细介绍及构建,有兴趣的读者可以查看作者写作的《图解数据结构与算法》一书。

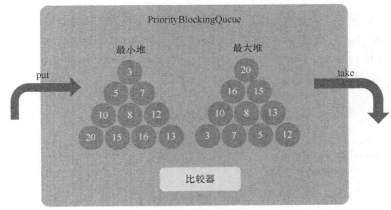

图 8.9 优先级阻塞队列

PriorityBlockingQueue 的使用可以分为两种情况:不提供自定义比较器和提供自定义比较器。我们先看不提供自定义比较器的情况,在这种情况下保存到队列中的对象必须实现 Comparable 接口。该接口定义了 compare 方法,队列则可以通过该方法来确定对象的优先级值。也就是说阻塞队列会通过该方法来维护二叉堆结构。该方法能比较当前对象与其他对象的大小,它可以返回负整数、0、正整数,分别表示当前对象小于、等于、大于指定的对象。

在代码清单 8.18 中定义了 Money 类,它实现了 Comparable 接口,compareTo 方法只返回 1 和-1。在 main 方法中将 1、100、2、20、10、5、50 这 7 张不同面额的币种依次加入到优先级队列中,然后输出为 "1|2|5|10|20|50|100|"。可以看到最终是按照币种面额从小到大输出的。

代码清单 8.18 不提供自定义比较器的情况

```
1.    public class PriorityBlockingQueueDemo {
2.        static class Money implements Comparable<Money> {
```

```
3.      private int value;
4.      Money(int v) {
5.          this.value = v;
6.      }
7.      public int compareTo(Money o) {
8.          return this.value > o.value ? 1 : -1;
9.      }
10. }
11.
12. public static void main(String[] args) throws InterruptedException {
13.     PriorityBlockingQueue<Money> queue = new PriorityBlockingQueue<Money>();
14.     int[] values = { 1, 100, 2, 20, 10, 5, 50 };
15.     for (int v : values)
16.         queue.put(new Money(v));
17.     for (int i = 0; i < values.length; i++)
18.         System.out.print(queue.take().value+"|");
19.     }
20. }
```

代码清单 8.19 提供了自定义比较器,此时我们的保存对象可以不必实现 Comparable,因为可以通过比较器对两个对象进行比较。比较器需要实现 Comparator 接口,并且重写里面的 compare 方法,这里只返回-1 和 1,分别表示第一个对象小于或大于第二个对象。最终程序输出结果与代码清单 8.18 一样。

代码清单 8.19 提供了自定义比较器的情况

```
1.  public class PriorityBlockingQueueDemo2 {
2.      static class Money {
3.          private int value;
4.          Money(int v) {
5.              this.value = v;
6.          }
7.      }
8.
9.      static class MoneyComparator implements Comparator<Money> {
10.         public int compare(Money o1, Money o2) {
11.             return o1.value < o2.value ? -1 : 1;
12.         }
13.     }
14.
15.     public static void main(String[] args) throws InterruptedException {
16.         PriorityBlockingQueue<Money> queue = new PriorityBlockingQueue<Money>(11, new MoneyComparator());
17.         int[] values = { 1, 100, 2, 20, 10, 5, 50 };
18.         for (int v : values)
19.             queue.put(new Money(v));
20.         for (int i = 0; i < values.length; i++)
21.             System.out.print(queue.take().value+"|");
22.     }
23. }
```

二叉堆有最大堆和最小堆两种。它是一棵完全二叉树，该树中的某个节点的值总是不大于（或不小于）其左右子节点的值，可以通过图 8.10 来理解。另外，为什么会使用数组来保存二叉堆呢？因为利用完全二叉树的性质，我们可以通过数组来表示完全二叉树。数组下标与完全二叉树节点存在映射关系，比如父节点可以通过 Math.floor((index-1)/2) 来获取，比如左子节点可以通过 2*index+1 来获取，比如右子节点可以通过 2*index+2 来获取。数组方式能简化实现及减少开销，从而避免使用额外的指针来实现树结构。

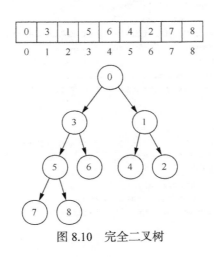

图 8.10　完全二叉树

二叉堆的性质包括以下两点：
- 二叉堆的树根节点的值是所有堆节点中的最小（大）值；
- 二叉堆树中每个节点的子树也都是最小（大）堆。

那么二叉堆有什么作用呢？以最小堆为例，它能保证堆顶元素为最小，而如果使用数组则无法达到该效果。如果数组要访问最小值，则需要遍历查找最小值，时间复杂度至少为 $O(N)$，而使用最小堆访问最小值的时间复杂度为 $O(1)$。当然，天下没有免费的午餐，我们需要做额外的工作去维护最小堆的结构，这也需要复杂度开销。最小堆的优势是通过动态维护使得最小值的获取代价很小，实际上维护的时间复杂度为 $O(\log N)$。而数组则无法做到，如果数组想要维护顺序性，则需要的时间复杂度至少为 $O(N)$。这样一来，最小（大）堆的优势就凸现出来了。

在了解了二叉堆结构后，我们来分析 PriorityBlockingQueue 类的实现原理，如代码清单 8.20 所示。先看相关属性和构造函数，DEFAULT_INITIAL_CAPACITY=11 是默认的数组大小，MAX_ARRAY_SIZE=Integer.MAX_VALUE-8 表示数组最大长度，queue 变量用来保存二叉堆，size 变量记录当前二叉堆包含的元素个数，comparator 是在构建二叉堆时使用的比较器，lock 是入队和出队时使用的锁，notEmpty 条件是阻塞队列非空条件，allocationSpinLock 是数组动态扩容时使用的自旋锁。static 块用于对 allocationSpinLock 进行 CAS 操作。构造函数主要分 3 种，第一种是使用默认数组长度 11 和不指定比较器，第二种是指定数组长度和不指定比较器，第三种是指定数组长度和指定比较器。

代码清单 8.20　PriorityBlockingQueue 类的实现

```java
1.  public class PriorityBlockingQueue<E> implements BlockingQueue<E> {
2.
3.      private static final int DEFAULT_INITIAL_CAPACITY = 11;
4.      private static final int MAX_ARRAY_SIZE = Integer.MAX_VALUE - 8;
5.      private transient Object[] queue;
6.      private transient int size;
7.      private transient Comparator<? super E> comparator;
8.      private final ReentrantLock lock = new ReentrantLock();
9.      private final Condition notEmpty = lock.newCondition();
10.     private transient volatile int allocationSpinLock;
11.     private static final VarHandle ALLOCATIONSPINLOCK;
12.
13.     static {
14.         try {
15.             MethodHandles.Lookup l = MethodHandles.lookup();
16.             ALLOCATIONSPINLOCK = l.findVarHandle(PriorityBlockingQueue.class, "allocationSpinLock",
17.                     int.class);
18.         } catch (ReflectiveOperationException e) {
19.             throw new ExceptionInInitializerError(e);
20.         }
21.     }
22.
23.     public PriorityBlockingQueue() {
24.         this(DEFAULT_INITIAL_CAPACITY, null);
25.     }
26.
27.     public PriorityBlockingQueue(int initialCapacity) {
28.         this(initialCapacity, null);
29.     }
30.
31.     public PriorityBlockingQueue(int initialCapacity, Comparator<? super E> comparator) {
32.         if (initialCapacity < 1)
33.             throw new IllegalArgumentException();
34.         this.comparator = comparator;
35.         this.queue = new Object[Math.max(1, initialCapacity)];
36.     }
37.
38.     ......
39.
40. }
```

二叉堆结构的维护是 PriorityBlockingQueue 的核心工作，在入队时需要向上检测是否符合二叉堆性质，而在出队时则要向下检测是否符合二叉堆性质。我们先看无比较器的情况，此时将使用实现了 Comparable 接口的对象的 compareTo 方法来判断对象值的大小。前面说过，二叉堆的父节点可以通过 Math.floor((index-1)/2) 来查找，左子节点通过 2*index+1 来查找，右子节点通过 2*index+2 来查找。siftUpComparable 方法用于入队时向上检测，它通过 while 循环不断向上检测，不断向上获取父节点并比较大小。如果当前对象的值大于等于父节点的值，则停止向上检测，因为此时已经符合最小堆的性质。siftDownComparable 方法用于出队时向下检测，在最小的元素出队后，数组的最后一个元素将从根节点往下寻找自己的位置。该方向通过 while 循环不断向下检测，选择左子节点和右子节点中较小的那个与当前对象比较。如果当前节点比左右两个子节点都小，则不必继续向下检测，因为此时已经符合最小堆的性质。siftUpComparable 方法和 siftDownComparable 方法的代码如代码清单 8.21 所示。

代码清单 8.21 siftUpComparable 方法和 siftDownComparable 方法

```
1.   private static <T> void siftUpComparable(int k, T x, Object[] es) {
2.       Comparable<? super T> key = (Comparable<? super T>) x;
3.       while (k > 0) {
4.           int parent = (k - 1) >>> 1;
5.           Object e = es[parent];
6.           if (key.compareTo((T) e) >= 0)
7.               break;
8.           es[k] = e;
9.           k = parent;
10.      }
11.      es[k] = key;
12.  }
13.
14.  private static <T> void siftDownComparable(int k, T x, Object[] es, int n) {
15.      Comparable<? super T> key = (Comparable<? super T>) x;
16.      int half = n >>> 1;
17.      while (k < half) {
18.          int child = (k << 1) + 1;
19.          Object c = es[child];
20.          int right = child + 1;
21.          if (right < n && ((Comparable<? super T>) c).compareTo((T) es[right]) > 0)
22.              c = es[child = right];
23.          if (key.compareTo((T) c) <= 0)
24.              break;
25.          es[k] = c;
26.          k = child;
27.      }
28.      es[k] = key;
29.  }
```

对于有比较器的情况，它与无比较器的逻辑基本一样，唯一不同的地方在于比较时使用的是比较器的 compare 方法。siftUpUsingComparator 和 siftDownUsingComparator 分别为向上检测和向下检测的方法，如代码清单 8.22 所示。可将代码清单 8.22 与代码清单 8.21 中的 siftUpComparable 和 siftDownComparable 方法进行对比。

代码清单 8.22　siftUpUsingComparator 方法和 siftDownUsingComparator 方法

```java
1.   private static <T> void siftUpUsingComparator(int k, T x, Object[] es,
2.           Comparator<? super T> cmp) {
3.       while (k > 0) {
4.           int parent = (k - 1) >>> 1;
5.           Object e = es[parent];
6.           if (cmp.compare(x, (T) e) >= 0)
7.               break;
8.           es[k] = e;
9.           k = parent;
10.      }
11.      es[k] = x;
12.  }
13.
14.  private static <T> void siftDownUsingComparator(int k, T x, Object[] es, int n,
15.          Comparator<? super T> cmp) {
16.      int half = n >>> 1;
17.      while (k < half) {
18.          int child = (k << 1) + 1;
19.          Object c = es[child];
20.          int right = child + 1;
21.          if (right < n && cmp.compare((T) c, (T) es[right]) > 0)
22.              c = es[child = right];
23.          if (cmp.compare(x, (T) c) <= 0)
24.              break;
25.          es[k] = c;
26.          k = child;
27.      }
28.      es[k] = x;
29.  }
```

PriorityBlockingQueue 的 put 和 offer 操作其实是相同的（见代码清单 8.23），它们都不提供超时机制，因为二叉堆是无界的，所以不会因为无法入队而导致阻塞。下面来看一下 offer 方法，它不允许 null 加入到阻塞队列中，必须先成功获取锁才能往下执行，如果数组的空间不足以存放数据，则调用 tryGrow 方法进行动态扩容。接着判断是否定义了比较器，如果没有则通过 siftUpComparable 方法维护最小堆性质，如果有则通过 siftUpUsingComparator 方法维护最小堆性质。接着将阻塞队列长度 size 加 1 并通知 notEmpty 条件，最终释放锁。

代码清单 8.23　put 和 offer 操作

```java
1.   public void put(E e) {
2.       offer(e);
3.   }
4.
5.   public boolean offer(E e, long timeout, TimeUnit unit) throws InterruptedException {
6.       return offer(e);
7.   }
```

```java
8.
9.    public boolean offer(E e) {
10.       if (e == null)
11.           throw new NullPointerException();
12.       final ReentrantLock lock = this.lock;
13.       lock.lock();
14.       int n, cap;
15.       Object[] es;
16.       while ((n = size) >= (cap = (es = queue).length))
17.           tryGrow(es, cap);
18.       try {
19.           final Comparator<? super E> cmp;
20.           if ((cmp = comparator) == null)
21.               siftUpComparable(n, e, es);
22.           else
23.               siftUpUsingComparator(n, e, es, cmp);
24.           size = n + 1;
25.           notEmpty.signal();
26.       } finally {
27.           lock.unlock();
28.       }
29.       return true;
30.   }
```

tryGrow 方法用于数组扩容操作，如代码清单 8.24 所示。由于扩容与入队和出队操作能够并发地进行，所以这里先将入队时获取的锁释放掉，以便出队和入队操作并发进行。然后尝试通过 CAS 算法将 allocationSpinLock 变量设置为 1，即表示成功获取扩容锁。旧容量如果小于 64，则增量为 oldCap+2，否则增量为 oldCap>>1（新容量不能超过最大容量限制）。然后按新容量创建 Object 数组并最终将 allocationSpinLock 变量设置为 0。需要注意的是，newArray==null 表示其他线程正在扩容，此时就调用 Thread.yield 放弃 CPU 时间片。最后重新获取 lock 锁，通过 System.arraycopy 方法将旧数组的元素都复制到新数组中。

代码清单 8.24　tryGrow 方法

```java
1.    private void tryGrow(Object[] array, int oldCap) {
2.        lock.unlock();
3.        Object[] newArray = null;
4.        if (allocationSpinLock == 0 && ALLOCATIONSPINLOCK.compareAndSet(this, 0, 1)) {
5.            try {
6.                int newCap = oldCap + ((oldCap < 64) ? (oldCap + 2) : (oldCap >> 1));
7.                if (newCap - MAX_ARRAY_SIZE > 0) {
8.                    int minCap = oldCap + 1;
9.                    if (minCap < 0 || minCap > MAX_ARRAY_SIZE)
10.                       throw new OutOfMemoryError();
11.                   newCap = MAX_ARRAY_SIZE;
12.               }
13.               if (newCap > oldCap && queue == array)
14.                   newArray = new Object[newCap];
```

```
15.        } finally {
16.            allocationSpinLock = 0;
17.        }
18.    }
19.    if (newArray == null)
20.        Thread.yield();
21.    lock.lock();
22.    if (newArray != null && queue == array) {
23.        queue = newArray;
24.        System.arraycopy(array, 0, newArray, 0, oldCap);
25.    }
26. }
```

PriorityBlockingQueue 的 take 和 poll 操作的差别就在于一个不支持超时机制，而另一个支持，如代码清单 8.25 所示。前面说过，入队不支持超时机制，但出队则支持等待超时，即如果阻塞队列为空时执行出队操作，则要等待。先看 take 方法，它在成功获取锁后通过 dequeue 方法从最小堆中取出最小元素。如果取出的是 null，则调用 notEmpty 条件的 await 方法进入等待状态。poll 方法与 take 方法的逻辑一样，不同的地方在于 notEmpty 条件的 awaitNanos 方法，该方法在超时时会解除等待状态。

代码清单 8.25　take 和 poll 操作的差别

```
1.  public E take() throws InterruptedException {
2.      final ReentrantLock lock = this.lock;
3.      lock.lockInterruptibly();
4.      E result;
5.      try {
6.          while ((result = dequeue()) == null)
7.              notEmpty.await();
8.      } finally {
9.          lock.unlock();
10.     }
11.     return result;
12. }
13.
14. public E poll(long timeout, TimeUnit unit) throws InterruptedException {
15.     long nanos = unit.toNanos(timeout);
16.     final ReentrantLock lock = this.lock;
17.     lock.lockInterruptibly();
18.     E result;
19.     try {
20.         while ((result = dequeue()) == null && nanos > 0)
21.             nanos = notEmpty.awaitNanos(nanos);
22.     } finally {
23.         lock.unlock();
24.     }
25.     return result;
26. }
```

dequeue 方法用于从最小堆中取出最小元素,然后拿最后一个元素来补充。它需要不断向下检测调整,如代码清单 8.26 所示。最小元素是数组的第一个元素 queue[0],然后将阻塞队列长度减 1 并取出数组最后一个元素 es[(n=--size)],然后将数组最后一个元素 es[n]设为 null,接着判断是否定义了比较器,如果没有则通过 siftDownComparable 方法向下检测最小堆性质,如果有则通过 siftDownUsingComparator 方法向下检测最小堆性质。dequeue 方法大概逻辑为,取出数组第一个元素(最小元素),然后将数组的最后一个元素补充到数组第一个元素的位置,最后不断向下检测并调整元素,直到满足最小堆性质。

代码清单 8.26　dequeue 方法

```
1.   private E dequeue() {
2.       final Object[] es;
3.       final E result;
4.       if ((result = (E) ((es = queue)[0])) != null) {
5.           final int n;
6.           final E x = (E) es[(n = --size)];
7.           es[n] = null;
8.           if (n > 0) {
9.               final Comparator<? super E> cmp;
10.              if ((cmp = comparator) == null)
11.                  siftDownComparable(0, x, es, n);
12.              else
13.                  siftDownUsingComparator(0, x, es, n, cmp);
14.          }
15.      }
16.      return result;
17.  }
```

8.5　延迟阻塞队列

延迟阻塞队列 DelayQueue 是一种具有延迟机制的阻塞队列,如图 8.11 所示。所谓延迟,是指队列中的元素指定了延时时长,只有当延时时间到期后才能出队。有很多方法可以监控哪些元素的延时到期(比如可以启动一个线程来监控),但 JDK 的实现使用了更优雅的方式,它通过 PriorityQueue 优先级队列(它与上节讲解的 PriorityBlockingQueue 非常相似)来维护最短延时的元素,进而只需检测二叉堆的第一个元素即可。由于它的实现基于 PriorityQueue,所以它是无界的队列。此外,入队的元素必须要实现 Delayed 接口,由该接口指定延迟时长。

下面通过代码清单 8.27 来了解延迟阻塞队列的功能。代码中定义了一个缓存对象 Cache,该对象指定了延迟时间,阻塞队列中的 Cache 对象只有当延迟时间到期后才能被取出。Cache 必须实现 Delayed 接口,需要实现 compareTo 方法和 getDelay 方法,前者用于二叉堆比较,后者则用于获取剩余的延迟时间(当其返回 0 或负数时表示延迟时间已到期)。Cache 对象包含 data 和 endTime 属性,data 表示保存的数据,endTime 表示延迟超时时间点。

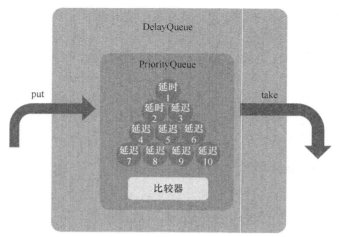

图 8.11 延迟阻塞队列

代码清单 8.27 延迟队列的功能

```
1.  public class DelayQueueDemo {
2.      static class Cache implements Delayed {
3.          private String data;
4.          private long endTime;
5.
6.          Cache(String data, long t) {
7.              this.data = data;
8.              this.endTime = t;
9.          }
10.
11.         public int compareTo(Delayed o) {
12.             Cache c = (Cache) o;
13.             return this.endTime - c.endTime > 0 ? 1 : (this.endTime - c.endTime < 0 ? -1 : 0);
14.         }
15.
16.         public long getDelay(TimeUnit unit) {
17.             return unit.convert(endTime - System.currentTimeMillis(), TimeUnit.MILLISECONDS);
18.         }
19.     }
20.
21.     public static void main(String[] args) throws InterruptedException {
22.         DelayQueue<Cache> queue = new DelayQueue<Cache>();
23.         String[] datas = { "data1", "data2", "data3" };
24.         for (int i = 0; i < datas.length; i++) {
25.             final int index = i;
26.             new Thread(() -> {
27.                 queue.put(new Cache(datas[index], System.currentTimeMillis() + (index + 1) * 3000));
28.             }).start();
```

```
29.         }
30.         for (int i = 0; i < datas.length; i++)
31.             new Thread(() -> {
32.                 try {
33.                     System.out.println(queue.take().data);
34.                 } catch (InterruptedException e) {
35.                 }
36.             }).start();
37.     }
38. }
```

main 方法中创建了一个 DelayQueue 对象，然后分别启动 3 个线程将 data1、data2、data3 加入到延迟队列中，延迟超时时间点为当前时间加上(index+1)*3000ms。最后再启动 3 个线程从延迟队列中执行 take 操作。由于刚开始还没有任何一个 Cache 对象到达延迟超时时间点，所以 3 个线程都被阻塞。大概 3s 后先输出 data1，接着 3s 后输出 data2，再 3s 后输出 data3。

8.5.1 优先级队列

DelayQueue 的内部使用了优先级队列 PriorityQueue 来实现二叉堆结构，大家是不是感觉这个优先级队列很熟悉？没错，它类似上节讲解的优先级阻塞队列 PriorityBlockingQueue。它们之间就差了"阻塞"两字。PriorityBlockingQueue 在入队和出队时可能会产生阻塞，而 PriorityQueue 则不会，它只是普通的二叉堆结构。

由于前面已经深入分析过 PriorityBlockingQueue，这里只是简单讲解一下 PriorityQueue 的实现，如代码清单 8.28 所示。我们对 PriorityQueue 的属性和构造函数都很熟悉，包括当前二叉堆长度、默认长度、最大长度、比较器以及用于保存二叉堆的 Object[]数组。可以看到属性中已经不包含任何锁对象。

代码清单 8.28　PriorityQueue 的实现

```
1.  public class PriorityQueue<E> {
2.      int size;
3.      transient Object[] queue;
4.      private static final int DEFAULT_INITIAL_CAPACITY = 11;
5.      private final Comparator<? super E> comparator;
6.      private static final int MAX_ARRAY_SIZE = Integer.MAX_VALUE - 8;
7.
8.      public PriorityQueue() {
9.          this(DEFAULT_INITIAL_CAPACITY, null);
10.     }
11.
12.     public PriorityQueue(int initialCapacity) {
13.         this(initialCapacity, null);
14.     }
15.
```

```
16.    public PriorityQueue(int initialCapacity, Comparator<? super E> comparator) {
17.        if (initialCapacity < 1)
18.            throw new IllegalArgumentException();
19.        this.queue = new Object[initialCapacity];
20.        this.comparator = comparator;
21.    }
22.
23.    ......
24.
25. }
```

主要的方法包括入队操作的 offer 方法、出队操作的 poll 方法和获取第一个元素的 peek 方法，如代码清单 8.29 所示。offer 方法主要先判断容量是否足够，如果不够则调用 grow 方法进行扩容，然后根据比较器向上检测二叉堆性质，如果没有比较器则根据 Comparable 接口的 compareTo 方法向上检测二叉堆性质。poll 方法直接取出第一个元素，如果成功取出，则进一步向下检测维护二叉堆结构；如果失败则返回 null。peek 方法只获取第一个元素而不执行出队操作。

代码清单 8.29 入队操作的 offer 方法、出队操作的 poll 方法和获取第一个元素的 peek 方法

```
1.  public boolean offer(E e) {
2.      if (e == null)
3.          throw new NullPointerException();
4.      int i = size;
5.      if (i >= queue.length)
6.          grow(i + 1);
7.      if (comparator != null)
8.          siftUpUsingComparator(i, e, queue, comparator);
9.      else
10.         siftUpComparable(i, e, queue);
11.     size = i + 1;
12.     return true;
13. }
14.
15. private void grow(int minCapacity) {
16.     int oldCapacity = queue.length;
17.     int newCapacity = oldCapacity
18.         + ((oldCapacity < 64) ? (oldCapacity + 2) : (oldCapacity >> 1));
19.     if (minCapacity < 0)
20.         throw new OutOfMemoryError();
21.     newCapacity = (minCapacity > MAX_ARRAY_SIZE) ? Integer.MAX_VALUE : MAX_ARRAY_SIZE;
22.     queue = Arrays.copyOf(queue, newCapacity);
23. }
24.
25. public E poll() {
26.     final Object[] es;
27.     final E result;
```

```
28.     if ((result = (E) ((es = queue)[0])) != null) {
29.         final int n;
30.         final E x = (E) es[(n = --size)];
31.         es[n] = null;
32.         if (n > 0) {
33.             final Comparator<? super E> cmp;
34.             if ((cmp = comparator) == null)
35.                 siftDownComparable(0, x, es, n);
36.             else
37.                 siftDownUsingComparator(0, x, es, n, cmp);
38.         }
39.     }
40.     return result;
41. }
```

8.5.2 DelayQueue 的阻塞与唤醒

DelayQueue 的阻塞与唤醒时机与其他阻塞队列存在不同。其他阻塞队列是当阻塞队列为空时进行出队操作或者当阻塞队列满时进行入队操作，都会导致阻塞发生，然后在队列非空非满时唤醒。由于 DelayQueue 是无界的，所以入队不会发生阻塞。出队时有两种情况会发生阻塞：队列为空时进行出队操作；队列第一个元素的延迟超时时间点还未到时进行出队操作。

第一种情况如图 8.12 所示。线程 1 获得锁后执行出队操作，但此时队列中没有任何一个元素，所以线程 1 只能进入阻塞状态并放弃锁。然后线程 2 获得锁并将元素 1 入队，此时线程 2 通知唤醒线程 1 并释放锁。最后线程 1 再次获得锁并执行出队操作将元素 1 取出。

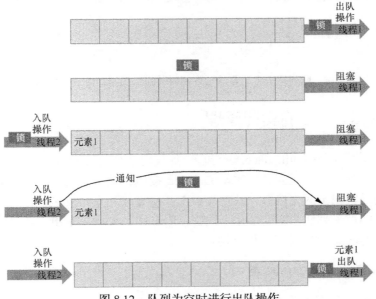

图 8.12　队列为空时进行出队操作

第二种情况如图 8.13 所示。原先队列中就有 3 个元素，线程 1 获取锁后执行出队操作。由于元素 1（延迟时间最小的元素）还未到延迟超时时间点，所以线程 1 进入限时阻塞并释放锁。接着线程 2 继续执行出队操作，此时元素 1 仍然未到时间点，于是线程 2 永久阻塞并释放锁（如果没有其他线程唤醒，它将永远阻塞）。假如线程 1 阻塞到达延迟超时时间点，那么它将自动解除阻塞，获取锁后将元素 1 出队，此时它将通知线程 2 并释放锁。最后线程 2 解除阻塞状态，获取锁后将元素 2 出队。

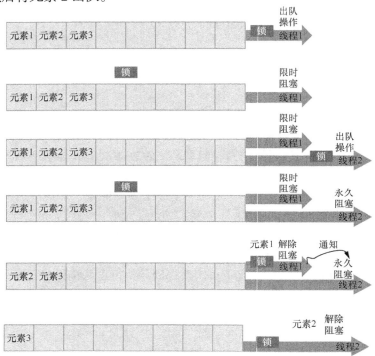

图 8.13　队列第一个元素的延迟超时时间点还未到时进行出队操作

8.5.3　DelayQueue 的实现原理

DelayQueue 类位于 java.util.concurrent 包下，如代码清单 8.30 所示。它实现了 BlockingQueue 接口，保存到延迟队列中的元素必须实现 Delayed 接口，该接口又继承了 Comparable 接口，所以实际上是同时实现了两个接口，对应着 getDelay 方法和 compareTo 方法。leader 是出队操作时由于延迟时间点未到而阻塞的第一个线程，lock 和 available 是出入队时使用的锁和条件，还有一个 PriorityQueue 对象是二叉堆结构。

代码清单 8.30　DelayQueue 类

```
1.    public class DelayQueue<E extends Delayed> implements BlockingQueue<E> {
2.        private Thread leader;
```

```
3.      private final transient ReentrantLock lock = new ReentrantLock();
4.      private final Condition available = lock.newCondition();
5.      private final PriorityQueue<E> q = new PriorityQueue<E>();
6.
7.      public DelayQueue() {
8.      }
9.
10.     ......
11.
12. }
13. public interface Delayed extends Comparable<Delayed> {
14.     long getDelay(TimeUnit unit);
15. }
```

入队操作主要包括 put 和 offer 方法，该类中 put 方法间接调用了 offer 方法，如代码清单 8.31 所示。offer 方法需要先成功获取锁，然后调用 PriorityQueue 对象的 offer 方法将元素加入到二叉堆中。尽管 PriorityQueue 是非线程安全的，但这里的锁却能保证只有一个线程能访问。通过 q.peek()==e 判断当前元素是否为第一个元素（即队列只有当前元素），如果是则将 leader 置为 null 并向 available 条件发送信号，因为可能有线程阻塞在出队操作。

代码清单 8.31　put 和 offer 方法

```
1.  public void put(E e) {
2.      offer(e);
3.  }
4.
5.  public boolean offer(E e) {
6.      final ReentrantLock lock = this.lock;
7.      lock.lock();
8.      try {
9.          q.offer(e);
10.         if (q.peek() == e) {
11.             leader = null;
12.             available.signal();
13.         }
14.         return true;
15.     } finally {
16.         lock.unlock();
17.     }
18. }
19.
20. public boolean offer(E e, long timeout, TimeUnit unit) {
21.     return offer(e);
22. }
```

take 方法对应出队操作，成功获取锁后通过一个 for(;;) 循环来控制阻塞与唤醒，并通过 q.peek() 获取二叉堆的第一个元素。如果第一个元素为空则表示队列为空，此时调用 available.await() 进入阻塞状态，等待其他线程执行入队操作，如代码清单 8.32 所示。如果第一个元素不为空，则通过 first.getDelay(NANOSECONDS)获取第一个元素的延迟时长，如果小于等于 0 则可以将其从二叉堆中取出来并返回；如果大于 0 则需要等待。如果 leader 不为空，则调用 available.await()永久等待并释放锁；如果 leader 为空，则将当前线程赋值给 leader，然后调用 available.awaitNanos(delay)阻塞指定时长并释放锁。注意，await 和 awaitNanos 都会自动释放锁。最终的 if(leader==null&&q.peek()!=null)条件用于说明被唤醒的线程要继续唤醒下一个阻塞的线程。

代码清单 8.32　take 方法

```
1.   public E take() throws InterruptedException {
2.       final ReentrantLock lock = this.lock;
3.       lock.lockInterruptibly();
4.       try {
5.           for (;;) {
6.               E first = q.peek();
7.               if (first == null)
8.                   available.await();
9.               else {
10.                  long delay = first.getDelay(NANOSECONDS);
11.                  if (delay <= 0L)
12.                      return q.poll();
13.                  first = null;
14.                  if (leader != null)
15.                      available.await();
16.                  else {
17.                      Thread thisThread = Thread.currentThread();
18.                      leader = thisThread;
19.                      try {
20.                          available.awaitNanos(delay);
21.                      } finally {
22.                          if (leader == thisThread)
23.                              leader = null;
24.                      }
25.                  }
26.              }
27.          }
28.      } finally {
29.          if (leader == null && q.peek() != null)
30.              available.signal();
31.          lock.unlock();
32.      }
33.  }
```

take 方法是不具有超时机制的阻塞出队操作。此外还有不阻塞和具有超时机制阻塞的两种出队操作，这两种方式对应两个 poll 方法。不阻塞的 poll 方法先获取锁，然后通过 PriorityQueue 的 peek 方法获取二叉堆第一个元素。如果该元素为空或延迟时间未到，则返回 null。如果该元素不为空且延迟时间已到则，调用 PriorityQueue 的 poll 出队，最后释放锁。

8.5 延迟阻塞队列

下面看具有超时机制的阻塞出队操作，如代码清单 8.33 所示。成功获取锁后通过 for(;;) 循环控制阻塞、唤醒和超时，然后通过 q.peek() 获取二叉堆的第一个元素。如果第一个元素为空，则表示队列为空，而且超时时间到期则返回 null；如果没到超时时间，则调用 available.await() 进入阻塞状态，等待其他线程执行入队操作。如果第一个元素不为空，则通过 first.getDelay (NANOSECONDS)获取第一个元素的延迟时长。如果小于等于 0，则可以将其从二叉堆中取出来并返回，如果超时时间到期，则返回 null。如果超时时间大于延迟时间或者 leader 不为空，则调用 available.awaitNanos(nanos)按超时时间进行阻塞并释放锁。如果超时时间小于延迟时间或者 leader 为空，则将当前线程赋值给 leader，然后调用 available.awaitNanos(delay)按延迟时间阻塞并释放锁。阻塞结束后会执行 nanos-=delay-timeLeft 来计算新的超时时间。

代码清单 8.33 具有超时机制的阻塞队列操作

```java
1.  public E poll() {
2.      final ReentrantLock lock = this.lock;
3.      lock.lock();
4.      try {
5.          E first = q.peek();
6.          return (first == null || first.getDelay(NANOSECONDS) > 0) ? null : q.poll();
7.      } finally {
8.          lock.unlock();
9.      }
10. }
11.
12. public E poll(long timeout, TimeUnit unit) throws InterruptedException {
13.     long nanos = unit.toNanos(timeout);
14.     final ReentrantLock lock = this.lock;
15.     lock.lockInterruptibly();
16.     try {
17.         for (;;) {
18.             E first = q.peek();
19.             if (first == null) {
20.                 if (nanos <= 0L)
21.                     return null;
22.                 else
23.                     nanos = available.awaitNanos(nanos);
24.             } else {
25.                 long delay = first.getDelay(NANOSECONDS);
26.                 if (delay <= 0L)
27.                     return q.poll();
28.                 if (nanos <= 0L)
29.                     return null;
30.                 first = null;
31.                 if (nanos < delay || leader != null)
32.                     nanos = available.awaitNanos(nanos);
33.                 else {
34.                     Thread thisThread = Thread.currentThread();
35.                     leader = thisThread;
```

```
36.                    try {
37.                        long timeLeft = available.awaitNanos(delay);
38.                        nanos -= delay - timeLeft;
39.                    } finally {
40.                        if (leader == thisThread)
41.                            leader = null;
42.                    }
43.                }
44.            }
45.        }
46.    } finally {
47.        if (leader == null && q.peek() != null)
48.            available.signal();
49.        lock.unlock();
50.    }
51. }
```

8.6 链表阻塞的双向队列

前面介绍的阻塞队列都属于单向队列。所谓单向，就是只能在队列的一端入队，在另一端出队。阻塞队列只能分别在固定的一端进行 put 操作和 take 操作，这就保证了队列的先进先出。双向队列则具有更大的灵活性，它可以同时在两端进行 put 和 take 操作，同时它也打破了先进先出的规则，如图 8.14 所示。

图 8.14　单向队列与双向队列的对比

链表阻塞双向队列 LinkedBlockingDeque 实现的是 BlockingDeque 接口，该接口主要的核心方法如代码清单 8.34 所示。这些方法都支持中断，其中 put 和 take 方法对应入队和出队操作，putFirst 和 putLast 分别表示从队列头和队列尾入队，takeFirst 和 takeLast 分别表示从队列头和队列尾出队。offer 和 poll 这两个方法分别提供了超时机制的入队和出队操作。

代码清单 8.34　BlockingDeque 接口的核心方法

```
1. public interface BlockingDeque<E> {
2.     void put(E e) throws InterruptedException;
3.     void putFirst(E e) throws InterruptedException;
4.     void putLast(E e) throws InterruptedException;
```

```
 5.         E take() throws InterruptedException;
 6.         E takeFirst() throws InterruptedException;
 7.         E takeLast() throws InterruptedException;
 8.     public boolean offer(E e, long timeout, TimeUnit unit) throws InterruptedException;
 9.         boolean offerFirst(E e, long timeout, TimeUnit unit) throws InterruptedException;
10.         boolean offerLast(E e, long timeout, TimeUnit unit) throws InterruptedException;
11.     public E poll(long timeout, TimeUnit unit) throws InterruptedException;
12.         E pollFirst(long timeout, TimeUnit unit) throws InterruptedException;
13.         E pollLast(long timeout, TimeUnit unit) throws InterruptedException;
14.     }
```

LinkedBlockingDeque 采用了双向链表作为存储数据结构。双向链表由于有 prev 和 next 两个方向的指针，所以在某个节点能方便地往前或往后查找，如图 8.15 所示。LinkedBlockingDeque 能够在队列头和队列尾进行入队（对应 putFirst 和 putLast 方法），同样地，出队也能够在队列的头尾进行（分别对应 takeFirst 和 takeLast 方法）。LinkedBlockingDeque 的队列是无界的，而且只提供非公平模式。

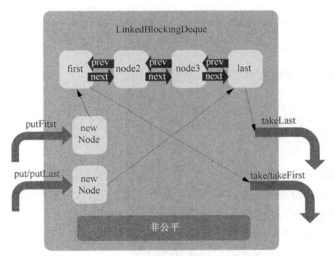

图 8.15　链表阻塞双向队列

假如有这样一个场景，常规数据按照先进先出的顺序执行，紧急数据则要优先执行。即在正常情况下，常规数据入队到队尾而从队头出队，紧急数据则入队到队头。在代码清单 8.35 中，先创建了一个 LinkedBlockingDeque 对象，然后用 3 个线程分别将 data1、data2、data3 入队到队尾。休眠 2s 后开始通过 take 方法执行出队操作，当将 data1 取出后刚好有紧急数据 urgency_data 需要优先处理，则要通过 putFirst 将其入队到队头，这样就能让紧急数据优先被处理，然后再处理 data2 和 data3。

代码清单 8.35　常规数据按照先进先出的顺序执行，紧急数据优先执行

```
1.   public class LinkedBlockingDequeDemo {
2.       static BlockingDeque<String> blockingDeque = new LinkedBlockingDeque<String>();
```

```
3.
4.      public static void main(String[] args) throws InterruptedException {
5.          String[] datas = { "data1", "data2", "data3" };
6.          String urgency = "urgency_data";
7.          for (int i = 0; i < 3; i++) {
8.              final int index = i;
9.              new Thread(() -> {
10.                 try {
11.                     blockingDeque.put(datas[index]);
12.                 } catch (InterruptedException e) {
13.                 }
14.             }).start();
15.             Thread.sleep(500);
16.         }
17.         Thread.sleep(2000);
18.         System.out.println("take " + blockingDeque.take() + " from deque");
19.         blockingDeque.putFirst("put " + urgency + " into deque");
20.         System.out.println("take " + blockingDeque.take() + " from deque");
21.         System.out.println("take " + blockingDeque.take() + " from deque");
22.         System.out.println("take " + blockingDeque.take() + " from deque");
23.     }
24. }
```

代码清单 8.35 执行后的输出结果如下所示。

```
1.  take data1 from deque
2.  take put urgency_data into deque from deque
3.  take data2 from deque
4.  take data3 from deque
```

在继续分析 LinkedBlockingDeque 的实现原理前，需要先了解一下双向链表的入队和出队操作。双向链表有 first 和 last 这两个特殊的引用，分别指向头部和尾部。此外，节点间通过 prev 和 next 引用连接起来。开始时，first 和 last 不指向任何对象，在创建 Node1 节点后由于只有一个节点，此时 first 和 last 都指向它。继续创建 Node2 并入队，此时 Node1 的 next 指向 Node2，而 Node2 的 prev 指向 Node1，first 还是指向 Node1，last 指向 Node2。类似地，再将 Node3 进行入队。最后，将 Node4 加入到队列头，让 first 指向 Node4 且维护对应的 prev 和 next 引用。双向链表的入队和出队操作的整个过程如图 8.16 所示。

出队操作默认是从队头取出节点并维护 prev 和 next 引用的指向，它根据 first 获取到队列的第一个节点，然后将 first 指向下一个节点。如图 8.17 所示，先将队头的 Node4 出队，first 指向 Node1 的同时将 Node1 的 prev 引用置为 null。类似地，将 Node1 进行出队。

LinkedBlockingDeque 类定义了一个 Node 内部类，item 用于保存数据，prev 和 next 分别指向前驱节点和后继节点，如代码清单 8.36 所示。first 和 last 分别指向头节点和尾节点，count 表示双向队列的当前长度，capacity 表示双向队列的容量大小，lock 是入队和出队时使用的锁，notEmpty 和 notFull 表示队列非空时和非满时的条件。该类的构造函数非常简单，只需指定容量大小；若不指定，则使用 Integer.MAX_VALUE 作为默认大小。

8.6 链表阻塞的双向队列

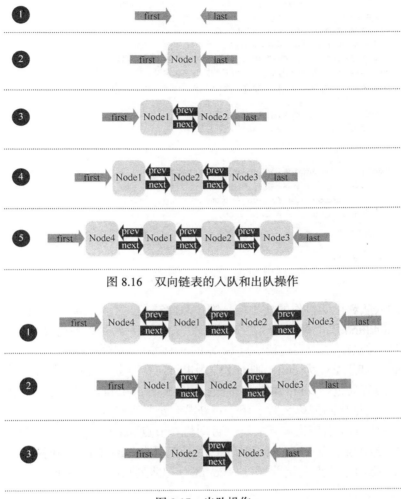

图 8.16 双向链表的入队和出队操作

图 8.17 出队操作

代码清单 8.36　LinkedBlockingDeque 类

```
1.   public class LinkedBlockingDeque<E> implements BlockingDeque<E> {
2.       static final class Node<E> {
3.           E item;
4.           Node<E> prev;
5.           Node<E> next;
6.
7.           Node(E x) {
8.               item = x;
9.           }
10.      }
11.
12.      transient Node<E> first;
```

```
13.     transient Node<E> last;
14.     private transient int count;
15.     private final int capacity;
16.     final ReentrantLock lock = new ReentrantLock();
17.     private final Condition notEmpty = lock.newCondition();
18.     private final Condition notFull = lock.newCondition();
19.
20.     public LinkedBlockingDeque() {
21.         this(Integer.MAX_VALUE);
22.     }
23.
24.     public LinkedBlockingDeque(int capacity) {
25.         if (capacity <= 0)
26.             throw new IllegalArgumentException();
27.         this.capacity = capacity;
28.     }
29.
30.     ......
31.
32. }
```

双向队列的入队和出队操作的核心就是维护双向链表，下面看看如何将新节点链接到双向链表的头和尾。linkFirst 用于将新节点链接到链表头，它先判断是否超过容量，如果 first 为 null，则将 first 引用指向新节点，否则将新节点 node 的 next 指向原来的头结点 first，接着将 first 引用指向新节点 node。如果 last 为 null，则 last 引用也指向新节点，否则将原来的头结点的 prev 指向新节点 node，接着通过 ++count 维护队列长度，最后向 notEmpty 发送信号，唤醒等待的线程。

linkLast 方法（见代码清单 8.37）用于将新节点链接到链尾，它先判断是否超过容量，如果 last 为 null，则将 last 引用指向新节点，否则将新节点 node 的 prev 指向原来的尾节点 last，接着将 last 引用指向新节点 node。如果 first 为 null，则 first 引用也指向新节点，否则将原来的尾节点的 next 指向新节点，接着通过 ++count 维护队列长度，最后向 notEmpty 发送信号，唤醒等待的线程。

代码清单 8.37　linkLast 方法

```
1.  private boolean linkFirst(Node<E> node) {
2.      if (count >= capacity)
3.          return false;
4.      Node<E> f = first;
5.      node.next = f;
6.      first = node;
7.      if (last == null)
8.          last = node;
9.      else
10.         f.prev = node;
11.     ++count;
12.     notEmpty.signal();
13.     return true;
```

```
14.    }
15.
16.    private boolean linkLast(Node<E> node) {
17.        if (count >= capacity)
18.            return false;
19.        Node<E> l = last;
20.        node.prev = l;
21.        last = node;
22.        if (first == null)
23.            first = node;
24.        else
25.            l.next = node;
26.        ++count;
27.        notEmpty.signal();
28.        return true;
29.    }
```

双向链表的删除对应着 unlinkFirst 和 unlinkLast 这两个方法，前者删除队头，后者删除队尾，如代码清单 8.38 所示。unlinkFirst 的逻辑为，直接获取 first 指向的节点，如果它为 null，则返回 null，否则通过 f.next 获取节点 n 作为新的队头，并将 first 指向 n。注意，f.item = null 和 f.next = f 是为了让垃圾回收器更好地回收取出后的节点，实际上我们只需要 item，其他对象都可以被回收。如果 n==null，则说明取出的是最后一个节点，此时将 last 指向 null，并将 n 的 prev 指向 null。接着通过--count 将队列长度减 1，最后向 notFull 发送信号唤醒等待的线程。unlinkLast 方法类似，此处不再赘述。关于在双向链表的两端进行新增节点和删除节点的操作，可以结合前面的示意图理解。

代码清单 8.38　unlinkFirst 和 unlinkLast 方法

```
1.     private E unlinkFirst() {
2.         Node<E> f = first;
3.         if (f == null)
4.             return null;
5.         Node<E> n = f.next;
6.         E item = f.item;
7.         f.item = null;
8.         f.next = f;
9.         first = n;
10.        if (n == null)
11.            last = null;
12.        else
13.            n.prev = null;
14.        --count;
15.        notFull.signal();
16.        return item;
17.    }
18.
19.    private E unlinkLast() {
20.        Node<E> l = last;
```

```
21.     if (l == null)
22.         return null;
23.     Node<E> p = l.prev;
24.     E item = l.item;
25.     l.item = null;
26.     l.prev = l;
27.     last = p;
28.     if (p == null)
29.         first = null;
30.     else
31.         p.next = null;
32.     --count;
33.     notFull.signal();
34.     return item;
35. }
```

put、putFirst 和 putLast 是没有超时机制的入队操作，如代码清单 8.39 所示。可以看到 put 间接调用了 putLast，所以实际上 put 就是入队到队尾。putFirst 先创建 Node 对象，然后成功获取锁后用 while 循环调用 linkFirst 方法，将新节点入队。注意容量满了后会执行 notFull.await() 进入等待，其他线程在容量不满时会唤醒它。putLast 也类似，调用的是 linkLast 方法。

代码清单 8.39　没有超时机制的入队操作

```
1.  public void put(E e) throws InterruptedException {
2.      putLast(e);
3.  }
4.
5.  public void putFirst(E e) throws InterruptedException {
6.      if (e == null)
7.          throw new NullPointerException();
8.      Node<E> node = new Node<E>(e);
9.      final ReentrantLock lock = this.lock;
10.     lock.lock();
11.     try {
12.         while (!linkFirst(node))
13.             notFull.await();
14.     } finally {
15.         lock.unlock();
16.     }
17. }
18.
19. public void putLast(E e) throws InterruptedException {
20.     if (e == null)
21.         throw new NullPointerException();
22.     Node<E> node = new Node<E>(e);
23.     final ReentrantLock lock = this.lock;
24.     lock.lock();
25.     try {
26.         while (!linkLast(node))
27.             notFull.await();
```

```
28.         } finally {
29.             lock.unlock();
30.         }
31. }
```

take、takeFirst 和 takeLast 是没有超时机制的出队操作，如代码清单 8.40 所示。其中 take 间接调用了 takeFirst，所以实际上 take 就是对队头进行出队操作。takeFirst 先获取锁，然后通过 while 调用 unlinkFirst 方法获取节点数据。如果队列为空，则执行 notEmpty.await() 进入等待，当队列不为空时会唤醒它。takeLast 也类似，调用的是 unlinkLast。

代码清单 8.40　没有超时机制的出队操作

```
1.  public E take() throws InterruptedException {
2.      return takeFirst();
3.  }
4.
5.  public E takeFirst() throws InterruptedException {
6.      final ReentrantLock lock = this.lock;
7.      lock.lock();
8.      try {
9.          E x;
10.         while ((x = unlinkFirst()) == null)
11.             notEmpty.await();
12.         return x;
13.     } finally {
14.         lock.unlock();
15.     }
16. }
17.
18. public E takeLast() throws InterruptedException {
19.     final ReentrantLock lock = this.lock;
20.     lock.lock();
21.     try {
22.         E x;
23.         while ((x = unlinkLast()) == null)
24.             notEmpty.await();
25.         return x;
26.     } finally {
27.         lock.unlock();
28.     }
29. }
```

offer、offerFirst 和 offerLast 是具有超时机制的入队操作，如代码清单 8.41 所示。offer 间接调用了 offerLast，所以默认是入队到队尾。offerFirst 先创建 Node 对象，接着成功获取锁后通过 while 循环调用 linkFirst 方法入队。可以看到，不同的地方在于增加了 if(nanos<=0L) 判断，当入队等待超时时，返回 false。awaitNanos 方法用于在等待指定时间后自动唤醒。offerLast 方法也类似。

代码清单 8.41　具有超时机制的入队操作

```java
1.  public boolean offer(E e, long timeout, TimeUnit unit) throws InterruptedException {
2.      return offerLast(e, timeout, unit);
3.  }
4.
5.  public boolean offerFirst(E e, long timeout, TimeUnit unit) throws InterruptedException {
6.      if (e == null)
7.          throw new NullPointerException();
8.      Node<E> node = new Node<E>(e);
9.      long nanos = unit.toNanos(timeout);
10.     final ReentrantLock lock = this.lock;
11.     lock.lockInterruptibly();
12.     try {
13.         while (!linkFirst(node)) {
14.             if (nanos <= 0L)
15.                 return false;
16.             nanos = notFull.awaitNanos(nanos);
17.         }
18.         return true;
19.     } finally {
20.         lock.unlock();
21.     }
22. }
23.
24. public boolean offerLast(E e, long timeout, TimeUnit unit) throws InterruptedException {
25.     if (e == null)
26.         throw new NullPointerException();
27.     Node<E> node = new Node<E>(e);
28.     long nanos = unit.toNanos(timeout);
29.     final ReentrantLock lock = this.lock;
30.     lock.lockInterruptibly();
31.     try {
32.         while (!linkLast(node)) {
33.             if (nanos <= 0L)
34.                 return false;
35.             nanos = notFull.awaitNanos(nanos);
36.         }
37.         return true;
38.     } finally {
39.         lock.unlock();
40.     }
41. }
```

poll、pollFirst 和 pollLast 是具有超时机制的出队操作，如代码清单 8.42 所示。poll 间接调用了 pollFirst，所以默认是从队头出队。pollFirst 方法先获取锁，然后通过 while 调用 unlinkFirst 方法获取节点数据。if(nanos<=0L)说明已经超时，直接返回 null。pollLast 方法也类似。

代码清单 8.42 具有超时机制的出队操作

```
1.    public E poll(long timeout, TimeUnit unit) throws InterruptedException {
2.        return pollFirst(timeout, unit);
3.    }
4.
5.    public E pollFirst(long timeout, TimeUnit unit) throws InterruptedException {
6.        long nanos = unit.toNanos(timeout);
7.        final ReentrantLock lock = this.lock;
8.        lock.lockInterruptibly();
9.        try {
10.           E x;
11.           while ((x = unlinkFirst()) == null) {
12.               if (nanos <= 0L)
13.                   return null;
14.               nanos = notEmpty.awaitNanos(nanos);
15.           }
16.           return x;
17.       } finally {
18.           lock.unlock();
19.       }
20.   }
21.
22.   public E pollLast(long timeout, TimeUnit unit) throws InterruptedException {
23.       long nanos = unit.toNanos(timeout);
24.       final ReentrantLock lock = this.lock;
25.       lock.lockInterruptibly();
26.       try {
27.           E x;
28.           while ((x = unlinkLast()) == null) {
29.               if (nanos <= 0L)
30.                   return null;
31.               nanos = notEmpty.awaitNanos(nanos);
32.           }
33.           return x;
34.       } finally {
35.           lock.unlock();
36.       }
37.   }
```

第 9 章

锁

9.1 可重入锁

Java 中的可重入锁（ReentrantLock）是一种并发控制工具，它的功能类似于 Java 内置的 synchronized 语法。可重入锁是指一个线程可以多次对某个锁进行加锁操作，比如程序在多层调用中进行多次加锁。而对于不可重入锁来说，进行两次以及以上的加锁会导致死锁的产生。

ReentrantLock 是一种独占锁，在独占模式下线程只能逐一使用锁，即任意时刻最多只能有一个线程持有锁。在图 9.1 中，线程 1 首先成功调用 lock 方法进行加锁操作从而持有该锁，其他两个线程只能等待线程 1 释放锁。因为锁是可重入的，所以线程 1 可多次调用 lock 方法对锁进行加锁操作。当线程 1 调用 unlock 方法释放锁后，线程 2 调用 lock 方法成功获得锁。接着线程 2 也调用 unlock 方法释放锁。然后线程 3 调用 lock 方法成功获得锁，它也可以多次调用 lock 方法对该锁进行加锁操作。

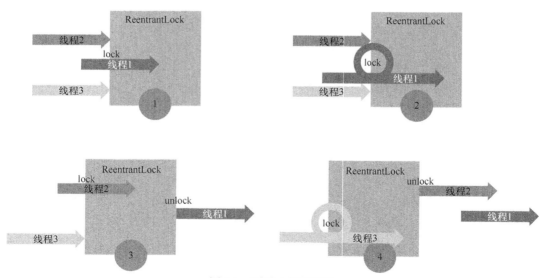

图 9.1 可重入锁的演示

ReentrantLock 类的四要素为公平/非公平模式、lock 方法、unlock 方法以及 newCondition 方法。公平/非公平模式表示多个线程在同时去获取锁时是否按照先到先得的顺序获得锁，如果是则为公平模式，否则为非公平模式。lock 方法用于获取锁，假如锁已被其他线程占有则进入等待状态。unlock 方法用于释放锁。newCondition 方法用于返回一个新创建的 Condition 对象，这个对象支持线程的阻塞和唤醒功能。每个 Condition 对象内部都有一个等待队列，具体的实现由 AQS 同步器提供（第 5 章讲过具体的实现）。

9.1.1 非公平模式的实现

ReentrantLock 类的内部是基于 AQS 同步器实现的，如图 9.2 所示。不管是公平模式还是非公平模式，ReentrantLock 都是基于 AQS 的独占模式，只是在获取锁的操作逻辑上有些差异。ReentrantLock 的默认模式为非公平模式，我们先看非公平模式的实现。

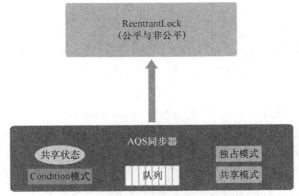

图 9.2　ReentrantLock 类的实现

ReentrantLock 类的方法较多，为了方便理解，这里分析几个主要的核心方法，如代码清单 9.1 所示。它提供了两个构造函数，无参的构造函数默认使用非公平模式。如果要指定模式，则可以为有参的构造函数传入一个 boolean 参数，当其为 true 时表示公平模式。其中 NonfairSync 为非公平同步器，而 FairSync 为公平同步器。lock 方法用于加锁操作，它间接调用 AQS 同步器的 acquire 方法获取独占锁。unlock 方法用于释放锁操作，它间接调用 AQS 同步器的 release 方法释放独占锁。newCondition 方法用于创建新的 Condition 对象并返回。

代码清单 9.1　ReentrantLock 类的核心方法

```
1.  public class ReentrantLock implements Lock, java.io.Serializable {
2.      private final Sync sync;
3.
4.      public ReentrantLock() {
5.          sync = new NonfairSync();
6.      }
```

```
7.
8.     public ReentrantLock(boolean fair) {
9.         sync = fair ? new FairSync() : new NonfairSync();
10.    }
11.
12.    public void lock() {
13.        sync.acquire(1);
14.    }
15.
16.    public void unlock() {
17.        sync.release(1);
18.    }
19.
20.    public Condition newCondition() {
21.        return sync.newCondition();
22.    }
23.
24.    ......
25.
26. }
```

ReentrantLock 类内部的 Sync 子类是公平模式 FairSync 类和非公平模式 NonfairSync 类的抽象父类。ReentrantLock 属于独占模式，所以非公平模式的锁获取方法为 nonfairTryAcquire。tryRelease 方法提供了释放锁的操作（公平模式和非公平模式都通过该方法释放锁）。newCondition 方法用于创建 AQS 同步器的 ConditionObject 对象，供 ReentrantLock 类的 newCondition 方法调用。

我们进一步分析 nonfairTryAcquire 方法，如代码清单 9.2 所示。该方法主要的逻辑是，获取 AQS 同步器的状态变量，其值只能为 0 或 1，分别表示未被加锁和已被加锁。如果状态变量为 0，则通过 compareAndSetState 方法进行 CAS 修改，将状态变量改为 1，还要通过 setExclusiveOwnerThread 方法将当前线程设为锁持有线程。如果状态变量为 1 且当前线程为锁持有线程，则表示正在进行锁重入操作，这时要将状态变量累加 1，而且如果重入的次数超过 int 类型的范围则抛出 Maximum lock count exceeded 错误。

继续分析 tryRelease 方法（见代码清单 9.2）。该方法主要的逻辑为，首先保证当前线程为锁持有线程，如果不是则抛出 IllegalMonitorStateException 异常。然后将 AQS 同步器的状态变量减 1，如果状态变量为 0，则表示锁已经完全释放成功，并且通过 setExclusiveOwnerThread 方法将持有锁线程设为空。如果状态变量不为 0，那么只需要将状态变量减 1 即可。

代码清单 9.2　nonfairTryAcquire 方法和 tryRelease 方法

```
1.  abstract static class Sync extends AbstractQueuedSynchronizer {
2.      final boolean nonfairTryAcquire(int acquires) {
3.          final Thread current = Thread.currentThread();
4.          int c = getState();
5.          if (c == 0) {
6.              if (compareAndSetState(0, acquires)) {
```

```
7.                 setExclusiveOwnerThread(current);
8.                 return true;
9.             }
10.         } else if (current == getExclusiveOwnerThread()) {
11.             int nextc = c + acquires;
12.             if (nextc < 0)
13.                 throw new Error("Maximum lock count exceeded");
14.             setState(nextc);
15.             return true;
16.         }
17.         return false;
18.     }
19.
20.     protected final boolean tryRelease(int releases) {
21.         int c = getState() - releases;
22.         if (Thread.currentThread() != getExclusiveOwnerThread())
23.             throw new IllegalMonitorStateException();
24.         boolean free = false;
25.         if (c == 0) {
26.             free = true;
27.             setExclusiveOwnerThread(null);
28.         }
29.         setState(c);
30.         return free;
31.     }
32.
33.     final ConditionObject newCondition() {
34.         return new ConditionObject();
35.     }
36. }
```

非公平模式的 NonfairSync 类仅需实现 tryAcquire 方法，它直接调用父类 Sync 的 nonfairTryAcquire 方法，如代码清单 9.3 所示。非公平就体现在这个方法中，这里的 tryAcquire 实际上就是使用了闯入策略，即准备获取锁的线程会先尝试去获取锁，失败了才会进入队列。这就是不公平的来源。

代码清单 9.3　tryAcquire 方法（非公平模式）

```
1. static final class NonfairSync extends Sync {
2.     protected final boolean tryAcquire(int acquires) {
3.         return nonfairTryAcquire(acquires);
4.     }
5. }
```

9.1.2　公平模式的实现

公平模式与非公平模式的主要差异就在于获取锁时的机制。非公平模式通过闯入策略来打破公平，而公平模式则是所有线程完全通过队列来实现公平机制。下面看公平模式的 FairSync

类。公平与非公平的差异主要在于 tryAcquire 方法，就是 if (!hasQueuedPredecessors() && compareAndSetState(0, acquires))那行代码，如代码清单 9.4 所示。它会检查是否已经存在等待队列，如果已经有等待队列，则返回 false（其实就是说队列已经有其他线程，直接放弃闯入操作）。这表示让 AQS 同步器将当前线程放入等待队列，这种放弃闯入操作的做法则意味着公平。

代码清单 9.4 tryAcquire 方法（公平模式）

```
1.    static final class FairSync extends Sync {
2.        protected final boolean tryAcquire(int acquires) {
3.            final Thread current = Thread.currentThread();
4.            int c = getState();
5.            if (c == 0) {
6.                if (!hasQueuedPredecessors() && compareAndSetState(0, acquires)) {
7.                    setExclusiveOwnerThread(current);
8.                    return true;
9.                }
10.           } else if (current == getExclusiveOwnerThread()) {
11.               int nextc = c + acquires;
12.               if (nextc < 0)
13.                   throw new Error("Maximum lock count exceeded");
14.               setState(nextc);
15.               return true;
16.           }
17.           return false;
18.       }
19.   }
```

9.1.3 公平模式的 3 个示例

公平模式的第一个示例非常简单，如代码清单 9.5 所示。3 个线程启动后都执行加锁操作，独占锁的特性保证了 3 个线程依次进行加锁和释放锁的操作。

代码清单 9.5 公平模式的实现示例 1

```
1.    public class ReentrantLockDemo {
2.        static ReentrantLock lock = new ReentrantLock();
3.
4.        public static void main(String[] args) {
5.            Thread thread1 = new Thread(() -> {
6.                try {
7.                    lock.lock();
8.                    System.out.println("thread1 got the lock");
9.                    Thread.sleep(2000);
10.                   lock.unlock();
11.                   System.out.println("thread1 release the lock");
12.               } catch (InterruptedException e) {
```

```
13.            }
14.        });
15.        Thread thread2 = new Thread(() -> {
16.            try {
17.                lock.lock();
18.                System.out.println("thread2 got the lock");
19.                Thread.sleep(2000);
20.                lock.unlock();
21.                System.out.println("thread2 release the lock");
22.            } catch (InterruptedException e) {
23.            }
24.        });
25.        Thread thread3 = new Thread(() -> {
26.            try {
27.                lock.lock();
28.                System.out.println("thread3 got the lock");
29.                Thread.sleep(2000);
30.                lock.unlock();
31.                System.out.println("thread3 release the lock");
32.            } catch (InterruptedException e) {
33.            }
34.        });
35.        thread1.start();
36.        thread2.start();
37.        thread3.start();
38.    }
39. }
```

代码清单 9.5 某次运行的输出结果如下。

```
1. thread2 got the lock
2. thread2 release the lock
3. thread1 got the lock
4. thread1 release the lock
5. thread3 got the lock
6. thread3 release the lock
```

线程 2 首先调用 lock 方法成功获取锁，睡眠 2s 后释放锁。接着线程 1 又成功获取锁，它也睡眠 2s 后释放锁。最后线程 3 成功获取锁，并在睡眠 2s 后释放锁。注意，这里使用了 ReentrantLock 的无参构造函数，这意味着使用了非公平模式，所以 3 个线程在竞争锁时都使用了闯入策略。

接着看第二个示例，如代码清单 9.6 所示。该示例主要为了说明 ReentrantLock 的可重入性。

代码清单 9.6 公平模式的实现示例 2

```
1. public class ReentrantLockDemo2 {
2.     static ReentrantLock lock = new ReentrantLock();
3.
4.     public static void main(String[] args) {
```

```java
5.        Thread thread1 = new Thread(() -> {
6.            try {
7.                lock.lock();
8.                System.out.println("thread1 got the lock");
9.                lock.lock();
10.               System.out.println("thread1 got the lock again");
11.               System.out.println("lock times : " + lock.getHoldCount());
12.               Thread.sleep(2000);
13.               lock.unlock();
14.               System.out.println("thread1 release the lock");
15.               System.out.println("lock times : " + lock.getHoldCount());
16.               lock.unlock();
17.               System.out.println("thread1 release the lock again");
18.               System.out.println("lock times : " + lock.getHoldCount());
19.           } catch (InterruptedException e) {
20.           }
21.       });
22.       thread1.start();
23.   }
24. }
```

在代码清单 9.6 中，线程 1 调用 lock 方法成功获取锁后再一次调用 lock 方法获取锁，也就是锁重入操作。此时调用 getHoldCount 方法输出该锁被加锁的次数，输出为 2。睡眠 2s 后调用 unlock 方法进行一次锁释放操作，此时再调用 getHoldCount 方法得知锁被加锁的次数为 1。最后再调用一次 unlock 方法完全释放该锁，此时锁被加锁的次数为 0。

```
1. thread1 got the lock
2. thread1 got the lock again
3. lock times : 2
4. thread1 release the lock
5. lock times : 1
6. thread1 release the lock again
7. lock times : 0
```

第三个示例（见代码清单 9.7）主要看 ReentrantLock 的 Condition 的使用。通过 ReentrantLock 的 newCondition 方法能得到一个 Condition 对象，这个对象就能实现等待与唤醒操作。ReentrantLock 与 Condition 对象的 await、signalAll 方法各自对应着 synchronized 与 object 对象的 wait、notifyAll 方法。

代码清单 9.7　公平模式的实现示例 3

```java
1. public class ReentrantLockDemo3 {
2.     static ReentrantLock lock = new ReentrantLock();
3.     static Condition con = lock.newCondition();
4.
5.     public static void main(String[] args) throws InterruptedException {
6.         Thread thread1 = new Thread(() -> {
7.             try {
```

```
8.              lock.lock();
9.              System.out.println("thread1 waiting for signal");
10.             con.await();
11.             System.out.println("thread1 release the lock");
12.             lock.unlock();
13.         } catch (InterruptedException e) {
14.         }
15.     });
16.     Thread thread2 = new Thread(() -> {
17.         try {
18.             lock.lock();
19.             System.out.println("thread2 waiting for signal");
20.             con.await();
21.             System.out.println("thread2 release the lock");
22.             lock.unlock();
23.         } catch (InterruptedException e) {
24.         }
25.     });
26.     thread1.start();
27.     thread2.start();
28.     Thread.sleep(2000);
29.     lock.lock();
30.     con.signalAll();
31.     lock.unlock();
32. }
33. }
```

在代码清单 9.7 中，线程 1 和线程 2 都通过 lock 方法依次获得锁，但加锁后都调用了 Condition 对象的 await 方法进入等待。这里要注意，await 方法会将锁释放后再进行等待，所以两个线程都能依次获得锁。而当在主线程中调用 Condition 对象的 signalAll 方法唤醒所有等待的线程时，线程 1 和线程 2 都会依次被唤醒。唤醒后又会把原来的线程放到等待队列中重新去获取锁，所以还需要 unlock 方法进行锁释放操作，如下所示。

```
1. thread1 waiting for signal
2. thread2 waiting for signal
3. thread1 release the lock
4. thread2 release the lock
```

9.2 读写锁

Java 语法层面的 synchronized 锁和 JDK 提供的 ReentrantLock 都属于独占锁，也就是说，这些锁最多只能由一个线程所持有，其他线程得排队依次获取锁。但在有些场景下，为了提高并发会引入共享锁来与独占锁共同对外构成一个锁，这就叫读写锁（ReadWriteLock）。之所以叫读写锁，主要是因为它在使用中考虑了读和写这两种场景，读操作不会改变数据，可以多线程进行读操作，而写操作会改变数据，只能有一个线程进行写操作。读写锁在内部维护了一对锁（读锁和写锁），它通过将锁进行分离得到更高的并发性能。

在图 9.3 中，存在一个读写锁对象，其内部包含了读锁和写锁两个对象。假如存在 5 个线程，其中线程 1 和线程 2 想要获取读锁，则两个线程可以同时获取到读锁。但是写锁不可以共享，它是独占锁。比如线程 3、线程 4 和线程 5 都想要持有写锁，则只能一个一个线程轮着持有。

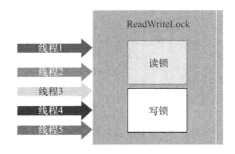

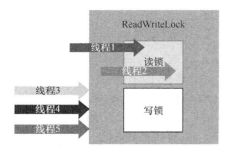
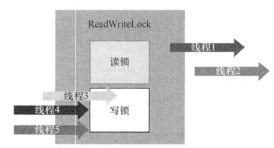

图 9.3　读写锁的演示

9.2.1　读写锁的性质

读写锁具有如下性质。
- 多个线程可以同时持有读锁，某个线程成功获取读锁后，其他线程仍然能成功获取读锁，即使该线程不释放读锁（见图 9.4 的左图）。
- 在某个线程持有读锁的情况下，其他线程不能持有写锁，除非持有读锁的线程全部都释放读锁（见图 9.4 的右图）。
- 在某个线程持有写锁的情况下，其他线程不能持有写锁或读锁。在某个线程成功获取写锁后，其他所有尝试获取读锁和写锁的线程都将进入等待状态，只有当该线程释放写锁后其他线程才能继续往下执行（见图 9.5 的左图）。

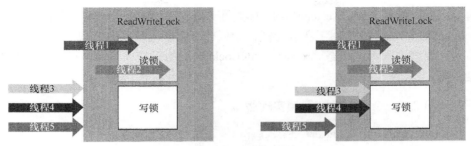

图 9.4 多个线程可同时持有读锁

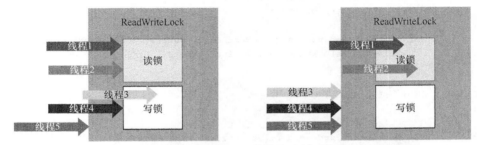

图 9.5 某个线程持有写锁的情况下其他线程就不能持有写锁或读锁

- 要获取读锁,则需要满足两个条件:目前没有线程持有写锁和目前没有线程请求获取写锁(见图 9.5 的右图)。
- 要获取写锁,则需要满足两个条件:目前没有线程持有写锁和目前没有线程持有读锁(见图 9.6)。

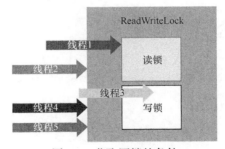

图 9.6 获取写锁的条件

9.2.2 简单的实现版本

为了加深读者对读写锁的理解,在分析 JDK 实现的读写锁之前,我们先来看一个简单的读写锁实现版本,如代码清单 9.8 所示。其中 3 个整型变量分别表示持有读锁的线程数、持有写锁的线程数以及请求获取写锁的线程数,4 个方法分别对应读锁、写锁各自的获取与释放操作。acquireReadLock 方法用于获取读锁,如果持有写锁的线程数或请求写锁的线程数大于 0,

则让线程进入等待状态。releaseReadLock 方法用于释放读锁，将读锁线程数减 1 并唤醒其他线程。acquireWriteLock 方法用于获取写锁，如果持有读锁的线程数量或持有写锁的线程数量大于 0，则让线程进入等待状态。releaseWriteLock 方法用于释放写锁，将写锁线程数减 1 并唤醒其他线程。

代码清单 9.8　一个简单的读写锁实现版本

```
1.  public class SimpleReadWriteLock {
2.      private int readLockNum = 0;
3.      private int writeLockNum = 0;
4.      private int writeRequests = 0;
5.
6.      public synchronized void acquireReadLock() throws InterruptedException {
7.          while (writeLockNum > 0 || writeRequests > 0) {
8.              wait();
9.          }
10.         readLockNum++;
11.     }
12.
13.     public synchronized void releaseReadLock() {
14.         readLockNum--;
15.         notifyAll();
16.     }
17.
18.     public synchronized void acquireWriteLock() throws InterruptedException {
19.         writeRequests++;
20.         while (readLockNum > 0 || writeLockNum > 0) {
21.             wait();
22.         }
23.         writeRequests--;
24.         writeLockNum++;
25.     }
26.
27.     public synchronized void releaseWriteLock() throws InterruptedException {
28.         writeLockNum--;
29.         notifyAll();
30.     }
31. }
```

9.2.3　读写锁的升级与降级

在某些场景下，我们希望某个已经拥有读锁的线程能够获得写锁，并将原来的读锁释放，这就涉及读锁升级为写锁的操作。读写锁的升级操作需要满足一定的条件，这个条件就是某个线程必须是唯一拥有读锁的线程，否则将无法成功升级。在图 9.7 中，线程 2 已经持有读锁，而且它是唯一的一个持有读锁的线程，所以它可以成功获得写锁。

与锁升级相对应的是锁降级。锁降级就是某个已经拥有写锁的线程希望能够获得读锁，并将原来的写锁释放。锁降级操作几乎没有什么风险，因为写锁是独占锁，持有写锁的线程肯定是唯一的，而且读锁也肯定不存在持有线程，所以写锁可以直接降级为读锁。在图 9.8 中，线程 3 持有写锁，此时其他线程不可能持有读锁和写锁，所以可以安全地将写锁降级为读锁。

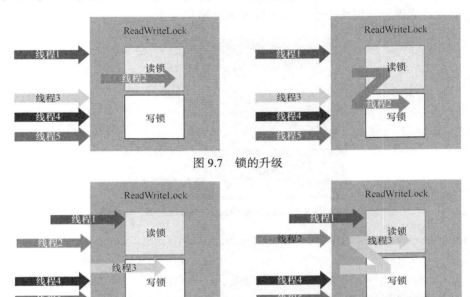

图 9.7 锁的升级

图 9.8 锁的降级

9.2.4 读写锁的实现思想

JDK 提供的读写锁 ReentrantReadWriteLock 实现了 ReadWriteLock 接口，它仅仅定义了 readLock 和 writeLock 这两个方法，分别表示获取读锁和获取写锁，如代码清单 9.9 所示。ReentrantReadWriteLock 类包含的属性和方法较多，为了做到简洁明了，我们将剔除非核心源码，只对核心功能进行分析。

代码清单 9.9　ReadWriteLock 接口

```
1.    public interface ReadWriteLock {
2.        Lock readLock();
3.        Lock writeLock();
4.    }
```

ReentrantReadWriteLock 类的三要素为公平/非公平模式、读锁和写锁。其中公平/非公平模式表示多个线程同时去获取锁时是否按照先到先得的顺序获得锁，如果是则为公平模式，否则为非公平模式。

总体来说，ReentrantReadWriteLock 类的内部包含了 ReadLock 内部类和 WriteLock 内部类，

分别对应读锁和写锁，这两种锁都提供了公平模式和非公平模式，如图 9.9 所示。不管是公平模式还是非公平模式，不管是读锁还是写锁，都是基于 AQS 同步器来实现的。实现的主要难点在于只使用一个 AQS 同步器来实现读锁和写锁，这就要求读锁和写锁共用同一个共享状态变量。下面会具体讲解如何用一个状态变量来供读锁和写锁使用。

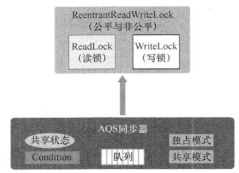

图 9.9　ReentrantReadWriteLock 类

ReentrantReadWriteLock.ReadLock 和 ReentrantReadWriteLock.WriteLock 分别为读锁和写锁，如代码清单 9.10 所示。Sync 对象表示 ReentrantReadWriteLock 类的同步器，它基于 AQS 同步器，而 FairSync 类和 NonfairSync 类分别表示公平模式和非公平模式的同步器，可以看到默认情况下使用的是非公平模式。

代码清单 9.10　ReentrantReadWriteLock.ReadLock 和 ReentrantReadWriteLock.WriteLock

```
1.  public class ReentrantReadWriteLock implements ReadWriteLock, java.io.Serializable {
2.      private final ReentrantReadWriteLock.ReadLock readerLock;
3.      private final ReentrantReadWriteLock.WriteLock writerLock;
4.      final Sync sync;
5.
6.      public ReentrantReadWriteLock() {
7.          this(false);
8.      }
9.
10.     public ReentrantReadWriteLock(boolean fair) {
11.         sync = fair ? new FairSync() : new NonfairSync();
12.         readerLock = new ReadLock(this);
13.         writerLock = new WriteLock(this);
14.     }
15.
16.     public ReentrantReadWriteLock.WriteLock writeLock() {
17.         return writerLock;
18.     }
19.
20.     public ReentrantReadWriteLock.ReadLock readLock() {
21.         return readerLock;
22.     }
```

```
23.
24.        ......
25.
26.  }
```

ReadLock 与 WriteLock 都属于 ReentrantReadWriteLock 的内部类，它们都实现了 Lock 接口，这里主要关注 lock、unlock 和 newCondition 这 3 个核心方法，分别表示对读锁和写锁的加锁操作、释放锁操作和创建 Condition 对象操作，如代码清单 9.11 所示。可以看到这些方法都间接调用了 ReentrantReadWriteLock 的同步器的方法。需要注意的是，读锁不支持创建 Condition 对象。

代码清单 9.11　ReadLock 与 WriteLock 类中的核心方法

```
1.  public static class ReadLock implements Lock, java.io.Serializable {
2.      private final Sync sync;
3.
4.      protected ReadLock(ReentrantReadWriteLock lock) {
5.          sync = lock.sync;
6.      }
7.
8.      public void lock() {
9.          sync.acquireShared(1);
10.     }
11.
12.     public void unlock() {
13.         sync.releaseShared(1);
14.     }
15.
16.     public Condition newCondition() {
17.         throw new UnsupportedOperationException();
18.     }
19. }
20.
21. public static class WriteLock implements Lock, java.io.Serializable {
22.     private final Sync sync;
23.
24.     protected WriteLock(ReentrantReadWriteLock lock) {
25.         sync = lock.sync;
26.     }
27.
28.     public void lock() {
29.         sync.acquire(1);
30.     }
31.
32.     public void unlock() {
33.         sync.release(1);
34.     }
35.
36.     public Condition newCondition() {
```

```
37.        return sync.newCondition();
38.    }
39. }
```

9.2.5 读写锁的共用状态变量

ReentrantReadWriteLock 实现的难点在于读锁和写锁都共用一个共享变量，下面介绍具体是如何共用的。我们知道 AQS 同步器的共享状态是整型的（32 位），那么最简单的共用方式就是读锁和写锁分别使用 16 位。其中高 16 位用于读锁的状态，而低 16 位则用于写锁的状态，如图 9.10 所示。但是这样设计后，如果要获取读锁和写锁的状态值，则需要一些额外的计算（比如移位和逻辑与操作）。

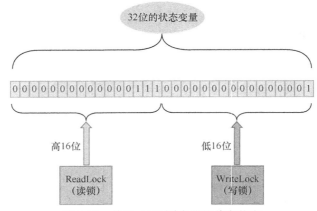

图 9.10　共用 AQS 同步器的共享状态

ReentrantReadWriteLock 中共用状态变量的逻辑如代码清单 9.12 所示。其中 SHARED_SHIFT 表示移动的位数为 16；SHARED_UNIT 表示读锁在每次加锁时对应的状态值大小，将 1 左移 16 位刚好对应高 16 位的 1；MAX_COUNT 表示读锁能执行加锁操作的最大次数，值为 16 个 1（二进制）；EXCLUSIVE_MASK 表示写锁的掩码，值为 16 个 1（二进制）。sharedCount 方法用于获取读锁（高 16 位）的状态值（左移 16 位就能得到）。exclusiveCount 方法用于获取写锁（低 16 位）的状态值（通过掩码就能得到）。

代码清单 9.12　ReentrantReadWriteLock 中共用状态变量的逻辑

```
1.  abstract static class Sync extends AbstractQueuedSynchronizer {
2.
3.      static final int SHARED_SHIFT = 16;
4.      static final int SHARED_UNIT = (1 << SHARED_SHIFT);
5.      static final int MAX_COUNT = (1 << SHARED_SHIFT) - 1;
6.      static final int EXCLUSIVE_MASK = (1 << SHARED_SHIFT) - 1;
7.
```

```
8.      static int sharedCount(int c) {
9.          return c >>> SHARED_SHIFT;
10.     }
11.
12.     static int exclusiveCount(int c) {
13.         return c & EXCLUSIVE_MASK;
14.     }
15.     ......
16.     ......
17.
18. }
```

9.2.6 读写锁的公平/非公平模式

ReentrantReadWriteLock 的默认模式为非公平模式,其内部类 Sync 是公平模式 FairSync 类和非公平模式 NonfairSync 类的抽象父类。因为 ReentrantReadWriteLock 的读锁使用了共享模式,而写锁使用了独占模式,所以该父类将不同模式下的公平机制抽象成 readerShouldBlock 和 writerShouldBlock 这两个抽象方法,然后子类就可以各自实现不同的公平模式,如代码清单 9.13 所示。换句话说,ReentrantReadWriteLock 的公平机制由这两个方法来决定。

代码清单 9.13　ReentrantReadWriteLock 的公平机制

```
1. abstract static class Sync extends AbstractQueuedSynchronizer {
2.     ......
3.     abstract boolean readerShouldBlock();
4.     abstract boolean writerShouldBlock();
5.     ......
6. }
```

公平模式的 FairSync 类的 readerShouldBlock 和 writerShouldBlock 方法都直接返回 hasQueuedPredecessors 方法的结果,如代码清单 9.14 所示。它是 AQS 同步器的方法,用于判断当前线程的前面是否有排队的线程。如果有,就要让当前线程也加入排队队列中,这样按照队列顺序获取锁也就保证了公平性。

非公平模式的 NonfairSync 类的 writerShouldBlock 方法直接返回 false,表示不让当前线程进入排队队列中,而是直接进行锁的获取竞争,如代码清单 9.14 所示。readerShouldBlock 方法则调用 apparentlyFirstQueuedIsExclusive 方法,它是 AQS 同步器的方法,用于判断头节点的下一个节点线程是否在请求获取独占锁(写锁)。如果是,则让其他线程先获取写锁,而自己则去排队。如果不是,则说明下一个节点线程是请求共享锁(读锁),此时直接与之竞争读锁。

代码清单 9.14　公平模式的 FairSync 类和非公平模式的 NonfairSync 类

```
1. static final class FairSync extends Sync {
2.
```

```
 3.        final boolean writerShouldBlock() {
 4.            return hasQueuedPredecessors();
 5.        }
 6.
 7.        final boolean readerShouldBlock() {
 8.            return hasQueuedPredecessors();
 9.        }
10.    }
11.
12.    static final class NonfairSync extends Sync {
13.
14.        final boolean writerShouldBlock() {
15.            return false;
16.        }
17.
18.        final boolean readerShouldBlock() {
19.            return apparentlyFirstQueuedIsExclusive();
20.        }
21.    }
```

9.2.7 写锁的实现

写锁（WriteLock）有 lock 和 unlock 两个核心方法，它们都间接调用了 ReentrantReadWriteLock 内部同步器的对应方法。在同步器中需要重写 tryAcquire 方法和 tryRelease 方法，分别用于获取写锁和释放写锁操作。

先看 tryAcquire 方法的逻辑（见代码清单 9.15）。它获取状态值并通过 exclusiveCount 方法得到低 16 位的写锁状态值。c!=0 时有两种情况：高 16 位的读锁状态不为 0；低 16 位的写锁状态不为 0。w==0 时表示还有线程持有读锁，直接返回 false 表示获取写锁失败。如果持有写锁的线程为当前线程，则表示写锁重入操作，此时需要将状态变量进行累加，还需要校验写锁重入状态值，确保其不能超过 MAX_COUNT。writerShouldBlock 方法用于判断是否需要将当前线程放入排队队列中，compareAndSetState 方法用于对状态变量进行累加操作，如果 CAS 失败，也需要将当前线程放入排队队列中。对于非公平模式，这里的 CAS 操作就是指闯入操作，即线程先尝试一次竞争写锁。最后通过 setExclusiveOwnerThread 方法设置当前线程持有写锁，该方法只是简单的 set 方法。

继续看 tryRelease 方法的逻辑（见代码清单 9.15）。它首先用 isHeldExclusively 方法检查当前线程是否为写锁持有线程，如果是，则将状态值减去释放的值，并通过 exclusiveCount 得到低 16 位的写锁状态值。如果其值为 0，则表示已经没有重入，可以彻底释放锁。调用 setExclusiveOwnerThread (null) 来确保没有线程持有写锁。最后设置新的状态值。

代码清单 9.15　tryAcquire 方法和 tryRelease 方法的逻辑

```
 1.    abstract static class Sync extends AbstractQueuedSynchronizer {
```

```
2.
3.          ......
4.
5.      protected final boolean tryAcquire(int acquires) {
6.          Thread current = Thread.currentThread();
7.          int c = getState();
8.          int w = exclusiveCount(c);
9.          if (c != 0) {
10.             if (w == 0 || current != getExclusiveOwnerThread())
11.                 return false;
12.             if (w + exclusiveCount(acquires) > MAX_COUNT)
13.                 throw new Error("Maximum lock count exceeded");
14.             setState(c + acquires);
15.             return true;
16.         }
17.         if (writerShouldBlock() || !compareAndSetState(c, c + acquires))
18.             return false;
19.         setExclusiveOwnerThread(current);
20.         return true;
21.     }
22.
23.     protected final boolean tryRelease(int releases) {
24.         if (!isHeldExclusively())
25.             throw new IllegalMonitorStateException();
26.         int nextc = getState() - releases;
27.         boolean free = exclusiveCount(nextc) == 0;
28.         if (free)
29.             setExclusiveOwnerThread(null);
30.         setState(nextc);
31.         return free;
32.     }
33.
34.         ......
35.
36. }
```

9.2.8 读锁的实现

写锁 ReadLock 同样有 lock 和 unlock 两个核心方法，它们间接调用了 ReentrantReadWriteLock 内部同步器的对应方法。在同步器中需要重写 tryAcquireShared 方法和 tryReleaseShared 方法，分别用于获取读锁和释放读锁操作。

先看 tryAcquireShared 方法的逻辑（见代码清单 9.16）。它首先通过 getState 方法获取状态值，然后通过 exclusiveCount 方法获取低 16 位的写锁状态。如果不为 0，则表示有其他线程持有写锁而且当前线程没有持有写锁，此时尝试获取读锁失败并返回-1，即将当前线程放到排队队列。注意，这里如果当前线程持有写锁，则可以继续获取读锁。继续通过 sharedCount 得到高 16 位的读锁，然后尝试用 CAS 设置新的状态值，如果成功，则返回 1，表示成功获取读锁。如果不成功，则继续调用 fullTryAcquireShared 方法。

继续看 fullTryAcquireShared 方法的逻辑（见代码清单 9.16）。它包括一个自旋操作，首先获取状态值，如果写锁不为 0 且当前线程不为持有写锁线程，则返回-1，表示尝试获取读锁失败，将当前线程加入排队队列中。如果写锁的状态为 0，则表示没有线程持有写锁，然后继续通过 readerShouldBlock 方法判断是否需要将该线程加入到排队队列中。如果需要，则返回 -1，AQS 同步器会将其加入到排队队列中。此外，读锁的状态值不能等于 MAX_COUNT（即已经达到最大读锁数）。最后，通过 compareAndSetState 方法设置新的状态值。需要注意的是，在非公平模式下，如果排队队列中下一个线程是要获取写锁，则这个自旋操作也会被打破。

再看 tryReleaseShared 方法的逻辑（见代码清单 9.16）。它通过 for 循环实现自旋，自旋的逻辑就是不断计算新的状态值，然后通过 compareAndSetState 方法设置新的状态值。

代码清单 9.16　tryAcquireShared 方法、fullTryAcquireShared 方法和 tryReleaseShared 方法的逻辑

```
1.   abstract static class Sync extends AbstractQueuedSynchronizer {
2.       ......
3.       protected final boolean tryReleaseShared(int unused) {
4.           for (;;) {
5.               int c = getState();
6.               int nextc = c - SHARED_UNIT;
7.               if (compareAndSetState(c, nextc))
8.                   return nextc == 0;
9.           }
10.      }
11.
12.      protected final int tryAcquireShared(int unused) {
13.          Thread current = Thread.currentThread();
14.          int c = getState();
15.          if (exclusiveCount(c) != 0 && getExclusiveOwnerThread() != current)
16.              return -1;
17.          int r = sharedCount(c);
18.          if (!readerShouldBlock() && r < MAX_COUNT && compareAndSetState(c, c + SHARED_UNIT)) {
19.              return 1;
20.          }
21.          return fullTryAcquireShared(current);
22.      }
23.
24.      final int fullTryAcquireShared(Thread current) {
25.          for (;;) {
26.              int c = getState();
27.              if (exclusiveCount(c) != 0) {
28.                  if (getExclusiveOwnerThread() != current)
29.                      return -1;
30.              } else if (readerShouldBlock()) {
31.                  return -1;
32.              }
```

```
33.            if (sharedCount(c) == MAX_COUNT)
34.                throw new Error("Maximum lock count exceeded");
35.            if (compareAndSetState(c, c + SHARED_UNIT))
36.                return 1;
37.        }
38.    }
39.
40.    ......
41.
42. }
```

9.2.9 读写锁的使用示例

代码清单 9.17 是一个读写锁的使用示例。其中实例化了一个 ReentrantReadWriteLock 对象，然后通过它的读锁和写锁来控制对某个线程不安全的 TreeMap 对象的访问。可以看到 get 方法属于读取数据的操作，所以使用共享的读锁即可。而 put 和 clear 这两个方法涉及修改数据的操作，需要使用独占的写锁。

代码清单 9.17 读写锁的使用示例

```
1.  public class ReentrantReadWriteLockDemo {
2.      private final Map<String, String> m = new TreeMap<String, String>();
3.      private final ReentrantReadWriteLock rwl = new ReentrantReadWriteLock();
4.      private final Lock r = rwl.readLock();
5.      private final Lock w = rwl.writeLock();
6.
7.      public String get(String key) {
8.          r.lock();
9.          try {
10.             return m.get(key);
11.         } finally {
12.             r.unlock();
13.         }
14.     }
15.
16.     public String put(String key, String value) {
17.         w.lock();
18.         try {
19.             return m.put(key, value);
20.         } finally {
21.             w.unlock();
22.         }
23.     }
24.
25.     public void clear() {
26.         w.lock();
27.         try {
```

```
28.         m.clear();
29.     } finally {
30.         w.unlock();
31.     }
32. }
33. }
```

9.3 锁的条件机制

我们知道，线程的阻塞与唤醒一共有 suspend/resume、wait/notify、park/unpark 这 3 种模式（后面两种比较常用），其中 wait/notify 模式对应的是锁的条件机制，如图 9.11 所示。为什么这样说呢？因为它们都属于条件队列，强调的重点是将所有等待的线程组织起来作为等待集，这些等待集中的线程在等待对应的条件变为真。

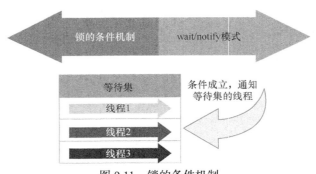

图 9.11　锁的条件机制

9.3.1　wait/notify 模式

一般在使用 wait/notify 模式时，并非只使用 wait 和 notify 方法，实际上会更多地使用 notifyAll 方法来替代 notify 方法。它们之间的区别主要在通知等待集中线程的数量，notify 只会通知其中一个，而 notifyAll 则会通知等待集中的所有线程。

下面先看 wait/notify 模式如何通过等待集来维护等待线程，从而能更好地理解条件队列的工作方式。在图 9.12 中，假设开始一共有 5 个线程一起竞争锁对象，此时等待集为空，然后线程 3 成功获得锁进而往下执行。接着线程 3 调用了锁对象的 wait 方法，使得该线程进入到等待集中并释放该锁。

释放锁后，其他 4 个线程开始竞争锁，其中线程 4 成功获得锁。它调用了锁对象的 notify 方法，从而将等待集中的线程 3 移除，并重新加入到锁竞争行列中。此时线程 4 还未释放锁，只有等到线程 4 释放锁后，其他线程才能继续获得锁。假设线程 4 释放锁后再次被线程 3 成功获得锁，那么它就能继续往下执行，它释放锁后线程 2 继续获得锁并执行，如图 9.13 所示。

9.3 锁的条件机制

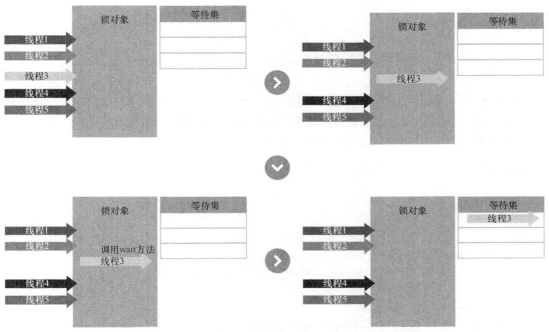

图 9.12 wait/notify 模式通过等待集维护等待线程

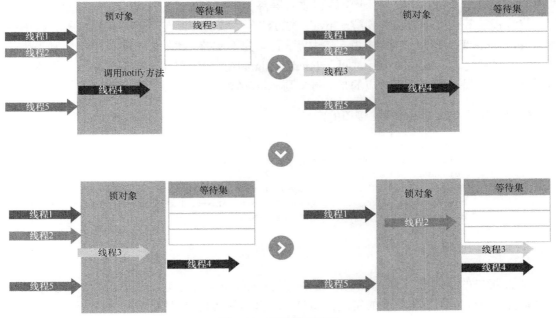

图 9.13 竞争锁的流程

9.3.2　Condition

在 JDK 1.5 之前，条件队列只能通过 wait/notify 模式实现，直到 JDK 1.5 时引入 Lock 接口后才有了另外一种选择方案。实际上，Lock 接口的 Condition（条件）机制就是为了替代 wait/notify 模式，Condition 由 Lock 接口定义。这是因为条件队列必须在加锁的情况下使用，而实现则由 AQS 的 Condition 队列提供支持。代码清单 9.18 是 Lock 接口的定义，我们要关注的是 newCondition 方法，它会返回 Condition 对象（也就是条件队列）。

代码清单 9.18　Lock 接口的定义

```
1.   public interface Lock {
2.       void lock();
3.       void unlock();
4.       boolean tryLock();
5.       boolean tryLock(long time, TimeUnit unit) throws InterruptedException;
6.       void lockInterruptibly() throws InterruptedException;
7.       Condition newCondition();
8.   }
```

Condition 也是一个接口，具体实现由 AQS 同步器负责，第 5 章已经深入分析过，这里不再赘述。Condition 接口的定义如代码清单 9.19 所示。从中可以看到主要定义了 await 和 signal 两类方法，其中 await 类方法用于将线程放入等待集中使之处于等待状态，而 signal 类方法则用于将线程从等待集中取出并唤醒。await 类方法分为支持中断与不支持中断两种，其中声明抛出 InterruptedException 异常的为支持中断，否则为不支持。此外还提供了多个超时等待的方法，传入的时间表示最大的等待时间。signal 与 signalAll 的区别在于唤醒的是一个等待线程还是所有等待线程，与它们相对应的是 notify 与 notifyAll。

代码清单 9.19　Condition 接口的定义

```
1.   public interface Condition {
2.       void await() throws InterruptedException;
3.       void awaitUninterruptibly();
4.       long awaitNanos(long nanosTimeout) throws InterruptedException;
5.       boolean await(long time, TimeUnit unit) throws InterruptedException;
6.       boolean awaitUntil(Date deadline) throws InterruptedException;
7.       void signal();
8.       void signalAll();
9.   }
```

Lock 接口的具体实现由 AQS 同步器内部的 Condition 来实现（见图 9.14），所以最终的核心内容就是 AQS 的 Condition 的实现，如果忘记具体的实现可以回到第 5 章中查看。

9.3 锁的条件机制

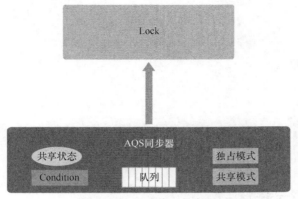

图 9.14 Lock 接口的实现

第 10 章 任务执行器

10.1 任务执行器接口

任务执行器 Executor 是一个接口，位于 java.util.concurrent 包下，它的作用主要是提供任务与执行机制（包括线程的使用和调度细节）之间的解耦。比如我们定义了一个任务，接下来是通过线程池来执行该任务呢，还是直接创建线程来执行该任务呢？通过 Executor 就能为任务提供不同的执行机制。执行器的实现方式各种各样，常见的执行器包括同步执行器、一对一执行器、线程池执行器、串行执行器等，如图 10.1 所示。下面分别介绍这 4 种执行器，以帮助我们更好地理解执行器的概念。

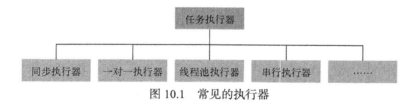

图 10.1　常见的执行器

10.1.1　同步执行器

同步执行器是最简单的执行器，提交给它的任务将由调用线程直接执行，不需要其他线程的帮助。在代码清单 10.1 中，DirectExecutor 继承了 Executor，它的 execute 方法直接调用了 Runnable 的 run 方法。在主线程中创建 DirectExecutor 对象后执行 MyTask 任务，实际上该任务将由调用线程（主线程）直接执行，最终输出 "executing task..."。

代码清单 10.1　同步执行器

```
1.   public class ExecutorTest1 {
2.
3.      public static void main(String[] args) {
4.          Executor executor = new DirectExecutor();
5.          executor.execute(new MyTask());
```

```
6.      }
7.
8.      static class MyTask implements Runnable {
9.          public void run() {
10.             System.out.println("executing task...");
11.         }
12.     }
13.
14.     static class DirectExecutor implements Executor {
15.         public void execute(Runnable r) {
16.             r.run();
17.         }
18.     }
19. }
```

10.1.2 一对一执行器

一对一执行器就是一个任务由一个线程负责,每个任务在提交给执行器时都将创建一个新的线程来执行该任务。在代码清单 10.2 中,ThreadPerTaskExecutor 的 execute 方法会新建一个线程来执行任务。

代码清单 10.2 一对一执行器

```
1.  public class ExecutorTest2 {
2.
3.      public static void main(String[] args) {
4.          Executor executor = new ThreadPerTaskExecutor();
5.          executor.execute(new MyTask());
6.      }
7.
8.      static class MyTask implements Runnable {
9.          public void run() {
10.             System.out.println("executing task...");
11.         }
12.     }
13.
14.     static class ThreadPerTaskExecutor implements Executor {
15.         public void execute(Runnable r) {
16.             new Thread(r).start();
17.         }
18.     }
19. }
```

10.1.3 线程池执行器

线程池执行器就是拥有线程池功能的执行器,任务在提交后将由线程池负责执行。下面来模拟实现一个简单的线程池,如代码清单 10.3 所示。ThreadPoolExecutor 类包含了任务队列

taskQueue 和 10 个工作线程 workers，构造函数中会创建 10 个线程并通过 while(true)循环不断从任务队列中取出任务并执行。而 execute 方法则是将任务添加到任务队列中，然后工作线程会执行任务队列中的任务。

代码清单 10.3　模拟实现一个简单的线程池

```java
public class ExecutorTest3 {

    public static void main(String[] args) {
        Executor executor = new ThreadPoolExecutor();
        executor.execute(new MyTask());
        executor.execute(new MyTask());
    }

    static class MyTask implements Runnable {
        public void run() {
            System.out.println("executing task...");
        }
    }

    static class ThreadPoolExecutor implements Executor {

        List<Runnable> taskQueue = new LinkedList<Runnable>();
        Thread[] workers = new Thread[10];

        public ThreadPoolExecutor() {
            for (int i = 0; i < workers.length; i++) {
                workers[i] = new Thread(() -> {
                    while (true) {
                        Runnable task = null;
                        synchronized (taskQueue) {
                            if (!taskQueue.isEmpty())
                                task = taskQueue.remove(0);
                        }
                        if (task != null)
                            task.run();
                    }
                });
                workers[i].start();
            }
        }

        public void execute(Runnable r) {
            synchronized (taskQueue) {
                taskQueue.add(r);
            }
        }
    }
}
```

10.1.4　串行执行器

串行执行器是一种具有串行功能的执行器，所有任务首先被加入到一个先进先出的队列中，然后内部的另外一个执行器会按照队列的顺序执行任务，且前一任务在执行完后负责启动后一任务的执行，这样就形成了串行。我们看一下串行执行器的简单实现，如代码清单 10.4 所示。

代码清单 10.4　串行执行器

```java
1.  public class ExecutorTest4 {
2.
3.      public static void main(String[] args) {
4.          Executor executor = new SerialExecutor();
5.          for (int i = 0; i < 5; i++)
6.              executor.execute(new MyTask("task" + i));
7.      }
8.
9.      static class MyTask implements Runnable {
10.         String name;
11.
12.         public MyTask(String name) {
13.             this.name = name;
14.         }
15.
16.         public void run() {
17.             System.out.println("executing " + name + " task...");
18.         }
19.     }
20.
21.     static class SerialExecutor implements Executor {
22.         final Queue<Runnable> tasks = new ArrayDeque<Runnable>();
23.         final Executor executor = new ThreadPerTaskExecutor();
24.         Runnable active;
25.
26.         public synchronized void execute(final Runnable r) {
27.             tasks.offer(new Runnable() {
28.                 public void run() {
29.                     try {
30.                         r.run();
31.                     } finally {
32.                         if ((active = tasks.poll()) != null)
33.                             executor.execute(active);
34.                     }
35.                 }
36.             });
37.             if (active == null)
38.                 if ((active = tasks.poll()) != null
```

```
39.                        executor.execute(active);
40.            }
41.        }
42. }
```

SerialExecutor 类中包含了一个任务队列和执行器，这里使用了 ThreadPerTaskExecutor 执行器。SerialExecutor 的 execute 方法负责将任务加入到队列中，而且还负责启动第一个任务的执行。finally 块主要负责启动下一个任务，从而形成环环相扣的执行顺序。

代码清单 10.4 执行后的输出结果如下。

```
1.  executing task0 task...
2.  executing task1 task...
3.  executing task2 task...
4.  executing task3 task...
5.  executing task4 task...
```

Executor 只是一个接口，它提供了任务和执行的解耦机制。以上分析了几种常见执行器的实现，大家在实际的工程中可以根据情况来设计任务执行器。

10.2 任务执行器的 ExecutorService 接口

Executor 是最简单的任务执行器接口，它仅仅定义了一个方法，即 void execute(Runnable command)。然而在实际工程中我们可能需要对任务进行某些控制，或者对任务执行器的生命周期进行管理，此时 Executor 接口就无法满足要求了。ExecutorService 接口应运而生，它是 Executor 接口的加强版。

ExecutorService 接口位于 java.util.concurrent 包下（见代码清单 10.5），该接口继承了 Executor 接口，并加强了对执行器的控制，且能执行有返回值的任务。ExecutorService 接口增加了一些新方法，比如关闭执行器的方法、查看执行器状态的方法、提交 Callable 任务的方法等。该接口包含的方法较多，由于我们的重点是理解执行器的设计思路，因此这里只列出了一些常用的方法。

- shutdown 方法：用于关闭执行器，调用该方法后停止接收新任务，但已经提交给执行器的任务将继续执行完毕。
- isShutdown 方法：判断执行器是否已关闭。
- isTerminated 方法：判断执行器是否已终止。
- awaitTermination 方法：阻塞等待执行器直到其终止。
- submit(Callable)方法：向执行器提交一个具有返回值的任务。

代码清单 10.5 ExecutorService 接口

```
1. public interface ExecutorService extends Executor {
```

```
2.      void shutdown();
3.      boolean isShutdown();
4.      boolean isTerminated();
5.      void awaitTermination() throws InterruptedException;
6.      <T> Future<T> submit(Callable<T> task);
7.  }
```

下面通过实现一个简易的任务执行器来加深读者对 ExecutorService 接口的理解。该任务执行器中主要包括任务队列和工作线程，工作线程不断从任务队列取出任务并执行，如图 10.2 所示。同时根据 ExecutorService 接口的定义，我们需要实现执行器的关闭操作及状态查询，还需要能执行具有返回值的任务。

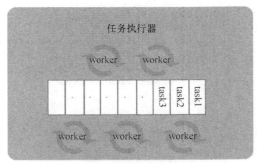

图 10.2　简易的任务执行器

我们定义一个 MyExecutorService 类，该类实现了 ExecutorService 接口，如代码清单 10.6 所示。先看看属性，isShutdown 和 isTerminated 分别表示是否关闭和是否已终止。taskQueue 是一个列表结构的任务队列。workers 是工作线程，这里创建了 5 个工作线程，count 之所以使用 AtomicInteger，是为了让执行器能在多线程中正确关闭。lock 用于实现线程阻塞的通知功能，因为 awaitTermination 方法会一直阻塞，直到执行器被终止。接着看构造函数，我们会在构造函数中创建工作线程并启动它们，启动后的线程不断从任务队列中取出任务并执行，它会根据 isShutdown 标识决定是否要跳出循环，跳出循环则意味着工作线程结束。由于希望在关闭任务管理器时能将原来在任务队列中的队列执行完，所以还加了 taskQueue.isEmpty() 作为判断条件。

代码清单 10.6　MyExecutorService 类

```
1.  public class MyExecutorService implements ExecutorService {
2.
3.      volatile boolean isShutdown = false;
4.      volatile boolean isTerminated = false;
5.
6.      List<Runnable> taskQueue = new LinkedList<Runnable>();
7.      AtomicInteger count = new AtomicInteger(5);
8.      Thread[] workers = new Thread[count.get()];
9.
```

```
10.     ReentrantLock lock = new ReentrantLock();
11.     Condition termination = lock.newCondition();
12.
13.     public MyExecutorService() {
14.         for (int i = 0; i < workers.length; i++) {
15.             workers[i] = new Thread(() -> {
16.                 while (true) {
17.                     Runnable task = null;
18.                     synchronized (taskQueue) {
19.                         if (!taskQueue.isEmpty()) {
20.                             task = taskQueue.remove(0);
21.                         }
22.                     }
23.                     if (task != null)
24.                         task.run();
25.                     if (taskQueue.isEmpty() && isShutdown) {
26.                         if (count.decrementAndGet() == 0) {
27.                             lock.lock();
28.                             isTerminated = true;
29.                             termination.signalAll();
30.                             lock.unlock();
31.                         }
32.                         break;
33.                     }
34.                 }
35.             });
36.             workers[i].start();
37.         }
38.     }
39.
40.     ...
41.
42. }
```

下面看各个方法的具体实现，如代码清单 10.7 所示。shutdown 直接将 isShutdown 置为 true。结合 execute 方法来看，当 isShutdown 为 true 时，则任务不会继续被添加到任务队列中，而且工作线程也会在任务队列为空时死亡。再看 awaitTermination 方法，它只需直接进入阻塞即可，在任务执行器终止时会通过 termination 条件唤醒阻塞的线程。submit 方法将传入的 Callable 对象封装成 FutureTask 对象，然后添加到任务队列中，最后返回 Future 对象。

代码清单 10.7　各个方法的具体实现

```
1.  public void shutdown() {
2.      isShutdown = true;
3.  }
4.
5.  public boolean isShutdown() {
6.      return isShutdown;
7.  }
```

```java
8.
9.     public boolean isTerminated() {
10.        return isTerminated;
11.    }
12.
13.    public void execute(Runnable r) {
14.        synchronized (taskQueue) {
15.            if (!isShutdown)
16.                taskQueue.add(r);
17.        }
18.    }
19.
20.    public void awaitTermination() throws InterruptedException {
21.        lock.lock();
22.        termination.await();
23.        lock.unlock();
24.    }
25.
26.    public <T> Future<T> submit(Callable<T> task) {
27.        RunnableFuture<T> ftask = new FutureTask<T>(task);
28.        execute(ftask);
29.        return ftask;
30.    }
```

根据实现的 MyExecutorService 类来编写第一个示例，如代码清单 10.8 所示。

代码清单 10.8　根据 MyExecutorService 类编写的示例 1

```java
1.  public class ExecutorServiceTest {
2.
3.      public static void main(String[] args) throws InterruptedException {
4.          ExecutorService executor = new MyExecutorService();
5.          for (int i = 0; i < 10; i++)
6.              executor.execute(new MyTask("task" + i));
7.      System.out.println("executor isShutdown = " + executor.isShutdown());
8.      System.out.println("executor isTerminated = " + executor.isTerminated());
9.          new Thread(() -> {
10.             try {
11.                 Thread.sleep(10000);
12.             } catch (InterruptedException e) {
13.             }
14.             executor.shutdown();
15.         }).start();
16.         executor.awaitTermination();
17.     System.out.println("executor isShutdown = " + executor.isShutdown());
18.     System.out.println("executor isTerminated = " + executor.isTerminated());
19.         System.out.println("executor has terminated.");
20.     }
21.
22.     static class MyTask implements Runnable {
23.         String name;
```

```
24.
25.        public MyTask(String name) {
26.            this.name = name;
27.        }
28.
29.        public void run() {
30.            try {
31.                Thread.sleep(1000);
32.            } catch (InterruptedException e) {
33.            }
34.            System.out.println("executing " + name + " task...");
35.        }
36.    }
37.
38. }
```

MyTask 对象睡眠 1s, 以下为代码清单 10.8 执行后的输出结果。由于执行器中只有 5 个工作线程, 所以最多只能有 5 个任务并发执行。主线程启动了另外一个线程, 在睡眠 10s 后对执行器进行 shutdown 操作。主线程调用 awaitTermination 方法后开始阻塞, 直到执行器终止后, 主线程才能继续往下执行。

```
1.  executor isShutdown = false
2.  executor isTerminated = false
3.  executing task3 task...
4.  executing task4 task...
5.  executing task2 task...
6.  executing task1 task...
7.  executing task0 task...
8.  executing task9 task...
9.  executing task8 task...
10. executing task5 task...
11. executing task7 task...
12. executing task6 task...
13. executor isShutdown = true
14. executor isTerminated = true
15. executor has terminated.
```

第二个示例用于展示具有可返回值的任务, 如代码清单 10.9 所示。MyTask 需要实现 Callable 接口, 然后在 call 方法中定义任务。将任务提交给执行器后调用 Future 的 get 方法将使主线程进入阻塞状态, 直到任务执行完毕返回结果, 最终输出 "receive result : task_result"。

代码清单 10.9　根据 MyExecutorService 类编写的示例 2

```
1.  public class ExecutorServiceTest2 {
2.      public static void main(String[] args) {
3.          MyExecutorService executor = new MyExecutorService();
4.          Future<String> future = executor.submit(new MyTask());
5.          try {
6.              String result = future.get();
```

```
7.                System.out.println("receive result : " + result);
8.            } catch (InterruptedException | ExecutionException e) {
9.                e.printStackTrace();
10.           }
11.           executor.shutdown();
12.       }
13.
14.       static class MyTask implements Callable<String> {
15.
16.           public String call() throws Exception {
17.               Thread.sleep(3000);
18.               return "task_result";
19.           }
20.
21.       }
22.   }
```

10.3 线程池任务执行器

JDK 提供的一种最常见的任务执行器是线程池任务执行器 ThreadPoolExecutor，从名字可以看出它的核心是线程池，该类位于 java.util.concurrent 包下，它实现了 ExecutorService 接口。图 10.3 为 ThreadPoolExecutor 核心组件示意图，BlockingQueue 是用于保存任务的阻塞队列，ThreadFactory 是线程工厂，RejectedExecutionHandler 是满负荷时的拒绝策略。此外，corePoolSize 是核心工作线程数量，而 maximumPoolSize 是最大的工作线程数量。

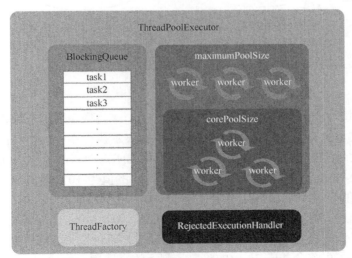

图 10.3　ThreadPoolExecutor 核心组件示意图

10.3.1 线程池任务执行器的运行状态

ThreadPoolExecutor 内部包含了 5 个运行状态：RUNNING、SHUTDOWN、STOP、TIDYING、TERMINATED。图 10.4 所示为 5 个状态及状态之间的转换示意图。

- RUNNING 状态：能正常接收新任务和正常处理队列中的任务。
- SHUTDOWN 状态：不接收新任务但正常处理队列中的任务。
- STOP 状态：不接收新任务且不处理队列中的任务，同时中断正在执行的任务。
- TIDYING 状态：所有任务都已被终止，没有工作线程，钩子方法 terminated()即将被执行。
- TERMINATED 状态：钩子方法 terminated()执行完毕。

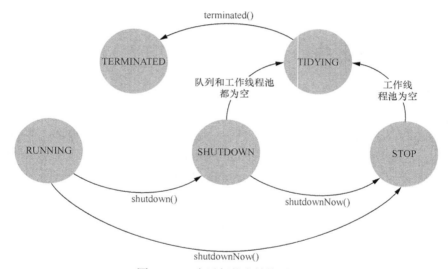

图 10.4　5 个运行状态转换示意图

为了提高使用率，ThreadPoolExecutor 只使用一个 int 类型来表示运行状态和工作线程数。其中高 3 位用来表示执行器的运行状态，而低 29 位则表示执行器中工作线程的数量，如图 10.5 所示。RUNNING 的值为-536870912（高 3 位为 111），SHUTDOWN 的值为 0（高 3 位为 000），STOP 的值为 536870912（高 3 位为 001），TIDYING 的值为 1073741824（高 3 位为 010），TERMINATED 的值为 1610612736（高 3 位为 011）。

对应上面的描述，代码清单 10.10 所示为相关的代码。AtomicInteger 类型的 ctl 变量用来表示运行状态和工作线程数，workerCountOf 方法通过掩码计算工作线程数量，ctlOf 方法用于将运行状态和工作线程数合并成 32 位的整数。剩下的方法用于比较判断运行状态的值，根据值的大小可以得出 TERMINATED > TIDYING > STOP >SHUTDOWN > RUNNING，所以可以根据值的大小来判断某个运行状态。

10.3 线程池任务执行器

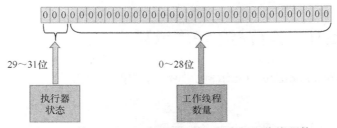

图 10.5 使用一个 int 类型表示运行状态和工作线程数

代码清单 10.10 使用一个 int 类型表示运行状态和工作线程数的相关代码

```java
1.  private final AtomicInteger ctl = new AtomicInteger(ctlOf(RUNNING, 0));
2.  private static final int COUNT_BITS = Integer.SIZE - 3;
3.  private static final int COUNT_MASK = (1 << COUNT_BITS) - 1;
4.  private static final int RUNNING    = -1 << COUNT_BITS;
5.  private static final int SHUTDOWN   =  0 << COUNT_BITS;
6.  private static final int STOP       =  1 << COUNT_BITS;
7.  private static final int TIDYING    =  2 << COUNT_BITS;
8.  private static final int TERMINATED =  3 << COUNT_BITS;
9.
10. private static int workerCountOf(int c) {
11.     return c & COUNT_MASK;
12. }
13.
14. private static int ctlOf(int rs, int wc) {
15.     return rs | wc;
16. }
17.
18. private static boolean runStateLessThan(int c, int s) {
19.     return c < s;
20. }
21.
22. private static boolean runStateAtLeast(int c, int s) {
23.     return c >= s;
24. }
25.
26. private static boolean isRunning(int c) {
27.     return c < SHUTDOWN;
28. }
29.
30. public boolean isShutdown() {
31.     return runStateAtLeast(ctl.get(), SHUTDOWN);
32. }
33.
34. public boolean isTerminated() {
35.     return runStateAtLeast(ctl.get(), TERMINATED);
36. }
```

10.3.2 线程池任务执行器的使用示例

下面看一个 ThreadPoolExecutor 示例，如代码清单 10.11 所示。

代码清单 10.11　ThreadPoolExecutor 示例

```java
1.   public class ThreadPoolExecutorDemo {
2.
3.       private static ExecutorService pool;
4.
5.       public static void main(String[] args) {
6.           pool = new ThreadPoolExecutor(1, 1, 1000, TimeUnit.MILLISECONDS,
7.                   new ArrayBlockingQueue<Runnable>(1), Executors.defaultThreadFactory(),
8.                   new ThreadPoolExecutor.AbortPolicy());
9.           for (int i = 0; i < 3; i++) {
10.              pool.execute(new ThreadTask());
11.          }
12.      }
13.  }
14.
15.  class ThreadTask implements Runnable {
16.      public void run() {
17.          System.out.println(Thread.currentThread().getName());
18.          try {
19.              Thread.sleep(500);
20.          } catch (InterruptedException e) {
21.              e.printStackTrace();
22.          }
23.      }
24.  }
```

该示例创建了一个 ThreadPoolExecutor 对象，构造方法中指定核心工作线程数和最大工作线程数都为 1，工作线程空闲超时时长为 1 000ms，使用容量为 1 的阻塞队列和默认线程工厂，使用 AbortPolicy 拒绝策略。

代码清单 10.11 的输出如下。可以看到有两个任务正常执行，而其中一个任务被拒绝并且抛出异常。这是因为阻塞队列容量为 1，执行器的最大工作线程数为 1，这就导致第三个任务因为无法放到阻塞队列中而被拒绝，而 AbortPolicy 的拒绝策略就是抛出 RejectedExecutionException 异常。

```
1.  pool-1-thread-1
2.  Exception in thread "main" java.util.concurrent.RejectedExecutionException: Task com.seaboat.
    thread.ThreadTask@161cd475 rejected from com.seaboat.thread.jdk.ThreadPoolExecutor@532760d8
3.      at com.seaboat.thread.jdk.ThreadPoolExecutor$AbortPolicy.rejectedExecution
    (ThreadPoolExecutor.java:302)
4.      at com.seaboat.thread.jdk.ThreadPoolExecutor.execute(ThreadPoolExecutor.java:195)
5.      at com.seaboat.thread.ThreadPoolExecutorDemo.main(ThreadPoolExecutorDemo.java:19
```

6.   pool-1-thread-1

10.3.3 线程池任务执行器的实现原理

下面分析 ThreadPoolExecutor 的实现，如代码清单 10.12 所示。先看构造函数，其参数分别为核心线程数、最大线程数、工作线程空闲超时时长、存储任务的阻塞队列、线程工厂、拒绝策略。各项参数的作用会在后面详细分析。

代码清单 10.12　ThreadPoolExecutor 的实现代码

```
1.   public class ThreadPoolExecutor implements ExecutorService {
2.
3.       private final BlockingQueue<Runnable> workQueue;
4.       private final ReentrantLock mainLock = new ReentrantLock();
5.       private final HashSet<Worker> workers = new HashSet<>();
6.       private final Condition termination = mainLock.newCondition();
7.       private volatile RejectedExecutionHandler handler = new AbortPolicy();
8.       private volatile ThreadFactory threadFactory = new DefaultThreadFactory();
9.       private volatile long keepAliveTime;
10.      private volatile int corePoolSize;
11.      private volatile int maximumPoolSize;
12.
13.      public ThreadPoolExecutor(int corePoolSize, int maximumPoolSize, long keepAliveTime, TimeUnit unit, BlockingQueue<Runnable> workQueue, ThreadFactory threadFactory, RejectedExecutionHandler handler) {
14.          this.corePoolSize = corePoolSize;
15.          this.maximumPoolSize = maximumPoolSize;
16.          this.workQueue = workQueue;
17.          this.keepAliveTime = unit.toNanos(keepAliveTime);
18.          this.threadFactory = threadFactory;
19.          this.handler = handler;
20.      }
21.
22.      ......
23.
24.  }
```

Worker 是工作线程，代码如代码清单 10.13 所示。它继承了 AbstractQueuedSynchronizer 同步器并实现了 Runnable 接口，所以它自己具备并发锁功能。构造 Worker 对象时会传入第一个任务，并且将同步状态设为-1，同时通过线程工厂创建一个新线程。run 方法的核心思想是通过 while 循环不断调用 getTask 方法以从阻塞队列中获取任务。它需要先加锁，然后判断是否需要中断当前工作线程，然后再调用 task 的 run 方法执行任务逻辑，最终将 task 赋值为 null 且释放锁。如果因为某些原因需要结束工作线程，则调用 processWorkerExit 方法执行退出工作。此外，注意 completedAbruptly 变量，它表示工作线程是正常退出还是因为异常而退出。

代码清单 10.13　Worker 工作线程

```
1.   private final class Worker extends AbstractQueuedSynchronizer implements Runnable {
2.       final Thread thread;
3.       Runnable firstTask;
4.
5.       Worker(Runnable firstTask) {
6.           setState(-1);
7.           this.firstTask = firstTask;
8.           this.thread = threadFactory.newThread(this);
9.       }
10.
11.      public void run() {
12.          Thread wt = Thread.currentThread();
13.          Runnable task = this.firstTask;
14.          this.firstTask = null;
15.          this.unlock();
16.          boolean completedAbruptly = true;
17.          try {
18.              while (task != null || (task = getTask()) != null) {
19.                  this.lock();
20.                  if ((runStateAtLeast(ctl.get(), STOP)
21.                          || (Thread.interrupted() && runStateAtLeast(ctl.get(), STOP)))
22.                          && !wt.isInterrupted())
23.                      wt.interrupt();
24.                  try {
25.                      task.run();
26.                  } finally {
27.                      task = null;
28.                      this.unlock();
29.                  }
30.              }
31.              completedAbruptly = false;
32.          } finally {
33.              processWorkerExit(this, completedAbruptly);
34.          }
35.      }
36.
37.      ......
38.
39.  }
```

Worker 具备并发锁的功能（见代码清单 10.14），与锁相关的主要是 tryAcquire、tryRelease、lock、tryLock 和 unlock 这 5 个方法。这些方法在前文中反复出现，这里不再赘述，如有疑惑可以重新回到第 5 章进行学习。

代码清单 10.14　Worker 具备并发锁的功能

```
1.   private final class Worker extends AbstractQueuedSynchronizer implements Runnable {
```

```
2.        ……
3.
4.
5.     protected boolean tryAcquire(int unused) {
6.         if (compareAndSetState(0, 1)) {
7.             setExclusiveOwnerThread(Thread.currentThread());
8.             return true;
9.         }
10.        return false;
11.    }
12.
13.    protected boolean tryRelease(int unused) {
14.        setExclusiveOwnerThread(null);
15.        setState(0);
16.        return true;
17.    }
18.
19.    public void lock() {
20.        acquire(1);
21.    }
22.
23.    public boolean tryLock() {
24.        return tryAcquire(1);
25.    }
26.
27.    public void unlock() {
28.        release(1);
29.    }
30. }
```

工作线程会不断循环以从任务队列获取任务，这部分的逻辑由 getTask 方法来定义，如代码清单 10.15 所示。首先获取 ctl 变量的值，该值包含了运行状态及工作线程数。然后判断运行状态是否为 STOP/TIDYING/TERMINATED，或者运行状态为非 RUNNING 但任务队列为空，如果是则通过 ctl.addAndGet(-1) 让工作线程数减 1，并返回 null 让工作线程退出。接着通过 wc > corePoolSize 判断是否需要超时控制。也就是说非核心工作线程都需要进行超时控制，timedOut 表示该工作线程等待任务超时，此时通过 ctl.compareAndSet(c, c - 1) 将工作线程数减 1 并返回 null。最后，非核心工作线程通过 workQueue.poll 获取阻塞队列的任务，而核心线程则通过 workQueue.take 获取任务 (这两者的差别在于超时与不超时)，获取的 Runnable 不为空则直接返回，为空则将 timedOut 标为 true，表示已经超时。

代码清单 10.15　getTask 方法

```
1.  private Runnable getTask() {
2.      boolean timedOut = false;
3.      for (;;) {
4.          int c = ctl.get();
5.          if (runStateAtLeast(c, SHUTDOWN) && (runStateAtLeast(c, STOP) || workQueue.
```

```
                isEmpty())) {
6.                  ctl.addAndGet(-1);
7.                  return null;
8.              }
9.              int wc = workerCountOf(c);
10.             boolean timed = wc > corePoolSize;
11.             if (((timed && timedOut)) && (wc > 1 || workQueue.isEmpty())) {
12.                 if (ctl.compareAndSet(c, c - 1))
13.                     return null;
14.                 continue;
15.             }
16.             try {
17.                 Runnable r = timed ? workQueue.poll(keepAliveTime, TimeUnit.NANOSECONDS)
18.                                    : workQueue.take();
19.                 if (r != null)
20.                     return r;
21.                 timedOut = true;
22.             } catch (InterruptedException retry) {
23.                 timedOut = false;
24.             }
25.         }
26.     }
```

有两种情况会导致工作线程退出 while 循环：通过 getTask 获取到的任务为 null 时；在执行任务的过程中发生异常时。这两种情况通过 completedAbruptly 变量来区别，最终都会执行 processWorkerExit 并退出工作线程。先看 processWorkerExit 方法的逻辑（见代码清单 10.16）。如果 completedAbruptly 为 true，则表示发生异常导致退出，此时需要通过 ctl.addAndGet(-1) 将工作线程数减 1。然后对 mainLock 加锁后通过 workers.remove(w) 将工作线程从工作线程集合中移除。接着调用 tryTerminate 方法尝试终止。最后通过 runStateLessThan(c, STOP) 判断运行状态是否为 SHUTDOWN 或 RUNNING，如果因为异常而导致工作线程退出，则需要调用 addWorker 给工作线程池增加一个线程。

代码清单 10.16　processWorkerExit 方法

```
1.  private void processWorkerExit(Worker w, boolean completedAbruptly) {
2.      if (completedAbruptly)
3.          ctl.addAndGet(-1);
4.      final ReentrantLock mainLock = this.mainLock;
5.      mainLock.lock();
6.      try {
7.          workers.remove(w);
8.      } finally {
9.          mainLock.unlock();
10.     }
11.     tryTerminate();
12.     int c = ctl.get();
13.     if (runStateLessThan(c, STOP))
14.         addWorker(null, false);
```

15.   }
```

addWorker 方法用于添加工作线程，可以为其传入第一个任务和是否核心线程的标识，如代码清单 10.17 所示。retry 块主要用于判断状态和将工作线程数加 1。它包含两个 for 循环，在应该停止的状态或超过核心工作线程数和最大工作线程数时都直接返回 false，并通过 ctl.compareAndSet(c, c + 1) 将工作线程数加 1。然后通过 new Worker(firstTask) 创建新的 Worker 对象，在成功获取 mainLock 后将该对象添加到 workers 集合中。最后调用 worker 对象中线程的 start 方法启动工作线程，并开始不断循环以从任务队列中获取任务进行处理。

**代码清单 10.17　addWorker 方法**

```
1.   private boolean addWorker(Runnable firstTask, boolean core) {
2.   retry: for (int c = ctl.get();;) {
3.   if (runStateAtLeast(c, SHUTDOWN)
4.   && (runStateAtLeast(c, STOP) || firstTask != null || workQueue.isEmpty()))
5.   return false;
6.   for (;;) {
7.   if (workerCountOf(c) >= ((core ? corePoolSize : maximumPoolSize) & COUNT_MASK))
8.   return false;
9.   if (ctl.compareAndSet(c, c + 1))
10.   break retry;
11.   c = ctl.get();
12.   if (runStateAtLeast(c, SHUTDOWN))
13.   continue retry;
14.   }
15.   }
16.   boolean workerStarted = false;
17.   boolean workerAdded = false;
18.   Worker w = new Worker(firstTask);
19.   final Thread t = w.thread;
20.   final ReentrantLock mainLock = this.mainLock;
21.   mainLock.lock();
22.   try {
23.   int c = ctl.get();
24.   if (isRunning(c) || (runStateLessThan(c, STOP) && firstTask == null)) {
25.   if (t.isAlive())
26.   throw new IllegalThreadStateException();
27.   workers.add(w);
28.   workerAdded = true;
29.   }
30.   } finally {
31.   mainLock.unlock();
32.   }
33.   if (workerAdded) {
34.   t.start();
35.   workerStarted = true;
36.   }
```
```

```
37.        return workerStarted;
38.    }
```

通过 execute 方法可以向 ThreadPoolExecutor 提交任务，如代码清单 10.18 所示。该方法的逻辑是先获取 ctl 的值并通过 workerCountOf 得到工作线程数，如果小于核心工作线程数，则调用 addWorker 添加核心工作线程。否则判断 ThreadPoolExecutor 是否处于 RUNNING 状态，如果是，则将任务添加到任务队列中。如果任务无法添加到任务队列中，则尝试调用 addWorker 增加非核心工作线程，此时若还不成功则会执行拒绝策略，即调用 RejectedExecutionHandler 的 rejectedExecution 方法。

代码清单 10.18　execute 方法

```java
1.  public void execute(Runnable command) {
2.      int c = ctl.get();
3.      if (workerCountOf(c) < corePoolSize) {
4.          if (addWorker(command, true))
5.              return;
6.          c = ctl.get();
7.      }
8.      if (isRunning(c) && workQueue.offer(command)) {
9.          ;
10.     } else if (!addWorker(command, false))
11.         handler.rejectedExecution(command, this);
12. }
```

shutdown 方法（见代码清单 10.19）用于关闭 ThreadPoolExecutor，它先获取 mainLock 锁，接着通过 for 循环和 compareAndSet 方法将运行状态修改为 SHUTDOWN，然后调用 interruptIdleWorkers 中断空闲的工作线程，最后执行 tryTerminate 方法并尝试终止执行器。

代码清单 10.19　shutdown 方法

```java
1.  public void shutdown() {
2.      final ReentrantLock mainLock = this.mainLock;
3.      mainLock.lock();
4.      try {
5.          for (;;) {
6.              int c = ctl.get();
7.              if (runStateAtLeast(c, SHUTDOWN)
8.                  || ctl.compareAndSet(c, ctlOf(SHUTDOWN, workerCountOf(c))))
9.                  break;
10.         }
11.         interruptIdleWorkers();
12.     } finally {
13.         mainLock.unlock();
14.     }
15.     tryTerminate();
16. }
```

interruptIdleWorkers 方法（见代码清单 10.20）用于中断工作线程池中的所有工作线程。它先获取 mainLock 锁，然后遍历所有 Worker 对象，再调用各 Worker 对象的 tryLock 方法尝试获取锁，如果能成功获取锁，则说明该工作线程不处于工作中，然后调用线程的 interrupt 方法将该线程的中断标识设置为 true，最后释放锁（两个锁都要被释放）。

代码清单 10.20　interruptIdleWorkers 方法

```
1.   private void interruptIdleWorkers() {
2.       final ReentrantLock mainLock = this.mainLock;
3.       mainLock.lock();
4.       try {
5.           for (Worker w : workers) {
6.               Thread t = w.thread;
7.               if (!t.isInterrupted() && w.tryLock()) {
8.                   try {
9.                       t.interrupt();
10.                  } catch (SecurityException ignore) {
11.                  } finally {
12.                      w.unlock();
13.                  }
14.              }
15.          }
16.      } finally {
17.          mainLock.unlock();
18.      }
19.  }
```

尝试终止执行器由 tryTerminate 方法来负责，如代码清单 10.21 所示。该方法的主体是一个 for 循环，它首先获取 ctl 的值并根据状态判断是否直接返回，如果处于 RUNNING 状态则直接返回，不做进一步处理。然后通过 workerCountOf 计算工作线程数，如果不为 0，则调用 interruptIdleWorkers 中断工作线程。接着加锁并通过 ctl.compareAndSet(c, ctlOf(TIDYING, 0)) 修改运行状态为 TIDYING，然后调用 terminated 方法（该方法是一个钩子方法）。最后将运行状态设置为 TERMINATED，并对 termination 条件进行唤醒操作。这里之所以要做唤醒操作，是因为 awaitTermination 方法是等待执行器关闭的方法，调用该方法后会调用 termination 条件的 await 方法进入等待状态，所以就需要在 tryTerminate 方法中去唤醒。

代码清单 10.21　tryTerminate 方法

```
1.   final void tryTerminate() {
2.       for (;;) {
3.           int c = ctl.get();
4.           if (isRunning(c) || runStateAtLeast(c, TIDYING)
5.               || (runStateLessThan(c, STOP) && !workQueue.isEmpty()))
6.               return;
7.           if (workerCountOf(c) != 0) {
8.               interruptIdleWorkers();
```

```
 9.            return;
10.        }
11.        final ReentrantLock mainLock = this.mainLock;
12.        mainLock.lock();
13.        try {
14.            if (ctl.compareAndSet(c, ctlOf(TIDYING, 0))) {
15.                try {
16.                    terminated();
17.                } finally {
18.                    ctl.set(ctlOf(TERMINATED, 0));
19.                    termination.signalAll();
20.                }
21.                return;
22.            }
23.        } finally {
24.            mainLock.unlock();
25.        }
26.    }
27. }
28.
29. public void awaitTermination() throws InterruptedException {
30.     final ReentrantLock mainLock = this.mainLock;
31.     mainLock.lock();
32.     try {
33.         while (runStateLessThan(ctl.get(), TERMINATED)) {
34.             termination.await();
35.         }
36.     } finally {
37.         mainLock.unlock();
38.     }
39. }
```

默认的线程工厂如代码清单 10.22 所示。其中主要实现 newThread 方法，在该方法内创建新的线程并做一些额外的工作。我们可以根据自己的需求去重写线程工厂类。

代码清单 10.22 默认线程工厂

```
 1. private static class DefaultThreadFactory implements ThreadFactory {
 2.     private static final AtomicInteger poolNumber = new AtomicInteger(1);
 3.     private final ThreadGroup group;
 4.     private final AtomicInteger threadNumber = new AtomicInteger(1);
 5.     private final String namePrefix;
 6.
 7.     DefaultThreadFactory() {
 8.         SecurityManager s = System.getSecurityManager();
 9.         group = (s != null) ? s.getThreadGroup() : Thread.currentThread().getThreadGroup();
10.         namePrefix = "pool-" + poolNumber.getAndIncrement() + "-thread-";
11.     }
12.
13.     public Thread newThread(Runnable r) {
```

```
14.         Thread t = new Thread(group, r, namePrefix + threadNumber.getAndIncrement (), 0);
15.         if (t.isDaemon())
16.             t.setDaemon(false);
17.         if (t.getPriority() != Thread.NORM_PRIORITY)
18.             t.setPriority(Thread.NORM_PRIORITY);
19.         return t;
20.     }
21. }
```

拒绝策略类需要实现 RejectedExecutionHandler 接口，具体的拒绝逻辑在 rejectedExecution 方法中实现，如代码清单 10.23 所示。我们所采用的 AbortPolicy 拒绝策略是直接抛出 RejectedExecutionException 异常，在实际使用中可以根据需求实现自己的拒绝策略类。

代码清单 10.23　拒绝策略类和拒绝逻辑的实现

```
1.  public static class AbortPolicy implements RejectedExecutionHandler {
2.      public AbortPolicy() {
3.      }
4.
5.      public void rejectedExecution(Runnable r, ThreadPoolExecutor e) {
6.          throw new RejectedExecutionException(
7.              "Task " + r.toString() + " rejected from " + e.toString());
8.      }
9.  }
```

第 11 章
其他并发工具

11.1 线程本地变量

线程本地变量 ThreadLocal 可用于实现线程与变量之间的绑定,也就是说每个线程都只能读写本线程所对应的变量。对于同一个 ThreadLocal 对象,每个线程在读写该对象时只能看到属于自己的变量,因此 ThreadLocal 也是一种线程安全的模式。ThreadLocal 的功能如图 11.1 所示。可以看到,一个 ThreadLocal 对象就是一个线程本地变量,该变量可以保存多个变量值,比如线程 1 对应变量值 1,其他两个线程也有自己的变量值。

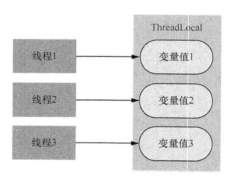

图 11.1　ThreadLocal 的功能

11.1.1　线程本地变量的使用示例

下面通过代码清单 11.1 来了解 ThreadLocal 的使用方法。首先创建一个 ThreadLocal 对象,由于 ThreadLocal 使用了泛型的声明方式,所以需要指定保存的数据类型,这里保存的是 String 类型。然后启动 5 个线程,每个线程都通过 ThreadLocal 对象的 set 方法设置要绑定该线程的变量值(要保存什么值就传入什么值),而当使用时则调用 ThreadLocal 对象的 get 方法(该方法无须传入参数值)。

代码清单 11.1　ThreadLocal 的使用示例

```
1.  public class ThreadLocalDemo {
2.
3.      static ThreadLocal<String> threadLocal = new ThreadLocal<String>();
4.
5.      public static void main(String[] args) {
6.
7.          for (int i = 0; i < 5; i++)
8.              new Thread(() -> {
9.                  threadLocal.set(Thread.currentThread().getName() + "的变量");
10.                 try {
11.                     Thread.sleep(2000);
12.                 } catch (InterruptedException e) {
13.                     e.printStackTrace();
14.                 }
15.                 System.out.println(Thread.currentThread().getName() + "--->" + threadLocal.get());
16.             }).start();
17.     }
18.
19. }
```

代码清单 11.1 执行后的输出结果如下。

```
1.  Thread-0--->Thread-0 的变量
2.  Thread-2--->Thread-2 的变量
3.  Thread-1--->Thread-1 的变量
4.  Thread-3--->Thread-3 的变量
5.  Thread-4--->Thread-4 的变量
```

代码清单 11.1 的执行效果如图 11.2 所示，其中 5 个线程都有各自对应的变量。

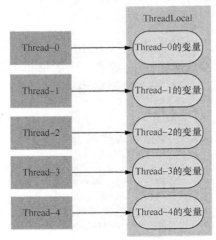

图 11.2　代码清单 11.1 的执行效果

11.1.2 线程本地变量的 3 个主要方法

ThreadLocal 有 3 个主要的方法，其名称和作用如下。
- set 方法：用于设置当前线程本地变量的值，传入的参数为要设置的值。比如 threadLocal.set("value")。
- get 方法：用于获取当前线程本地变量的值，无须传入任何参数。比如 String threadLocalValue = (String) threadLocal.get()。
- remove 方法：用于删除当前线程的本地变量，无须传入任何参数。比如 threadLocal.remove()。

在了解了 ThreadLocal 的功能后，我们来看一下 ThreadLocal 是如何实现的，以及变量与线程之间是如何绑定的。实际上，如果让我们自己来实现 ThreadLocal 功能，只需一个 Map 结构即可。其中 Map 的 key 是当前线程，而 Map 的 value 则是变量值。图 11.3 所示为 ThreadLocal 的设计思想。

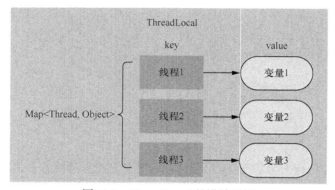

图 11.3　ThreadLocal 的设计思想

再看具体的模拟实现代码，如代码清单 11.2 所示。该模拟类提供了 set、get 和 remove 这 3 个方法，这 3 个方法都是间接地操作 Map 对象。注意 Map 对象的 key 值都是当前线程，由 Thread.currentThread() 来获取，这个 key 值不必由调用方传入。这样就实现了一个简单的 ThreadLocal。

代码清单 11.2　ThreadLocal 的模拟实现代码

```
1.  public class ThreadLocal<T> {
2.      static Map<Thread, Object> map = new Hashtable<Thread, Object>();
3.
4.      public void set(T obj) {
5.          map.put(Thread.currentThread(), obj);
6.      }
7.
```

```
 8.     public T get() {
 9.         return (T) map.get(Thread.currentThread());
10.     }
11.
12.     public void remove() {
13.         map.remove(Thread.currentThread());
14.     }
15. }
```

11.1.3　JDK 中线程本地变量的实现思想

上面的实现方式虽然很简单，而且符合我们的思考方式，但是它存在多线程并发的性能问题。这怎么理解呢？其实很明显，我们实现的 ThreadLocal 内部使用了一个 Map 对象，所有线程的操作都是针对该 Map 对象进行的操作，因此需要保证该对象访问的线程安全。这就需要额外的锁机制来保证，但与此同时也会带来性能问题。

JDK 提供的 ThreadLocal 的实现则比较巧妙，为了避免并发时涉及锁问题，它在每个线程对象中都放入一个 Map 对象，但它并没有直接使用 JDK 的 Map 类，而是自己实现了一个 key-value 数据结构。每个线程都操作自己的 Map 对象，这样就不存在并发问题。在图 11.4 中，线程 1 包含了一个 Map 对象，该 Map 对象的 key 是 ThreadLocal 对象，而 value 则是变量值。注意这里的 ThreadLocal 对象变成了 key，也就是说可能存在很多不同的 ThreadLocal 对象，在查找时需要传入对应的 ThreadLocal 对象。

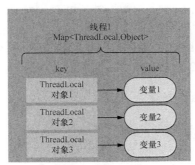
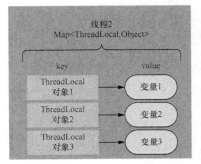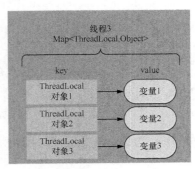

图 11.4　JDK 提供的 ThreadLocal 实现

11.1.4　JDK 中线程本地变量的实现源码

注意这里只分析核心内容，并非包括所有源码细节，并且为了达到简洁、清晰的效果，这里还删除或修改了少量源码。我们先来看 Thread 类与 ThreadLocal 类的关系，如图 11.5 所示。Thread 类中包含了一个 threadLocals 变量，它是 ThreadLocal.ThreadLocalMap 类型。该类型定义在 ThreadLocal 类里面，也就是一个内部类。ThreadLocalMap 内部类实现了一个 Map 结构，该类又包含了 Entry 内部类，ThreadLocal 对象和变量值则通过 Entry 来保存。

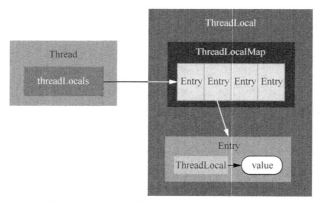

图 11.5　Thread 类与 ThreadLocal 类的关系

Thread 类里面声明了 threadLocals 变量，用于关联 ThreadLocal.ThreadLocalMap 对象（注意默认为 null），如代码清单 11.3 所示。

代码清单 11.3　threadLocals 变量

```
1.    public class Thread implements Runnable {
2.        ......
3.
4.        ThreadLocal.ThreadLocalMap threadLocals = null;
5.
6.        ......
7.    }
```

ThreadLocal 类的大体结构如代码清单 11.4 所示。其中提供了 3 个主要的方法，ThreadLocalMap 内部类实现了 Map 结构。Map 结构具体由 Entry 类实现，该类继承了 WeakReference 类，目的是为了避免内存泄漏。

代码清单 11.4　ThreadLocal 类的大体结构

```
1.    public class ThreadLocal<T> {
```

```
2.
3.     public ThreadLocal() {
4.     }
5.
6.     public void set(T value) {
7.
8.         ...
9.
10.    }
11.
12.    public T get() {
13.
14.        ...
15.
16.    }
17.
18.    public void remove() {
19.
20.        ...
21.
22.    }
23.
24.    static class ThreadLocalMap {
25.
26.        static class Entry extends WeakReference<ThreadLocal<?>> {
27.            Object value;
28.
29.            Entry(ThreadLocal<?> k, Object v) {
30.                super(k);
31.                value = v;
32.            }
33.        }
34.
35.        ...
36.
37.    }
38. }
```

对于多个线程与多个线程本地变量来说，它们的结构如图 11.6 所示。

ThreadLocalMap 类实际上就是一个 Map 结构的实现。Java 开发人员对 Map 再熟悉不过，而且由于 ThreadLocalMap 类的实现涉及很多细节，如果我们详细讲它繁琐的实现源码，会导致篇幅冗长，所以这里我们主要来看它的结构和操作。ThreadLocalMap 类使用数组来保存 key-value，数组的每个元素都对应着一个 key-value，所以新增、修改、删除等操作都是围绕着数组进行的，如图 11.7 所示。而且在保存之前会先用哈希算法计算线程对象的哈希值，这个哈希值是一个整型值，通过该值就能定位到数组中某个位置的元素，这样就能找到对应的 key-value 进行操作。

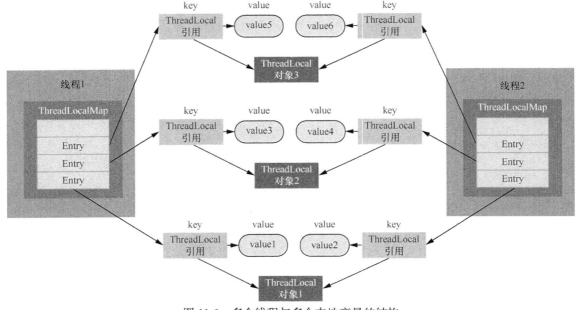

图 11.6 多个线程与多个本地变量的结构

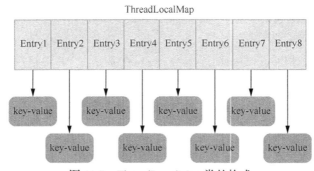

图 11.7 ThreadLocalMap 类的构成

ThreadLocal 类的 set 方法（见代码清单 11.5）的逻辑为：首先获取当前线程对象，然后通过 getMap 方法获取当前线程的 ThreadLocalMap（其实就是从 Thread 对象中获取），最后调用 ThreadLocalMap 对象的 set 方法保存 key-value。注意，如果 Thread 对象中的 ThreadLocalMap 对象为空，则需要调用 createMap 方法先创建 ThreadLocalMap 对象，然后关联到 Thread 对象中。

代码清单 11.5 set 方法

```
1.    public void set(T value) {
2.        Thread t = Thread.currentThread();
3.        ThreadLocalMap map = getMap(t);
4.        if (map != null) {
5.            map.set(this, value);
```

```
6.       } else {
7.           createMap(t, value);
8.       }
9.   }
10.
11.  ThreadLocalMap getMap(Thread t) {
12.      return t.threadLocals;
13.  }
14.
15.  void createMap(Thread t, T firstValue) {
16.      t.threadLocals = new ThreadLocalMap(this, firstValue);
17.  }
```

get 方法（见代码清单 11.6）的逻辑为：首先获取当前线程对象，然后通过 getMap 方法获取当前线程的 ThreadLocalMap 对象，如果该对象不为空，则调用 ThreadLocalMap 对象的 getEntry 方法获取 Entry。Entry 对象包含了我们要的 value。如果获取不到值，则还会执行 setInitialValue 方法，它会根据 ThreadLocal 对象的 initialValue 方法来设置初始值，默认是 null。如果想要设置一个初始值，则可以重写 initialValue 方法。

代码清单 11.6　get 方法

```
1.   public T get() {
2.       Thread t = Thread.currentThread();
3.       ThreadLocalMap map = getMap(t);
4.       if (map != null) {
5.           ThreadLocalMap.Entry e = map.getEntry(this);
6.           if (e != null) {
7.               T result = (T) e.value;
8.               return result;
9.           }
10.      }
11.      return setInitialValue();
12.  }
13.
14.  private T setInitialValue() {
15.      T value = initialValue();
16.      Thread t = Thread.currentThread();
17.      ThreadLocalMap map = getMap(t);
18.      if (map != null) {
19.          map.set(this, value);
20.      } else {
21.          createMap(t, value);
22.      }
23.      return value;
24.  }
25.
26.  protected T initialValue() {
27.      return null;
28.  }
```

remove 方法的逻辑很简单，它直接获取当前线程的 ThreadLocalMap 对象，然后调用该对象的 remove 方法删除对应的 key-value，如代码清单 11.7 所示。

代码清单 11.7　remove 方法

```
1.   public void remove() {
2.       ThreadLocalMap m = getMap(Thread.currentThread());
3.       if (m != null) {
4.           m.remove(this);
5.       }
6.   }
```

11.1.5　线程本地变量的内存泄漏

由于 Entry 继承了 WeakReference 类，所以可以指定对某个对象进行弱引用。在没有其他强引用的情况下，弱引用类型会被 JVM 的垃圾回收器回收。我们通过图 11.8 来理解如何产生内存泄漏。ThreadLocal 在创建后就会伴随 Thread 的整个生命周期，假如这个线程的生命周期很长，就可能导致内存泄漏。

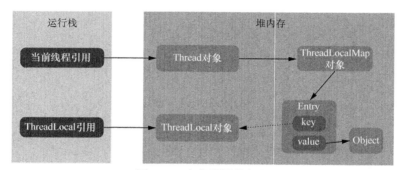

图 11.8　内存泄漏的产生

在运行栈的运行过程中，假如某个时刻 ThreadLocal 引用不再指向 ThreadLocal 对象，则该对象仅仅剩下一个弱引用，此时该对象就会被 JVM 回收，从而导致 Entry 的 key 为 null，而 key 为 null 时就会导致 ThreadLocalMap 无法再找到这个 Entry 的 value。一旦运行时间被拉长，value 将一直存在内存中而无法被回收，从而造成内存泄漏。整个引用关系为 Thread 对象→ThreadLocalMap 对象→Entry 对象→value。

是不是不继承 WeakReference 类，让它默认强引用就不会导致内存泄漏呢？肯定不是。在运行栈的运行过程中，假如某个时刻 ThreadLocal 引用不再指向 ThreadLocal 对象，则 ThreadLocal 对象因为存在强引用而不被 JVM 回收，此时除了 value 无法被回收外，ThreadLocal 对象也无法被回收，同样会产生内存泄漏问题。

综上所述，不管 Entry 有没有继承 WeakReference 类，都存在内存泄漏问题，如果我们不是手动去执行 remove 操作，就会导致内存泄漏。JDK 团队为什么又要继承 WeakReference

类呢？原因是他们想采取一些措施来尽量保证内存不泄漏。也就是说，在 ThreadLocalMap 类的 get、set、remove 方法中去执行一个清除操作，把 ThreadLocalMap 包含的所有 Entry 中 key 为 null 的 value 都清除掉，并且将对应的 Entry 也置为 null，以便被 JVM 回收。所以在使用 ThreadLocal 时要注意，当使用完 ThreadLocal 后，都要手动调用 remove 方法，以避免内存泄漏。

11.2 写时复制数组列表

写时复制（CopyOnWrite）是针对读多写少的场景提出的一种并发优化策略，它的核心思想是，如果多个线程对某资源只进行读操作，那么这些线程都访问同一个资源即可，直到某个线程要对资源进行写操作时才会复制一个资源副本，且资源副本在修改完后再作为新的资源，而原来的旧资源则被废弃。在图 11.9 中，线程 1、线程 2、线程 3 都对某资源进行读操作，当线程 4 准备对资源进行修改时，会复制一个资源副本，然后对资源副本进行修改，并最终将原来的资源废弃，而资源副本作为新资源供所有线程进行访问。

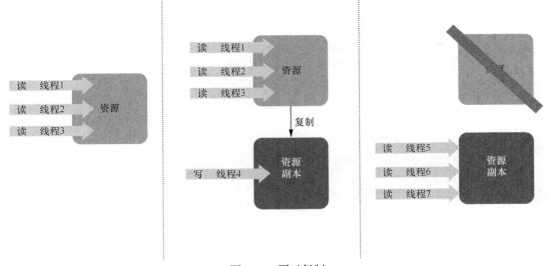

图 11.9 写时复制

CopyOnWriteArrayList 是一种具有写时复制功能的列表，从名字就可以看出它的实现是基于数组结构，所以该类其实就是具有 CopyOnWrite 功能的 ArrayList。该类内部包含了一个数组和一个锁对象，数组用于存放数据，锁用于修改操作，如图 11.10 所示。

下面通过代码清单 11.8 来了解 CopyOnWriteArrayList 的使用方法。这是一个黑名单列表场景，我们创建一个 CopyOnWriteArrayList 对象作为黑名单列表，并传入 name1、name2 和 name3 作为初始值。

图 11.10 CopyOnWriteArrayList 类

代码清单 11.8　CopyOnWriteArrayList 的使用方法

```
1.  public class CopyOnWriteArrayListDemo {
2.
3.      static CopyOnWriteArrayList<String> blackList = new CopyOnWriteArrayList<>(
4.              new String[] { "name1", "name2", "name3" });
5.
6.      public static void main(String[] args) throws InterruptedException {
7.
8.          new Thread(() -> {
9.              System.out.println("thread1->black list size is " + blackList.size());
10.             for (int i = 0; i < blackList.size(); i++)
11.                 System.out.print(blackList.get(i) + " ");
12.             try {
13.                 Thread.sleep(4000);
14.             } catch (InterruptedException e) {
15.             }
16.             System.out.println("\nthread1->black list size is " + blackList.size());
17.             for (int i = 0; i < blackList.size(); i++)
18.                 System.out.print(blackList.get(i) + " ");
19.         }).start();
20.
21.         Thread.sleep(1000);
22.
23.         new Thread(() -> {
24.             System.out.println("\nthread2->update the black ");
25.             blackList.add("name4");
26.             blackList.add("name5");
27.             blackList.set(0, "name_updated");
28.         }).start();
29.
30.         Thread.sleep(1000);
31.
32.         new Thread(() -> {
33.             System.out.println("thread3->black list size is " + blackList.size());
34.             for (int i = 0; i < blackList.size(); i++)
35.                 System.out.print(blackList.get(i) + " ");
36.         }).start();
37.
38.         Thread.sleep(3000);
```

```
39.        }
40.
41.    }
```

下面是代码清单 11.8 的输出结果，我们结合代码来看整个过程。首先，thread1 访问的黑名单大小为 3，分别为 name1、name2、name3，然后睡眠 4s。接着 thread2 对黑名单进行修改，增加了 name4 和 name5，并将黑名单的第一个元素改为 name_updated。再接着是 thread3 访问黑名单，此时黑名单大小为 5，分别为 name_updated、name2、name3、name4、name5。最后，thread1 休眠结束后也继续访问黑名单，此时的黑名单已被更新。

```
1.    thread1->black list size is 3
2.    name1 name2 name3
3.    thread2->update the black
4.    thread3->black list size is 5
5.    name_updated name2 name3 name4 name5
6.    thread1->black list size is 5
7.    name_updated name2 name3 name4 name5
```

CopyOnWriteArrayList 类位于 java.util.concurrent 包中，如代码清单 11.9 所示。该类原本实现了某些接口，为了方便讲解这里将它们都去掉了。先看属性和构造函数，它只包含 lock 和 array 这两个属性，lock 用于修改时加锁，而 array 则用来保存数据。该类提供了两种构造函数，不传参数时直接使用 new Object[0]，而传入数组时会将数组复制到 array 中。

代码清单 11.9　CopyOnWriteArrayList 类

```java
1.    public class CopyOnWriteArrayList<E> {
2.
3.        final transient Object lock = new Object();
4.        private transient volatile Object[] array;
5.
6.        public CopyOnWriteArrayList() {
7.            array = new Object[0];
8.        }
9.
10.       public CopyOnWriteArrayList(E[] toCopyIn) {
11.           array = Arrays.copyOf(toCopyIn, toCopyIn.length, Object[].class);
12.       }
13.
14.       ......
15.
16.   }
```

添加元素操作有两种方式：添加到末尾和添加到指定位置。在将元素添加到末尾时，获取锁后通过 Arrays.copyOf 方法创建新的数组并把原来数组的元素复制到新数组中。新数组的长度为原数组长度加 1，新元素放到新数组的最后位置上。在将元素添加到指定位置时，需要指定添加的位置 index，且 index 必须大于等于 0 而小于等于数组长度，然后创建一个新的数组并将新元素放到 index 位置，原数组被 index 分成的两部分分别复制到新数组中。最后将 array 引用指向新数组（等同于将原来数组废弃）。

为了帮助大家理解，图 11.11 给出了这两种方式的示意图，大家结合代码清单 11.10 就能很好地理解了。

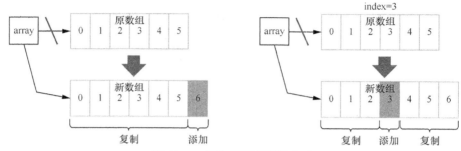

图 11.11　添加元素的两种方式

代码清单 11.10　添加元素的代码实现

```
1.  public boolean add(E e) {
2.      synchronized (lock) {
3.          Object[] es = array;
4.          int len = es.length;
5.          es = Arrays.copyOf(es, len + 1);
6.          es[len] = e;
7.          array = es;
8.          return true;
9.      }
10. }
11.
12. public void add(int index, E element) {
13.     synchronized (lock) {
14.         Object[] es = array;
15.         int len = es.length;
16.         if (index > len || index < 0)
17.             throw new IndexOutOfBoundsException("Index: " + index + ", Size: " + len);
18.         Object[] newElements;
19.         int numMoved = len - index;
20.         if (numMoved == 0)
21.             newElements = Arrays.copyOf(es, len + 1);
22.         else {
23.             newElements = new Object[len + 1];
24.             System.arraycopy(es, 0, newElements, 0, index);
25.             System.arraycopy(es, index, newElements, index + 1, numMoved);
26.         }
27.         newElements[index] = element;
28.         array = newElements;
29.     }
30. }
```

在删除元素时，需要指定索引或者指定元素。实际上这两种方式差不多，在删除指定元素时只需先查找对应元素的索引，然后统一按索引删除就好了。下面来看按索引的删除方式，通过图 11.12 再结合代码清单 11.11 就能很轻松地理解删除的过程。获取锁后先根据 index 计算两个要复制到新数组的区间，然后将这两部分分别复制到新数组中，新数组的长度为原数组长度减 1，最终将 array 引用指向新数组并返回被删除的元素。比如 index=3 时，第一个复制区间为 0～2，第二个复制区间为 4～5。将这两个区间的元素复制到新数组中，然后将 array 指向新数组并返回原数组中 index 为 3 的元素。

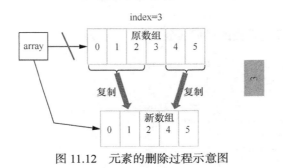

图 11.12　元素的删除过程示意图

代码清单 11.11　删除元素的代码实现

```
1.   public E remove(int index) {
2.       synchronized (lock) {
3.           Object[] es = array;
4.           int len = es.length;
5.           E oldValue = (E) array[index];
6.           int numMoved = len - index - 1;
7.           Object[] newElements;
8.           if (numMoved == 0)
9.               newElements = Arrays.copyOf(es, len - 1);
10.          else {
11.              newElements = new Object[len - 1];
12.              System.arraycopy(es, 0, newElements, 0, index);
13.              System.arraycopy(es, index + 1, newElements, index, numMoved);
14.          }
15.          array = newElements;
16.          return oldValue;
17.      }
18.  }
```

在设置元素时，需要指定索引和替换的元素，获取锁后先根据 index 得到对应的元素，然后判断要设置的元素是否与原来的元素相等，若相等则不用改动。如果不相等，则将原数组复制出来，然后将新值赋值到新数组所对应的索引位置，如图 11.13 所示。最后将 array 指向新数组并返回原数组中的 index 元素。另外的 get、size 方法很简单，此处不再展开讲解，如代码清单 11.12 所示。

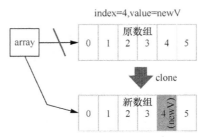

图 11.13　设置元素的操作示意图

代码清单 11.12　设置元素的代码实现

```
1.   public E set(int index, E element) {
2.       synchronized (lock) {
3.           Object[] es = array;
4.           E oldValue = (E) array[index];
5.           if (oldValue != element) {
6.               es = es.clone();
7.               es[index] = element;
8.               array = es;
9.           }
10.          return oldValue;
11.      }
12.  }
13.
14.  public E get(int index) {
15.      return (E) array[index];
16.  }
17.
18.  public int size() {
19.      return array.length;
20.  }
```

写时复制数组列表的核心思想是写时复制，也就是说只有在"增删改"时才会加锁并把原数组中的元素复制到一个新的数组进行修改。每次修改时基本上会重新复制一个新的数组，这就导致在写操作频繁和数组本身很大的场景下可能效率较低。写时复制数组列表适用于读多写少的场景，因为读是不加锁的。

第 12 章
C++模拟实现 Java 线程

本书前面的所有章节分析的都是 Java 线程，从层次结构上来看，可以将线程分为 Java 层线程、JVM 层线程、内核层线程。为了让大家更加具体、深入地理解不同层面上的线程概念，本章将讲解如何使用 C++语言来模拟 JVM 实现 Java 层面的线程。其中涉及的代码可通过异步社区获得。

12.1 模拟实现 Java 线程

我们先看整体的设计思路。由于 Java 被定义为一种跨平台语言，而且跨平台是通过 JVM 层实现的，所以很多概念都通过 JVM 层进行抽象，包括 Java 语言的线程，它需要 JVM 来提供具体的实现。模拟实现 Java 线程的整体设计思路如图 12.1 所示。我们在 Java 层用 Java 语言定义一个 Thread 类，该类表示 Java 层的线程。在 JVM 层则需要定义 JavaThread 类和 OSThread 类，这两个类都通过 C++进行定义，其中 JavaThread 类用于表示 Java 层的线程，而 OSThread 类则用于对不同操作系统底层线程的抽象。这里将基于 Linux 操作系统来模拟实现，Linux 系统提供了 pthread 库来操作线程。

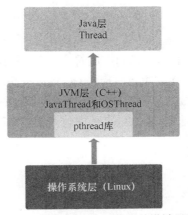

图 12.1　模拟实现 Java 线程的设计思路

最开始需要在 Java 层定义一个 com.seaboat.Thread 类，这个类就是模拟的线程类，对应着 Java 官方提供的 java.lang.Thread 类。JVM 与 Java 进行了约定，Thread 类中的 run 方法定义了线程要执行的任务，而 start 方法则用于启动线程。run 方法由用户在自定义线程类中进行重写，当用户调用 start 方法时则会调用本地的 start0 方法，该方法的实现在 com_seaboat_Thread.so 库中。此外还创建了一个 AtomicInteger 对象，用于生成线程 ID，在构造方法中会让线程 ID 不断加 1，如代码清单 12.1 所示。

代码清单 12.1　模拟的线程类

```
1.    package com.seaboat;
2.
3.    public class Thread {
4.        static {
5.            ai = new AtomicInteger();
6.            System.load("/root/seaboat/native_test/com_seaboat_Thread.so");
7.        }
8.        static AtomicInteger ai;
9.        public int threadId;
10.
11.       public Thread() {
12.           this.threadId = ai.incrementAndGet();
13.       }
14.
15.       public void run() {
16.       }
17.
18.       public void start() {
19.           start0();
20.       }
21.
22.       private native void start0();
23.
24.   }
```

为了实现 com_seaboat_Thread.so 库，我们要先定义一个头文件。它可以通过 javac -h ./ com/seaboat/Thread.java 命令让工具帮我们生成符合 JNI 调用的本地方法名。方法的命名也是 JVM 已经规定好的，这里的 start0 方法将对应 Java_com_seaboat_Thread_start0 方法。其中第一个参数 JNIEnv* 指针表示 Java 运行环境，通过它就能够调用 Java 语言定义的方法，而第二个参数 jobject 表示调用该本地方法的 Java 对象，如代码清单 12.2 所示。

代码清单 12.2　start0 方法

```
1.    #include <jni.h>
2.    /* Header for class com_seaboat_Thread */
3.
4.    #ifndef _Included_com_seaboat_Thread
5.    #define _Included_com_seaboat_Thread
```

```
6.   #ifdef __cplusplus
7.   extern "C" {
8.   #endif
9.
10.  /*
11.   * Class:     com_seaboat_Thread
12.   * Method:    start0
13.   * Signature: ()V
14.   */
15.  JNIEXPORT void JNICALL Java_com_seaboat_Thread_start0(JNIEnv*, jobject);
16.
17.  #ifdef __cplusplus
18.  }
19.  #endif
20.  #endif
```

接着在 JVM 层定义一个 JavaThread 类，该类用来封装 Java 层的线程对象，如代码清单 12.3 所示。该类主要实现了 3 个函数：构造函数、析构函数和执行 run 方法的函数。构造函数主要通过 GetJavaVM 函数获取 JavaVM 指针，并将 Java 层 Thread 对象封装成全局引用。析构函数则是将当前线程从 JVM 中分离。execRunMethod 函数会去调用 Java 层 Thread 对象的 run 方法，它先通过 AttachCurrentThread 函数将当前线程附加到 JVM 中，从而获得 Java 运行环境指针，然后通过 GetObjectClass 函数获取到 Java 层 Thread 对象，再通过 GetMethodID 函数获取该对象的 run 方法对象的 ID，进而通过 CallVoidMethod 函数执行该方法，执行完后删除 Thread 对象的全局引用。该类中的 JavaVM*和 jobject 都属于 JNI 的内容，这里不展开讲解，我们需要关注的是 Parker 和 interruptState 这两个属性。每个线程都有自己的 Parker 对象，通过它可以对线程进行阻塞和唤醒操作，而 interruptState 则表示该线程的中断状态，它是一个布尔型变量。

代码清单 12.3 定义 JavaThread 类

```
1.   class JavaThread {
2.   public:
3.       JavaVM *jvm;
4.       jobject jThreadObjectRef;
5.       Parker parker;
6.       bool interruptState = false;
7.       ~JavaThread();
8.       JavaThread(JNIEnv *env, jobject jThreadObject);
9.       void execRunMethod();
10.  };
11.
12.  JavaThread::JavaThread(JNIEnv *env, jobject jThreadObject) {
13.      env->GetJavaVM(&(this->jvm));
14.      this->jThreadObjectRef = env->NewGlobalRef(jThreadObject);
15.  }
16.
17.  JavaThread::~JavaThread() {
18.      jvm->DetachCurrentThread();
```

```
19.     }
20.
21.     void JavaThread::execRunMethod() {
22.         JNIEnv *env;
23.         if (jvm->AttachCurrentThread((void**) &env, NULL) != 0) {
24.             std::cout << "Failed to attach" << std::endl;
25.         }
26.         jclass cls = env->GetObjectClass(jThreadObjectRef);
27.         jmethodID runId = env->GetMethodID(cls, "run", "()V");
28.         if (runId != nullptr) {
29.             env->CallVoidMethod(jThreadObjectRef, runId);
30.         } else {
31.             cout << "No run method found in the Thread object!!" << endl;
32.         }
33.         env->DeleteGlobalRef(jThreadObjectRef);
34.     }
```

Parker 类（见代码清单 12.4）提供的函数为 park 和 unpark，分别表示阻塞和唤醒操作，如代码清单 12.5 所示。该类封装了 pthread 提供的线程阻塞、唤醒的相关函数，park 函数间接调用了 pthread_cond_timedwait 函数，通过它可以让线程阻塞指定的时间。如果超时，则会自动唤醒，而且操作前必须通过 pthread_mutex_lock 获取锁，操作完成后则通过 pthread_mutex_unlock 释放锁。unpark 函数间接调用了 pthread_cond_signal 函数，同样也是需要先获取锁，操作完后释放锁。

代码清单 12.4　Parker 类

```
1.  class Parker {
2.  private:
3.      pthread_mutex_t _mutex;
4.      pthread_cond_t _cond;
5.  public:
6.      Parker();
7.      void park(long millis);
8.      void unpark();
9.  };
```

代码清单 12.5　park 和 unpark 函数

```
1.  Parker::Parker() {
2.      pthread_mutex_init(&_mutex, NULL);
3.      pthread_cond_init(&_cond, NULL);
4.  }
5.
6.  void Parker::park(long millis) {
7.      struct timespec ts;
8.      struct timeval now;
9.      int status = pthread_mutex_lock(&_mutex);
10.     gettimeofday(&now, NULL);
11.     ts.tv_sec = time(NULL) + millis / 1000;
12.     ts.tv_nsec = now.tv_usec * 1000 + 1000 * 1000 * (millis % 1000);
```

```
13.        ts.tv_sec += ts.tv_nsec / (1000 * 1000 * 1000);
14.        ts.tv_nsec %= (1000 * 1000 * 1000);
15.        status = pthread_cond_timedwait(&_cond, &_mutex, &ts);
16.        if (status == 0) {
17.
18.        } else if (status == ETIMEDOUT) {
19.            // TODO: Time out.
20.        }
21.        status = pthread_mutex_unlock(&_mutex);
22.    }
23.
24.    void Parker::unpark() {
25.        int status = pthread_mutex_lock(&_mutex);
26.        status = pthread_cond_signal(&_cond);
27.        status = pthread_mutex_unlock(&_mutex);
28.    }
```

还要在 JVM 层定义一个 OSThread 类，该类负责封装 Linux 系统的 pthread 库所提供的线程，如代码清单 12.6 所示。它关联了 JavaThread 对象，构造函数中会传入 JavaThread 指针。call_os_thread 函数用于通过 pthread 库来创建操作系统线程，核心就是调用 pthread_create 函数。该函数的第三个和第四个参数分别表示线程执行的函数和对应的参数，也就是任务的定义。线程执行的具体任务由 OSThread::thread_entry_function 函数来定义，该函数负责调用 JavaThread 对象的 execRunMethod 函数，也就是 Java 层的 Thread 对象的 run 方法。

代码清单 12.6　OSThread 类

```
1.  class OSThread {
2.  private:
3.      JavaThread *javaThread;
4.  public:
5.      OSThread(JavaThread *javaThread);
6.      void call_os_thread();
7.      static void* thread_entry_function(void *args);
8.  };
9.
10. OSThread::OSThread(JavaThread *javaThread) {
11.     this->javaThread = javaThread;
12. }
13.
14. void OSThread::call_os_thread() {
15.     pthread_t tid;
16.     pthread_attr_t Attr;
17.     pthread_attr_init(&Attr);
18.     pthread_attr_setdetachstate(&Attr, PTHREAD_CREATE_DETACHED);
19.     std::cout << "creating linux thread!" << endl;
20.     if (pthread_create(&tid, &Attr, &OSThread::thread_entry_function,
21.             this->javaThread) != 0) {
22.         std::cout << "Error creating thread\n" << endl;
23.         return;
```

```
24.    }
25.    std::cout << "Started a linux thread! tid=" << tid << endl;
26.    pthread_attr_destroy(&Attr);
27. }
28.
29. void* OSThread::thread_entry_function(void *args) {
30.    JavaThread *javaThread = (JavaThread*) args;
31.    javaThread->execRunMethod();
32.    delete javaThread;
33.    return NULL;
34. }
```

上面已经定义好了 Java 层的 Thread 线程类以及 JVM 层的相关类，最后定义 start0 本地方法，如代码清单 12.7 所示。在方法外先创建一个 map<jint, JavaThread*>变量，它用于存放 JVM 层的所有线程对象，以便在不同方法的内部根据线程 ID 来获取 JavaThread*。这里简单使用 Map 结构管理线程对象，不考虑线程安全的问题。在方法内先创建 JavaThread 对象，然后获取 Java 层的线程 ID 值，对应 Thread 类的 threadId 属性，成功获取后以 threadId 为键且以 javaThread 为值增加到 map 结构中。接着继续创建 OSThread 对象，最后调用 OSThread 对象的 call_os_thread 函数。

代码清单 12.7　定义 start0 本地方法

```
1.  map<jint, JavaThread*> threads;
2.
3.  JNIEXPORT void JNICALL Java_com_seaboat_Thread_start0(JNIEnv *env,
4.          jobject jThreadObject) {
5.      std::cout << "creating a JavaThread object!" << endl;
6.      JavaThread *javaThread = new JavaThread(env, jThreadObject);
7.
8.      //将新线程保存到 map 中，方便后面根据 threadId 来获取 javaThread
9.      jclass cls = env->GetObjectClass(javaThread->jThreadObjectRef);
10.     jfieldID fID = env->GetFieldID(cls, "threadId", "I");
11.     jint threadId = env->GetIntField(javaThread->jThreadObjectRef, fID);
12.     threads.insert(map<jint, JavaThread*>::value_type(threadId, javaThread));
13.
14.     std::cout << "threadId = " << threadId << endl;
15.     std::cout << "creating a OSThread object!" << endl;
16.     OSThread osThread(javaThread);
17.     osThread.call_os_thread();
18.     return;
19. }
```

在所有代码编写好后，对 C++代码进行编译。我们使用 g++进行编译，具体命令为 g++ -fPIC -c -std=c++0x com_seaboat_Thread.cpp -I /usr/java/jdk1.8.0_111/include/ -I /usr/java/jdk1.8.0_111/include/linux/，它会编译生成 com_seaboat_Thread.o 目标文件。接着继续通过 g++ -shared com_seaboat_Thread.o -o com_seaboat_Thread.so 命令生成 com_seaboat_Thread.so 动态库文件。

最后在 Java 层创建一个线程测试类，测试代码如代码清单 12.8 所示。然后通过 java

com.seaboat.Thread 命令执行该类。

代码清单 12.8　线程测试类

```
1.  public class ThreadTest {
2.
3.      public static void main(String[] args) {
4.          new MyThread().start();
5.          Thread.sleep(100, 0);
6.      }
7.
8.      static class MyThread extends Thread {
9.          public void run() {
10.             System.out.println("simulates Java thread!");
11.             System.out.println("thread id is " + this.threadId);
12.         }
13.     }
14. }
```

代码清单 12.8 执行后的输出结果如下。

```
1.  creating a JavaThread object!
2.  threadId = 1
3.  creating a OSThread object!
4.  creating linux thread!
5.  Started a linux thread! tid=140586779625216
6.  calling sleep operation!
7.  simulates Java thread!
8.  thread id is 1
```

12.2　模拟实现 yield 语义

为了能更加深入地理解 yield 的语义，下面通过代码清单 12.9 来模拟它的实现。前面已经定义了线程的启动方法 start，现在继续定义 yield 本地方法。

代码清单 12.9　模拟 yield 的实现

```
1.  public class Thread {
2.      static {
3.          ai = new AtomicInteger();
4.          System.load("/root/seaboat/native_test/com_seaboat_Thread.so");
5.      }
6.      static AtomicInteger ai;
7.      public int threadId;
8.
9.      public Thread() {
10.         this.threadId = ai.incrementAndGet();
11.     }
```

```
12.
13.      public void run() {
14.      }
15.
16.      public void start() {
17.          start0();
18.      }
19.
20.      private native void start0();
21.
22.      public native void yield();
23.
24. }
```

然后通过 javac -h ./ com/seaboat/Thread.java 命令让工具帮我们生成头文件，对应的方法为 JNIEXPORT **void** JNICALL Java_com_seaboat_Thread_yield(JNIEnv*, jobject)。yield 本地方法对应的函数如代码清单 12.10 所示（实际上就是调用了 pthread 的 sched_yield() 函数）。

代码清单 12.10　yield 本地方法对应的函数

```
1.  JNIEXPORT void JNICALL Java_com_seaboat_Thread_yield(JNIEnv *env,
2.          jobject jThreadObject) {
3.      std::cout << "calling yield operation!" << endl;
4.      sched_yield();
5.      return;
6.  }
```

接着对 C++ 代码进行编译并生成 so 库。我们编写一个测试类（见代码清单 12.11），使用自己定义的 com.seaboat.Thread 线程类，MyThread 会不断执行 yield 方法以将 CPU 让给主线程。最终的输出包含了大量的"主线程"和"MyThread 放弃 CPU 时间"，但前者比后者输出的次数多得多，因为后者在每轮执行中都让出了 CPU 的使用权。

代码清单 12.11　测试类（模拟实现 yield 语义）

```
1.  public class YieldThreadTest {
2.
3.      public static void main(String[] args) {
4.          MyThread mt = new MyThread();
5.          mt.start();
6.          while (true) {
7.              System.out.println("主线程");
8.          }
9.      }
10.
11.     static class MyThread extends Thread {
12.         public void run() {
13.             while (true) {
14.                 System.out.println("MyThread 放弃 CPU 时间");
15.                 this.yield();
```

```
16.        }
17.      }
18.    }
19. }
```

12.3 模拟实现 sleep 操作

下面继续模拟 sleep 操作。在 com.seaboat.Thread 类中定义 sleep 本地方法，表示睡眠操作。为了方便 JVM 层的处理，该方法中额外添加了 threadId 参数（即传入当前线程的 ID），如代码清单 12.12 所示。这里规定主线程的 ID 为 0。

代码清单 12.12　在 com.seaboat.Thread 类中定义 sleep 本地方法

```
1.  public class Thread {
2.      static {
3.          ai = new AtomicInteger();
4.          System.load("/root/seaboat/native_test/com_seaboat_Thread.so");
5.      }
6.      static AtomicInteger ai;
7.      public int threadId;
8.
9.      public Thread() {
10.         this.threadId = ai.incrementAndGet();
11.     }
12.
13.     public void run() {
14.     }
15.
16.     public void start() {
17.         start0();
18.     }
19.
20.     private native void start0();
21.
22.     public native void yield();
23.
24.     public static native void sleep(long millis, int threadId);
25.
26. }
```

通过 javac -h ./ com/seaboat/Thread.java 命令让工具帮我们生成头文件，对应的方法为 JNIEXPORT void JNICALL Java_com_seaboat_Thread_sleep(JNIEnv*, jclass, jlong, jint)。

sleep 本地方法的核心实现如代码清单 12.13 所示。实际上它会分两种情况处理：睡眠时间为零时和睡眠时间为非零时。睡眠时间为零时则等同于 yield 方法（即间接调用 pthread 的 sched_yield 函数），而非零时则根据 threadId 判断是否为主线程。如果是主线程，则直接创建一个 Parker 对象并调用 park 函数将其阻塞。如果不是主线程，则通过前面讲到的 Map 来获取线程 ID 对应的 JavaThread 对象指针。每个 JavaThread 对象都有自己的 Parker 对象，调用相应 Parker 对象的 park 函数可使当前线程进入等待状态，传入的参数为阻塞的超时时长。

代码清单 12.13　sleep 本地方法的核心实现

```
1.    JNIEXPORT void JNICALL Java_com_seaboat_Thread_sleep(JNIEnv *env,jclass jc,
2.            jlong millis, jint threadId) {
3.        std::cout << "calling sleep operation!" << endl;
4.        if (millis == 0) {
5.            sched_yield();
6.        } else {
7.            //主线程
8.            if (threadId == 0) {
9.                Parker parker;
10.                parker.park(millis);
11.                return;
12.            }
13.            JavaThread *javaThread = threads.find(threadId)->second;
14.            javaThread->parker.park(millis);
15.        }
16.        return;
17.    }
```

然后对 C++代码进行编译并生成 so 库，接着我们写一个测试类。这个测试类很简单，就是让 MyThread 线程睡眠 3s，如代码清单 12.14 所示。注意这里实现的 sleep 方法不抛出 InterruptedException 异常。

代码清单 12.14　测试类（模拟实现 sleep 操作）

```
1.    public class SleepThreadTest {
2.    
3.        public static void main(String[] args) {
4.            new MyThread().start();
5.            Thread.sleep(4000, 0);
6.        }
7.    
8.        static class MyThread extends Thread {
9.            public void run() {
10.                System.out.println("当前线程睡眠 3000ms");
11.                Thread.sleep(3000, this.threadId);
12.                System.out.println("睡眠结束");
13.            }
14.        }
15.    }
```

代码清单 12.14 执行后的输出结果如下。

```
1.    creating a JavaThread object!
2.    threadId = 1
```

```
3.    creating a OSThread object!
4.    creating linux thread!
5.    Started a linux thread! tid=140632075712256
6.    calling sleep operation!
7.    当前线程睡眠 3000ms
8.    calling sleep operation!
9.    睡眠结束
```

12.4 模拟实现 synchronized 语义

Synchronized 在 JVM 中通过监视器（monitor）来实现。监视器最简单的实现方式就是互斥锁，下面我们来模拟实现 synchronized 语义，如代码清单 12.15 所示。先定义 monitorEnter 方法和 monitorExit 方法，分别表示进入 monitor 和退出 monitor。注意，这里因为我们不实现 JVM 指令执行引擎，所以改用这两个方法来替代 monitorenter、monitorexit 指令，虽然有差异，但仍然能够说明意义。在 Java 层面主动调用这两个方法，就能模拟在 JVM 中执行这两个指令的情况。

代码清单 12.15 模拟实现 synchronized 语义

```
1.    public class Thread {
2.        static {
3.            ai = new AtomicInteger();
4.            System.load("/root/seaboat/native_test/com_seaboat_Thread.so");
5.        }
6.        static AtomicInteger ai;
7.        public int threadId;
8.
9.        public Thread() {
10.           this.threadId = ai.incrementAndGet();
11.       }
12.
13.       public void run() {
14.       }
15.
16.       public void start() {
17.           start0();
18.       }
19.
20.       private native void start0();
21.
22.       public native void yield();
23.
24.       public static native void sleep(long millis, int threadId);
25.
26.       public static native void monitorEnter();
27.
28.       public static native void monitorExit();
```

```
29.
30. }
```

通过 javac -h ./ com/seaboat/Thread.java 命令让工具帮我们生成头文件,两个方法分别如下。

```
1.    JNIEXPORT void JNICALL Java_com_seaboat_Thread_monitorEnter(JNIEnv*,jclass);
2.
3.    JNIEXPORT void JNICALL Java_com_seaboat_Thread_monitorExit(JNIEnv*,jclass);
```

然后定义一个 Monitor 类,用于实现我们的监视器,如代码清单 12.16 所示。这个监视器中包含了一个 pthread 库提供的互斥锁,构造函数中通过 pthread_mutex_init 函数来初始化互斥锁。

代码清单 12.16　Monitor 类

```
1.    class Monitor {
2.    public:
3.        pthread_mutex_t _mutex;
4.        Monitor();
5.    };
6.
7.    Monitor::Monitor() {
8.        pthread_mutex_init(&_mutex, NULL);
9.    }
```

接下来看这两个本地方法的核心实现,如代码清单 12.17 所示。其中主要是通过 pthread 库的 pthread_mutex_lock 函数来加锁,然后通过 pthread_mutex_unlock 函数来释放锁,锁对象则保存在 monitor 对象中。这里为了方便只创建了一个 Monitor 对象,所有线程都共用它,但在实际的 JDK 中则是根据 synchronized 来指定使用哪个对象的 Monitor。

代码清单 12.17　两个本地方法的核心实现

```
1.    //为方便起见,这里假设所有线程都使用同一个monitor,实际JDK中由synchronized指定。
2.    Monitor monitor;
3.
4.    JNIEXPORT void JNICALL Java_com_seaboat_Thread_monitorEnter(JNIEnv *env,
5.        jclass jc) {
6.        pthread_mutex_lock(&(monitor._mutex));
7.        std::cout << "monitorEnter tid=" << pthread_self() << endl;
8.        return;
9.    }
10.
11.   JNIEXPORT void JNICALL Java_com_seaboat_Thread_monitorExit(JNIEnv *env,
12.       jclass jc) {
13.       pthread_mutex_unlock(&(monitor._mutex));
14.       std::cout << "monitorExit tid=" << pthread_self() << endl;
15.       return;
16.   }
```

然后对 C++代码进行编译并生成 so 库,接着我们写一个测试类,如代码清单 12.18 所示。主线程中同时启动了两个线程,因为它们都使用了 monitorEnter 和 monitorExit 方法的组合,所以在它们包括的范围内都属于同步块区域，具有互斥效果。其中 monitorEnter 与 monitorExit 的组合相当于 synchronized(obj){xxx}指定的同步块，只是一个是用两个方法来指定同步块，而另一个则是通过括号来指定同步块。

代码清单 12.18　测试类（模拟实现 synchronized 语义）

```
1.   public class SynchronizedThreadTest {
2.
3.       public static void main(String[] args) {
4.           new MyThread().start();
5.           new MyThread2().start();
6.           Thread.sleep(8000, 0);
7.       }
8.
9.       static class MyThread extends Thread {
10.          public void run() {
11.              Thread.monitorEnter();
12.              System.out.println("mythread sleeps 3s.");
13.              Thread.sleep(3000, this.threadId);
14.              Thread.monitorExit();
15.          }
16.      }
17.
18.      static class MyThread2 extends Thread {
19.          public void run() {
20.              Thread.monitorEnter();
21.              System.out.println("mythread2 thread sleeps 2s.");
22.              Thread.sleep(2000, this.threadId);
23.              Thread.monitorExit();
24.          }
25.      }
26.  }
```

从 Java 层到 JVM 层输出的完整信息如下所示。可以看到这两个线程是互斥的。

```
1.   creating a JavaThread object!
2.   threadId = 1
3.   creating a OSThread object!
4.   creating linux thread!
5.   Started a linux thread! tid=140273819399936
6.   creating a JavaThread object!
7.   threadId = 2
8.   creating a OSThread object!
9.   creating linux thread!
10.  Started a linux thread! tid=140273811007232
11.  calling sleep operation!
12.  monitorEnter tid=140273811007232
13.  mythread2 thread sleeps 2s.
14.  calling sleep operation!
15.  monitorExit tid=140273811007232
```

```
16.    monitorEnter tid=140273819399936
17.    mythread sleeps 3s.
18.    calling sleep operation!
19.    monitorExit tid=140273819399936
```

12.5 模拟实现 Interrupt 操作

中断机制为线程解除阻塞提供了方法。下面模拟中断相关的两个操作，如代码清单 12.19 所示。在 Thread 类中添加 isInterrupted 方法和 interrupt 方法，它们都是本地方法，分别用于获取 Java 线程的中断状态和对 Java 线程进行中断操作。

代码清单 12.19　模拟中断操作

```
1.    public class Thread {
2.        static {
3.            ai = new AtomicInteger();
4.            System.load("/root/seaboat/native_test/com_seaboat_Thread.so");
5.        }
6.        static AtomicInteger ai;
7.        public int threadId;
8.
9.        public Thread() {
10.            this.threadId = ai.incrementAndGet();
11.        }
12.
13.        public void run() {
14.        }
15.
16.        public void start() {
17.            start0();
18.        }
19.
20.        private native void start0();
21.
22.        public native void yield();
23.
24.        public static native void sleep(long millis, int threadId);
25.
26.        public static native void monitorEnter();
27.
28.        public static native void monitorExit();
29.
30.        public native boolean isInterrupted();
31.
32.        public native boolean interrupt();
33.
34.    }
```

12.5 模拟实现 Interrupt 操作

通过 javac -h ./ com/seaboat/Thread.java 命令让工具帮我们生成头文件，两个方法如下。

```
1.  JNIEXPORT bool JNICALL Java_com_seaboat_Thread_isInterrupted(JNIEnv*, jobject);
2.
3.  JNIEXPORT void JNICALL Java_com_seaboat_Thread_interrupt(JNIEnv*, jobject);
```

下面先看用于获取中断状态的 isInterrupted 方法的本地实现，如代码清单 12.20 所示。首先获取 Java 层的线程 ID 值（它对应 Thread 类的 threadId 属性），然后根据 threadId 从 map 中获取对应的 JavaThread 指针，最后获取 JavaThread 对象的 interruptState 属性并返回。需要注意的是，不能直接返回当前线程所对应的中断状态，因为有可能在某个线程中获取另外一个线程的中断状态，比如在线程 1 中查询线程 2 的中断状态。

代码清单 12.20　isInterrupted 方法的本地实现

```
1.  JNIEXPORT bool JNICALL Java_com_seaboat_Thread_isInterrupted(JNIEnv *env,
2.          jobject jThreadObject) {
3.      std::cout << "getting interrupt flag..." << endl;
4.      jclass cls = env->GetObjectClass(jThreadObject);
5.      jfieldID fID = env->GetFieldID(cls, "threadId", "I");
6.      jint threadId = env->GetIntField(jThreadObject, fID);
7.      JavaThread *javaThread = threads.find(threadId)->second;
8.      return javaThread->interruptState;
9.  }
```

继续看用于中断操作的 interrupt 方法的本地实现，如代码清单 12.21 所示。同样是首先获取 Java 层的线程 ID 值，然后根据 threadId 从 map 中获取对应的 JavaThread 指针，接着再获取对应的 parker 对象并调用它的 unpark 函数唤醒已阻塞的线程，最后还要把中断状态 interruptState 置为 true。

代码清点 12.21　interrupt 方法的本地实现

```
1.  JNIEXPORT void JNICALL Java_com_seaboat_Thread_interrupt(JNIEnv *env,
2.          jobject jThreadObject) {
3.      std::cout << "interrupt..." << endl;
4.      jclass cls = env->GetObjectClass(jThreadObject);
5.      jfieldID fID = env->GetFieldID(cls, "threadId", "I");
6.      jint threadId = env->GetIntField(jThreadObject, fID);
7.      JavaThread *javaThread = threads.find(threadId)->second;
8.      javaThread->interruptState = true;
9.      javaThread->parker.unpark();
10.     return;
11. }
```

然后对 C++ 代码进行编译并生成 so 库，接着我们写一个测试类，如代码清单 12.22 所示。在代码清单 12.22 中，主线程会先启动线程 thread1，该线程在启动后进入长时间的阻塞睡眠，然后启动线程 thread2。thread2 则会中断 thread1，最终 thread1 被唤醒。

代码清单 12.22　测试类（模拟实现 Interrupt 操作）

```
1.  public class InterruptThreadTest {
```

```java
2.
3.      static MyThread thread1 = new MyThread();
4.      static MyThread2 thread2 = new MyThread2();
5.
6.      public static void main(String[] args) {
7.          thread1.start();
8.          Thread.sleep(500, 0);
9.          thread2.start();
10.         Thread.sleep(100, 0);
11.     }
12.
13.     static class MyThread extends Thread {
14.         public void run() {
15.           System.out.println("MyThread1 中断状态为" + this.isInterrupted());
16.           Thread.sleep(500000, this.threadId);
17.           System.out.println("MyThread1 成功被中断" + this.isInterrupted());
18.         }
19.     }
20.
21.     static class MyThread2 extends Thread {
22.         public void run() {
23.             thread1.interrupt();
24.           System.out.println("MyThread1 中断状态为" + thread1.isInterrupted());
25.           System.out.println("MyThread2 中断状态为" + this.isInterrupted());
26.         }
27.     }
28. }
```

代码清单 12.22 执行后的输出结果如下。

```
1.  creating a JavaThread object!
2.  threadId = 1
3.  creating a OSThread object!
4.  creating linux thread!
5.  Started a linux thread! tid=140150227187456
6.  calling sleep operation!
7.  getting interrupt flag...
8.  MyThread1 中断状态为 false
9.  calling sleep operation!
10. creating a JavaThread object!
11. threadId = 2
12. creating a OSThread object!
13. creating linux thread!
14. Started a linux thread! tid=140149899917056
15. calling sleep operation!
16. interrupt...
17. getting interrupt flag...
18. MyThread1 成功被中断 true
19. getting interrupt flag...
20. MyThread1 中断状态为 true
21. getting interrupt flag...
22. MyThread2 中断状态为 false
```